Enrico Drioli, Lidietta Giorno, Francesca Macedonio (Eds.)
Membrane Engineering

Also of Interest

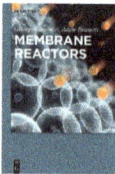

Membrane Reactors.
Barbieri, Brunetti, 2020
ISBN 978-3-11-033347-3, e-ISBN 978-3-11-033376-3

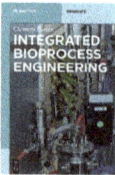

Integrated Bioprocess Engineering.
Posten, 2018
ISBN 978-3-11-031538-7, e-ISBN 978-3-11-031539-4

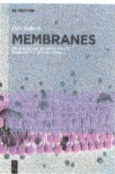

Membranes.
From Biological Functions to Therapeutic Applications
Jelinek, 2018
ISBN 978-3-11-045368-3, e-ISBN 978-3-11-045369-0

Membrane Systems.
For Bioartificial Organs and Regenerative Medicine
De Bartolo, Curcio, Drioli, 2017
ISBN 978-3-11-026798-3, e-ISBN 978-3-11-026801-0

Membrane Engineering

Edited by
Enrico Drioli, Lidietta Giorno, Francesca Macedonio

DE GRUYTER

Editors
Prof. Dr. Enrico Drioli
National Research Council of Italy
Institute for Membrane Technology (ITM-CNR)
at University of Calabria
Via P. Bucci, cubo 17/C
87036 Rende (CS)
Italy
e.drioli@itm.cnr.it

Dr. Lidietta Giorno
National Research Council of Italy
Institute for Membrane Technology (ITM-CNR)
at University of Calabria
Via P. Bucci, cubo 17/C
87036 Rende (CS)
Italy
l.giorno@itm.cnr.it

Dr. Francesca Macedonio
National Research Council of Italy
Institute for Membrane Technology (ITM-CNR)
at University of Calabria
Via P. Bucci, cubo 17/C
87036 Rende (CS)
Italy
f.macedonio@itm.cnr.it

ISBN 978-3-11-028140-8
e-ISBN (PDF) 978-3-11-028139-2
e-ISBN (EPUB) 978-3-11-038154-2

Library of Congress Control Number: 2018954783

Bibliographic information published by the Deutsche Nationalbibliothek
The Deutsche Nationalbibliothek lists this publication in the Deutsche Nationalbibliografie; detailed
bibliographic data are available on the Internet at http://dnb.dnb.de.

© 2019 Walter de Gruyter GmbH, Berlin/Boston
Typesetting: Integra Software Services Pvt. Ltd.
Printing and binding: CPI books GmbH, Leck
Cover image: Sebastian Kaulitzki / Shutterstock

www.degruyter.com

Preface

Not many years have passed from the days when Loeb and Sourirajan, with their preparation of asymmetric membranes, made the reverse osmosis process of industrial interest. It was the early 1960s when they discovered an effective method for significantly increasing the permeation flux of polymeric membranes without significantly changing their selectivity. This made the use of membranes possible in large-scale operations for desalting brackish water and seawater by reverse osmosis and for various other molecular separations in different industrial areas. Today, reverse osmosis is a well-recognized basic unit operation, accounting for more than 60% of all desalination plants, mainly due to its higher recovery factor, lower investment and total water cost compared to other conventional methodologies as well as to the continuing technological advances enabling reverse osmosis desalination to also treat high salinities raw water [1].

Composite polymeric membranes developed in the 1970s made the separation of components from gas streams commercially feasible. Billions of cubic meters of pure gases are now produced via selective permeation in polymeric membranes.

The combination of molecular separation with a chemical reaction, or membrane reactors, offers important new opportunities for improving the production efficiency in biotechnology and in the chemical industry. The availability, moreover, of new high-temperature-resistant membranes and of new membrane operations as membrane contactors (MCs) offers an important tool for the design of alternative production systems appropriate for sustainable growth.

The early membranologists have always been optimistic about the possibilities of membrane operations, but the scientific and technical results reached today are even superior to the expectations.

The basic properties of membrane operations make them ideal for industrial production: they are generally athermal and do not involve phase changes or chemical additives, they are simple in concept and operation, they are modular and easy to scale-up and they are low in energy consumption with a potential for more rational utilization of raw materials and recovery and reuse of by-products.

Membrane technologies, compared to those commonly used today, respond efficiently to the requirements of the so-called process intensification [2], because they permit drastic improvements in manufacturing and processing, substantially decreasing the equipment-size/production-capacity ratio, energy consumption and/ or waste production and resulting in cheaper, sustainable technical solutions.

Today, membrane technology has well-established uses in many industrial processes including water desalination, wastewater treatments, agro-food, gas separation, in artificial organs and in chemical and petrol-chemical industry. Membrane operations are already dominant technologies in molecular separations, but they are also becoming of interest as membrane reactors and MCs. Practically all of the typical unit operations of process engineering could be redesigned as membrane unit operations (membrane

https://doi.org/10.1515/9783110281392-201

distillation, membrane crystallizer, membrane reactors, membrane condensers and membrane gas separation).

The significant positive results achieved in various membrane systems are, however, still far from realizing the potentialities of this technology. There are still problems related to the pretreatment of streams, membrane life time, aging, fouling and sealing, thereby slowing down the growth of large-scale industrial use. A good understanding of the materials' properties and transport mechanisms and the creation of innovative functional materials with improved properties are key challenges for a further development of this technology, which requires further intensive research activities both at academic and industrial level. In addition, the design and optimization of the membrane process will lead to significant innovation toward large-scale diffusion of membrane technologies in various sectors, and the role of membrane engineering is crucial in this respect. Moreover, another interesting development for industrial membrane technologies is related to the possibility of integrating different types of these membrane operations in the same industrial cycle, with overall important benefits in terms of product quality, plant compactness, environmental impact and energetic aspects. It is known that existing nonmembrane-based equilibrium-driven separation technologies (e.g., distillation, extraction, absorption, adsorption, ion exchange and stripping), which represent the core of the traditional desalination, chemical and petrochemical industry, have significant shortcomings: high energy consumption, inherent operational difficulties, lack of flexibility and modularity, slower rates, need for hazardous chemicals, high capital costs, need for large equipment volume and footprint. These shortcomings are exacerbated by new separation demands (e.g., environmental pollution control laws). New membrane-based separation concepts and technologies (e.g., membrane distillation, membrane crystallization, membrane condenser, membrane dryer, membrane emulsifier) do not suffer from many such deficiencies and are poised to invade more and more the domain of traditional separation technologies. It is, then, realistic to affirm that new wide perspectives of membrane technologies and integrated membrane solutions for sustainable industrial growth are possible.

The purpose of this book is to present membrane science and technology, to provide a description of membrane structures and membrane processes in various fields, from mass separation to (bio)chemical reactors, energy conversion and storage.

The membranes used in the various applications differ widely in their structure, in their function and the way they are operated. Furthermore, membrane properties can be tailored and adjusted to specific separation tasks, and membrane processes are often technically simple and are equally well suited for large-scale continuous operations as for batch-wise treatment of very small quantities.

At the heart of every membrane processes, there is an interface, which is clearly materialized by a nanostructured/functionalized thin barrier that controls the exchange between two phases, not only by external forces and under the effect of fluid properties but also through the intrinsic characteristics of the membrane material itself. A membrane may be biological or synthetic, solid or liquid, homogeneous or heterogeneous, isotropic or anisotropic in its structure. A membrane can be a fraction of a micrometer or several millimeters thick. Its electrical resistance can vary from millions of ohm to a fraction of an ohm.

Another characteristic property of a membrane is its permselectivity, which is determined by differences in the transport rates of various components in the membrane matrix. The permeability of a membrane is a measure of the rate at which a given component is transported through the membrane under specific conditions of concentration, temperature, pressure and/or electric field. The transport rate of a component through a membrane is determined by the structure of the membrane, by the size of the permeating component, by the chemical nature and the electrical charge of the membrane material and permeating components and by the driving force, that is, concentration, pressure or electrical potential gradient across the membrane. The transport of certain components through a membrane may be facilitated by certain chemical compounds, coupled to the transport of other components, or activated by a chemical reaction occurring in the membrane. These phenomena are referred to as facilitated, coupled or active transport. Another important characteristic is the driving force acting on the permeating components. Some driving forces such as concentration, pressure or temperature gradients act equally on all components, in contrast to an electrical potential driving force, which is only effective with charged components. The use of different membrane structures and driving forces has resulted in a number of rather different membrane processes such as reverse osmosis, micro-, ultra- and nanofiltration, dialysis, electrodialysis, Donnan dialysis, pervaporation, gas separation, membrane reactors, MCs, membrane distillation, membrane emulsification, membrane crystallization, membrane condenser, membrane dryer and so on.

In Chapter 1, the common fundamentals of different membrane processes are described. In the first part, the main terms used in membrane processes such as membrane permeability, membrane permselectivity, selectivity and membrane rejection are defined. The basic thermodynamic relations relevant for the description of mass transport phenomena in membranes and membrane processes are treated and the mathematical relations used to describe the mass transport in membranes are discussed. In the second part, fundamental aspects of pressure-driven membrane processes, including transport mechanisms, polarization phenomena, fouling/biofouling/scaling problems, are discussed. The technical advantages and limitations of the various processes are also listed.

Pervaporation is described in Chapter 2. It is a membrane process in which the permeation of certain components through a membrane from a liquid feed mixture into a vapor phase is combined with the evaporation of these components. The driving force for the transport is the chemical potential gradient of the permeating components in the membrane. The membranes to be utilized in this application are described.

Ion-exchange membranes (IEMs) are described in Chapter 3. IEMs are charge-selective membranes characterized by the presence of ionizable groups linked by covalent bonds with the polymeric membrane matrix. Ions having the same charge of these ionizable fixed groups (or coions) are theoretically impermeable through the IEM by electrostatic repulsion, while ions with the opposite charge (or counterions) can be transported through the IEM under the effect of an electrochemical potential. The transport of ions through an IEM and its membrane permselctivity are illustrated. In addition, the concepts of polarization phenomena, limiting current and membranes and interfaces characterization by impedance spectroscopy are presented. Finally, the main application of IEMs (i.e., electrodialysis) is described.

Chapter 4 is dedicated to gas separation via membranes. The selective transport of gases and the separation of gases are strongly a function of the membrane material, and the gas transport is described based on a solution-diffusion mechanism. The different polymeric matrices currently used in this process are presented.

Among the large variety of membrane operations, MCs represent relatively new membrane-based devices that are gaining wide consideration. Their description is given in Chapter 5. MCs are systems in which microporous hydrophobic membranes are used not as selective barriers but to promote the mass transfer between phases on the basis of the principles of phase equilibrium. All traditional stripping, scrubbing, absorption and liquid–liquid extraction operations, as well as emulsification, crystallization and phase transfer catalysis, can be carried out according to this configuration. In the first part of the chapter, the description of the basic principles of this technology is introduced. The second part is dedicated to the description of membrane distillation, membrane crystallization, membrane dryer, membrane condenser and membrane emulsification.

Systems where a chemical or biochemical conversion is combined with a membrane separation process, that is, the membrane reactors, are described in Chapter 6. Both membrane reactors working at high temperature and those working at low temperature are illustrated. In addition, the two main configurations of both types of reactors are presented: "membrane reactor" is the one with the membrane catalytically inert, and it does not participate directly in the reaction but it simply acts as a barrier to reagents allowing selective separation of the product(s) or intermediate(s); on the contrary, "catalytic membrane reactor" is the one where the membrane not only separates but also contains the (bio)catalyst and participates directly in the reaction.

Chapter 7 is dedicated to the description of the various methods for membrane preparation and characterization. In the first part of the chapter, the preparation of porous (symmetric and asymmetric) membranes made from polymeric and inorganic material are discussed. Different methods for the preparation of composite membrane as well as for the modification of homogeneous dense membrane are illustrated. In the second part, the methods for the characterization of membrane structural properties are analyzed. Since porous, dense and composite membranes are very different in their properties and applications, a large number of different techniques are required for their characterization. Some of the most important membrane characterization procedures, distinguished by porous and dense membrane, are reviewed.

In Chapter 8, as a case study, the applications of membrane-based operations and integrated membrane systems in fruit juice processing are discussed. The chapter gives an outlook on the most relevant applications of membrane operations in fruit juice processing as an alternative to conventional methodologies. Special attention is paid to the combination of membrane operations in integrated systems, which can play a key role in redesigning the traditional flow sheet of fruit processing industry, with remarkable benefits in terms of product quality, plant compactness, environmental impact and energetic aspects.

In Chapter 9, the concept of blue energy as well as the principle, prospects and limitation of two major energy harvesting technologies from salinity (i.e., pressure-retarded osmosis and reverse electrodialysis) are discussed.

Finally, in Chapter 10, the use and limits of detailed atomistic simulations of transport phenomena in polymeric membranes (i.e., solubility, diffusivity and permselectivity) are illustrated through some case studies.

References

[1]   Fritzmann, C., Löwenberg, J., Wintgens, T., & Melin, T. State-of-the-art of reverse osmosis desalination. Desalination. (2007); 216: 1–76.
[2]   Drioli, E., Stankiewicz, A.I., & Macedonio, F. Membrane Engineering in Process Intensification – An Overview. J. Membr. Sci. (2011); 380: 1–8.

Contents

List of contributors

Aamer Ali
National Research Council of Italy
Institute for Membrane Technology (ITM-CNR)
at University of Calabria
Via P. Bucci, cubo 17/C
87036 Rende (CS)
Italy
a.aamer@itm.cnr.it

Giuseppe Barbieri
National Research Council of Italy
Institute for Membrane Technology (ITM-CNR)
at University of Calabria
Via P. Bucci, cubo 17/C
87036 Rende (CS)
Italy
g.barbieri@itm.cnr.it

Adele Brunetti
National Research Council of Italy
Institute for Membrane Technology (ITM-CNR)
at University of Calabria
Via P. Bucci, cubo 17/C
87036 Rende (CS)
Italy
a.brunetti@itm.cnr.it

Alfredo Cassano
National Research Council of Italy
Institute for Membrane Technology (ITM-CNR)
at University of Calabria
Via P. Bucci, cubo 17/C
87036 Rende (CS)
Italy
a.cassano@itm.cnr.it

Carmela Conidi
National Research Council of Italy
Institute for Membrane Technology (ITM-CNR)
at University of Calabria
Via P. Bucci, cubo 17/C
87036 Rende (CS)
Italy
c.conidi@itm.cnr.it

Enrico Drioli
National Research Council of Italy
Institute for Membrane Technology (ITM-CNR)
at University of Calabria
Via P. Bucci, cubo 17/C
87036 Rende (CS)
Italy
e.drioli@itm.cnr.it

and

University of Calabria
Department of Environmental and Chemical
Engineering
Rende
Italy
e.drioli@unical.it

and

Hanyang University
WCU Energy Engineering Department
Seoul
South Korea

and

Center of Excellence in Desalination
Technology
King Abdulaziz University
Jeddah
Saudi Arabia

Enrica Fontananova
National Research Council of Italy
Institute for Membrane Technology (ITM-CNR)
at University of Calabria
Via P. Bucci, cubo 17/C
87036 Rende (CS)
Italy
e.fontananova@itm.cnr.it

Juntae Jung
Department of Energy Engineering
Hanyang University
Seoul
South Korea
steelymoon@hanyang.ac.kr

https://doi.org/10.1515/9783110281392-202

Jeong F. Kim
Membrane Research Center
Korea Research Institute of Chemical
Technology (KRICT)
Daejeon
South Korea
Jeongkim0653@gmail.com

Francesca Macedonio
National Research Council of Italy
Institute for Membrane Technology (ITM-CNR)
at University of Calabria
Via P. Bucci, cubo 17/C
87036 Rende (CS)
Italy
f.macedonio@itm.cnr.it

and

University of Calabria
Department of Environmental and Chemical
Engineering
Rende
Italy
francesca.macedonio@unical.it

Elena Tocci
National Research Council of Italy
Institute for Membrane Technology (ITM-CNR)
at University of Calabria
Via P. Bucci, cubo 17/C
87036 Rende (CS)
Italy
e.tocci@itm.cnr.it

Francesca Macedonio, Enrico Drioli

1 Pressure-driven membrane processes

1.1 Definition

The performance of a membrane process is determined by a number of components and process parameters, such as membrane structure and material, membrane geometry, driving force and various engineering aspects that are relevant for designing process.

An interface is present at the core of every membrane process, which is clearly materialized by either a selective or a nonselective barrier that separates and/or contacts two adjacent phases and allows or promotes the exchange of matter, energy and information between the phases in a specific or nonspecific manner.

The concept of a selective and nonselective membrane is illustrated in Figure 1.1 that shows (a) a membrane, which is highly selective and capable of separating, for example, two enantiomers and (b) a membrane that acts as a barrier between two phases and does not allow the mixing of phases, however, it has no effect on the transportation of different components from one phase to the other.

The molecular mixture separated during the process is referred to as feed, the mixture containing the components retained by the membrane is called the retentate and the mixture composed of the components that permeates from the membrane is referred to as permeate (or filtrate in microfiltration [MF] and ultrafiltration [UF]) (Figure 1.2).

A membrane can be porous or nonporous, symmetric or asymmetric, homogeneous or heterogeneous. Dense membranes generally show low fluxes. To increase the flux the effective membrane thickness must be reduced to the maximum extent. This may be achieved through the preparation of asymmetric membranes. A particular class of asymmetric membrane is represented by composite membranes. A composite membrane consists of two different materials with a very selective membrane material being deposited as a thin layer on a more or less porous sublayer. The actual selectivity is determined by the thin top layer, whereas the porous sublayer merely serves as a support.

In porous membranes the separation is accomplished via mechanical sieving and particles are separated solely according to their dimensions. Membrane material is of crucial importance to chemical, thermal and mechanical stability. Nonporous membranes are capable of separating molecules of approximately the same size. Separation takes place through differences in solubility and/or diffusivity (Figure 1.3).

Membrane separation processes can be grouped according to the applied driving forces (Table 1.1) into the following: (1) hydrostatic pressure-driven processes such as reverse osmosis (RO), nano-, ultra- and MF or gas separation (GS); (2) concentration

Francesca Macedonio, Enrico Drioli, National Research Council of Italy, Institute for Membrane Technology (ITM-CNR), University of Calabria, Rende, Italy

https://doi.org/10.1515/9783110281392-001

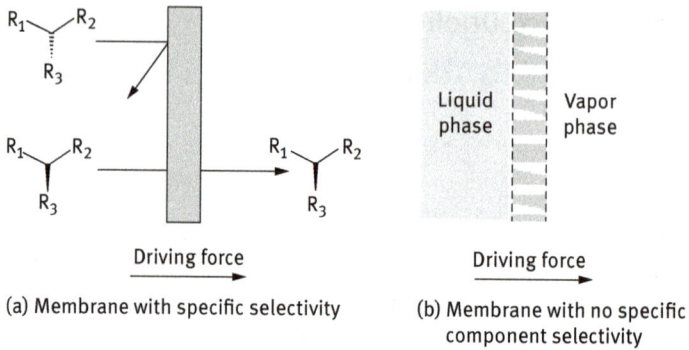

(a) Membrane with specific selectivity

(b) Membrane with no specific component selectivity

Figure 1.1: Schematic drawing illustrating (a) a membrane that is selective for the transport of different components and (b) a membrane that separates a liquid and a vapor phase and allows passage of vapor molecules; however, it is not selective for the transport of different components. Adapted from Strathmann et al. [1].

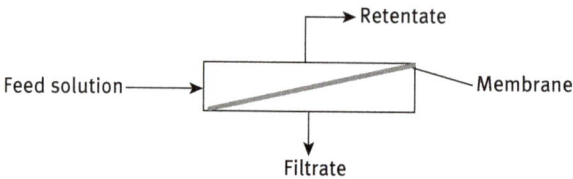

Figure 1.2: General scheme of a separation process.

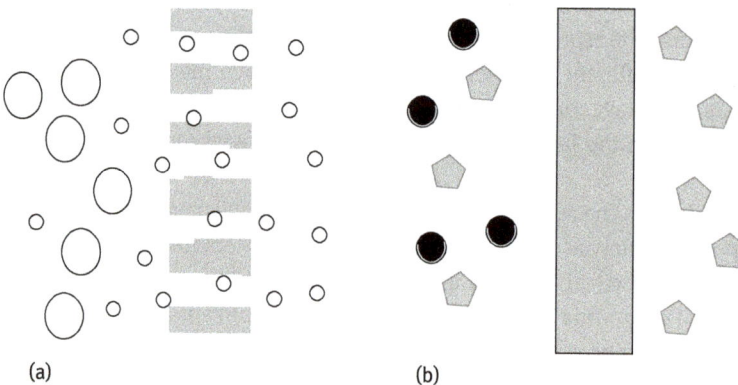

Figure 1.3: Schematic diagram of the most common transport mechanisms. (a) The components that permeate through the membrane are transported by convective flow through pores and the separation occurs because of size exclusion. (b) Transport through a membrane is based on the solution and diffusion of individual molecules in the nonporous membrane matrix. Adapted from Strathmann et al. [1].

Table 1.1: Basic membrane operations. Adapted from Strathmann et al. [1].

Process	Driving force	Mode of transport	Species passed	Species retained
Microfiltration	Pressure difference 100– 500 kPa	Size exclusion convection	Solvent (water) and dissolved solutes	Suspended solids, fine particulars and some colloids
Ultrafiltration	Pressure difference 100–800 kPa	Size exclusion convection	Solvent (water) and low molecular weight solutes (<1000 Da)	Macrosolutes and colloids
Nanofiltration	Pressure difference 0.3–3 MPa	Size exclusion Solution diffusion Donnan exclusion	Solvent (water),low molecular weight solutes and monovalent ions	Molecular weight compounds >200 Da multivalent ions
Reverse osmosis	Pressure difference 1–10 MPa	Solution diffusion mechanism	Solvent (water)	Dissolved and suspended solids
Gas separation	Pressure difference 0.1–10 MPa	Solution diffusion mechanism	Gas molecules having low molecular weight or high solubility diffusivity	Gas molecules having high molecular weight or low solubility diffusivity
Pervaporation	Chemical potential or concentration difference	Solution diffusion mechanism	High permeable solute or solvents	Less permeable solute or solvents
Electrodialysis	Electrical potential difference 1–2 V/cell pair	Donnan exclusion	Solutes (ions) Small quantity of solvent	Nonionic and macromolecular species
Dialysis	Concentration difference	Diffusion	Solute (ions and low molecular weight organic molecules) Small solvent quantity	Dissolved and suspended solids with MW > 1,000 Da
Membrane contactors	Chemical potential, concentration difference and temperature difference	Diffusion	Compounds soluble in the extraction solvent; volatiles	Compounds nonsoluble in the extraction solvent; nonvolatiles

(continued)

Table 1.1 (continued)

Process	Driving force	Mode of transport	Species passed	Species retained
Membrane-based solvent extraction	Chemical potential or concentration difference	Diffusion partition	Compounds soluble in the extraction solvent	Compounds nonsoluble in the extraction solvent
Membrane distillation	Temperature difference	Diffusion	Volatiles	Nonvolatiles
Supported liquid membranes	Concentration difference	Diffusion	Ions, low MW organics	Ions, less permeable organics
Membrane reactors	Various	Various	Permeable product	Nonpermeable reagents

gradient or chemical potential-driven processes, such as dialysis (D), Donnan dialysis, pervaporation and membrane contactors (MCs), such as membrane-based solvent extraction, membrane scrubbers and strippers, osmotic distillation; (3) electrical potential-driven processes such as electrodialysis (ED); and (4) temperature difference-driven membrane processes such as membrane distillation (MD).

A schematic diagram of the separation characteristics of the various membrane structures and related processes is given in Figure 1.4.

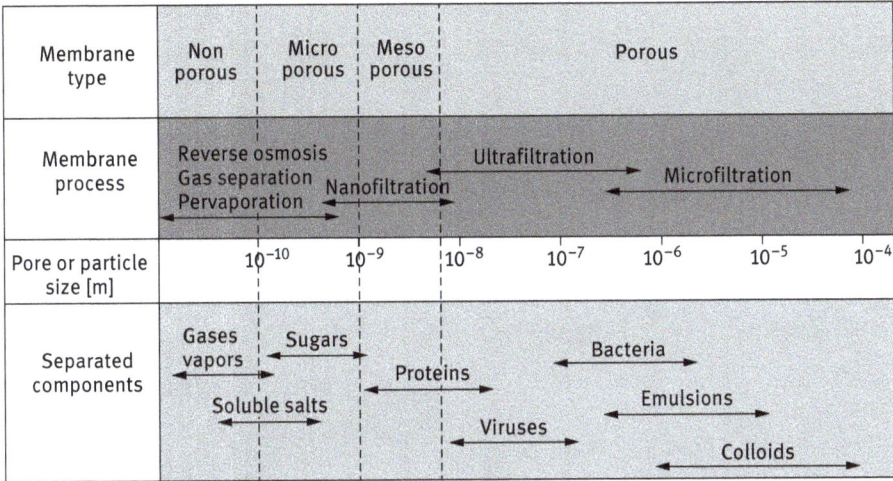

Membrane type	Non porous	Micro porous	Meso porous	Porous				
Membrane process	Reverse osmosis Gas separation Pervaporation	Nanofiltration		Ultrafiltration		Microfiltration		
Pore or particle size [m]	10^{-10}	10^{-9}	10^{-8}	10^{-7}	10^{-6}	10^{-5}	10^{-4}	
Separated components	Gases vapors Soluble salts	Sugars	Proteins Viruses	Bacteria	Emulsions	Colloids		

Figure 1.4: Membrane process characteristics.

The possible membrane geometries are reported in Figure 1.5 and are as follows:
1. *Spiral wound membrane* consists of consecutive layers of large membrane and support materials in an envelope shaped design rolled up around a perforated steel tube. This design is capable of maximizing the surface area in a minimum amount of space. It is less expensive but more sensitive to pollution because of its manufacturing process.
2. *Plate and frame module* is normally used for poor quality water. They are set up with a stack of membranes and support plates.
3. *Tubular membranes* are generally used for viscous or poor quality fluids; they are not self-supporting membranes. They are located inside of a tube, which is made of a special kind of microporous material. This material is the supporting layer for the membrane. Because the feed solution flows through the membrane core, the permeate passes through the membrane and is collected in the tubular housing. The main cause for this is that the attachment of the membrane to the supporting layer is very weak. Tubular membranes have a diameter of about 5 to 15 mm. Because of the size of the membrane surface, plugging of tubular membranes is not likely to occur. Therefore, these modules do not need a preliminary

pretreatment of water. The main drawback is that tubular membrane is not very compact and has a high cost per m^2 installed.

4. *Hollow fiber membrane* modules contain several small tubes or fibers (diameter of below 0.5 mm); consequently, the chances of plugging of a hollow fiber membrane are very high. The membranes can only be used for the treatment of water with a low suspended solid content. The packing density of a hollow fiber membrane is very high. Hollow fiber membranes are mostly used for NF and RO.

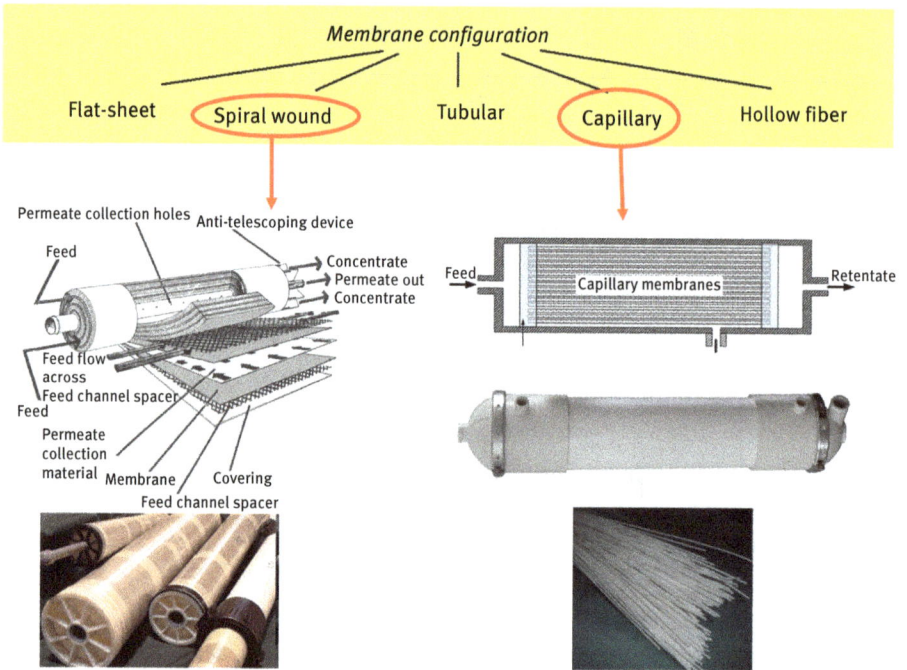

Figure 1.5: Possible membranes geometries.

Table 1.2. presents some general characteristics of the four basic membrane-module types.

Table 1.2: Qualitative comparison of membrane configurations.

	Tubular	Plate-and-frame	Spiral-wound	Hollow-fibre
Packing density	low	----------------------→		very high
Investment	high	←----------------------		low
Fouling tendency	low	----------------------→		very high
Ease to cleaning	good	←----------------------		poor

In the most generalized form, mass transportation through a membrane can be through the following procedures:

- Passive membrane transport (where a membrane is just a passive physical barrier without specific interactions between the permeating components and the membrane matrix)
- Carrier facilitated transport (where the transport of certain components is facilitated because of specific interactions with a "carrier" component in the membrane matrix. The carriers can be fixed to the membrane matrix or the carrier can be mobile and dissolved in a liquid located inside the membrane pores.

Two forms of flux coupling can be distinguished (Figure 1.6): (1) *cocurrent-coupled transport* when co- and counterions are transported in the same direction and (2) *countercurrent-coupled transport* where no counterions permeate through the membrane and coion fluxes are in the opposite direction; in case (1) (Figure 1.6a) the anion-exchange membrane is permeable for the anion A^-. The cation C^+ can permeate through the membrane only by a specific carrier and the cation B^+ cannot permeate through the membrane at all; in the case (2) (Figure 1.6b) the membrane is a negatively charged cation-exchange membrane through which the anion A^- cannot permeate. For electroneutrality, electrical charges of the component B^+ will be transported in the opposite direction against its concentration gradient.

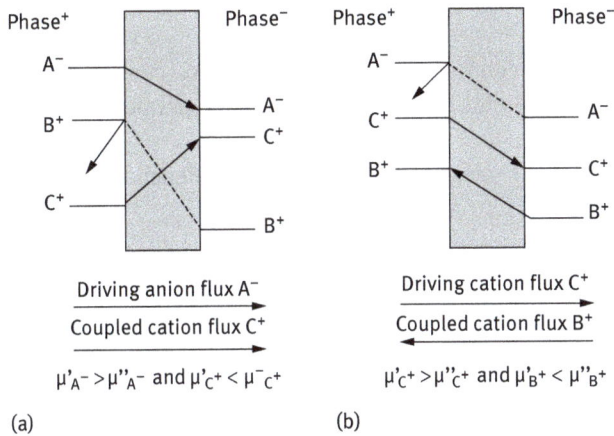

Figure 1.6: Schematic diagram illustrating (a) cocurrent-coupled transport in a selective cation carrier membrane and (b) countercurrent-coupled transport in a cation-exchange membrane. Adapted from Strathmann et al. [1].

In the most generalized form, the transmembrane flux can be expressed as follows:

$$J = -P\frac{dX}{dz} \tag{1.1}$$

where J is a flux, P is the phenomenological coefficient expressing the permeability of the membrane and $\frac{dX}{dz}$ is the driving force.

The flux through the membrane can be expressed by phenomenological equations:

1. Volume flux, J_v, expressed in volume per time, for example, m s^{-1}
2. Mass flux, J_m, expressed in mass per time, for example, kg m^{-2} s^{-1}
3. Molar flux, J_n, expressed in mole per time, that is, mol m^{-2} s^{-1}
4. Electrical flux, J_e, expressed in Faraday per time, that is, A m^{-2}

The different fluxes are conventionally described by simple linear relations between the flux and the driving force, such as

Darcy's law

$$J_v = - L_p \frac{dp}{dz} \tag{1.2}$$

Fick's law

$$J_i = - D_i \frac{dC_i}{dz} \tag{1.3}$$

Ohm's law

$$J_e = - \kappa \frac{d\varphi}{dz} \tag{1.4}$$

where p is the pressure, C_i is the concentration and D_i is the diffusion coefficient of a component i, φ is the electrical potential, L_p is the hydrodynamic permeability and κ is the electrical conductivity.

In membrane processes, driving forces and fluxes may be interdependent, giving rise to new effects.

The performance of a given membrane is determined not only by its flow but also by its selectivity. The membrane permselectivity $S^P_{A,B}$ between components A and B are defined as follows:

$$S^P_{A,B} = \frac{P_A}{P_B}$$

where P_A and P_B are the permeability of components A and B, respectively.

For nonporous membranes, permeability is derived from solubility and diffusivity as follows:

$$Permeability = Solubility \times Diffusivity$$

The selectivity of a membrane toward a mixture is generally expressed by one of the two parameters: (1) the separation factor α (for gas mixture and mixture of organic liquids) or (2) the rejection coefficient R (for dilute aqueous mixtures) as follows:

$$\alpha_{A/B} = \frac{A^{\text{Permeate}} B^{\text{Retentate}}}{A^{\text{Retentate}} B^{\text{Permeate}}}$$

$$R_A = \left(1 - \frac{C_A^{\text{Permeate}}}{C_A^{\text{Feed}}}\right)$$

1.2 Basic thermodynamic relations with relevance to membrane processes

The transport of heat, mass, electrical charges and individual components between two systems separated by a semipermeable membrane occurs only when the systems are not in equilibrium, that is, when a chemical or an electrical potential difference acts on the individual components in the system. The equilibrium condition in a closed system is given as follows:

$$(dG)_{p,T} = 0$$

For a closed system with no exchange of matter with the surrounding, Gibb's free energy G is given as follows:

$$dG = dU + d(pV) - d(TS) = Vdp - SdT$$

where U is the internal energy, S is the entropy, T is the temperature, p is the pressure, V is the volume of the system under consideration.

For an open system (i.e., a system with exchange of matter with the surrounding), the change of the Gibb's free energy is given as follows:

$$dG = Vdp - SdT + \sum_i \left(\frac{\partial G}{\partial n_i}\right)_{p,T,n_j} dn_i = Vdp - SdT + \sum_i \mu_i dn_i$$

where $\left(\frac{\partial G}{\partial n_i}\right)_{p,T,n_j} = \mu_i$ is the chemical potential.

Most transport processes take place because of a difference in chemical and/or electrical potential $\Delta\mu$. Under isothermal conditions, the electrochemical potential of a component i is given as follows:

$$\mu_i = \mu_i^o + RT\ln a_i + V_i P + z_i FE$$

where z, F and E are the charge number, Faraday constant and electrical potential, respectively. The subscript i refers to the ions in the solution.

The number of moles of a component transported from one system to the other through the membrane due to the differences of their chemical potentials in the two systems is given by the following equation:

$$\frac{dn_i{'}}{dt} = -\frac{dn_i{''}}{dt} = AJ_i$$

where n_i is the number of moles of the component i, $'$ and $''$ refer to the two systems separated by the membrane, J_i is the flux through the membrane from one system to the other, A is the membrane area and t is the time.

1.3 Osmotic equilibrium, osmotic pressure, osmosis and reverse osmosis

If two aqueous salt solutions of different concentrations are separated by a membrane that is permeable with the solvent, for example, water, but impermeable with the solute, for example, salt, a transport of water from the more dilute solution in the more concentrated solution is observed. This is the natural osmosis phenomenon.

Therefore, when a semipermeable membrane separates two solutions (the first one indicated by $'$, whereas the second one by $''$), three different situations can be distinguished depending on the concentrations and hydrostatic pressures in the two phases (Figure 1.7):

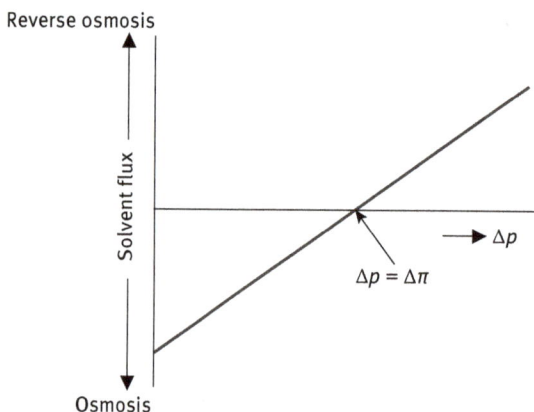

Figure 1.7: Solvent flux between two solutions of different concentrations through a strictly semipermeable membrane as a function of the hydrostatic pressure applied to the more concentrated solution.

a) Solution (1) and solution (2) have the same hydrostatic pressure but the solute concentration in solution (1) is higher than the one in solution (2). This situation is referred to as *osmosis* because there will be a flow of solvent from the more diluted solution (2) into the more concentrated solution (1) due to the higher osmotic pressure of solution (2).

b) The two solutions have different hydrostatic pressures; however, the difference in hydrostatic pressure is equal to the difference in the osmotic pressures between the two solutions acting in opposite directions. This situation is referred to as *osmotic equilibrium* and there will be no flow of solvent through the membrane, though the concentrations in the two solutions are different.

c) The two solutions have different hydrostatic pressures; however, the difference in hydrostatic pressure across the membrane is larger than that in the osmotic pressure and is acting in opposite directions. Thus, solvent will flow from the solution (1) with the higher solute concentration into the solution (2) with the lower solute concentration. This phenomenon is referred to as *RO*.

In order to allow the solvent (i.e., water) to pass through the membrane, the applied pressure Δp (between the concentrated side and the dilute side) must be higher than the osmotic pressure $\Delta \pi$.

The osmotic pressure of a solution π is proportional to the activity of the solvent in a solution (a_i^s) and is given by the following equation:

$$p^s - p^l = \pi = \frac{RT}{V_l} \ln a_i^s$$

For dilute solutions, the osmotic pressure can be expressed as follows:

$$\pi = RT \sum_i g_i C_i$$

where C_i is equal to the concentration of the individual components in the solution and g_i = van't Hoff coefficient.

1.4 Pressure-driven membrane operations

Pressure-driven membrane operations can be divided into four overlapping categories of increasing selectivity: (1) MF, (2) UF, (3) nanofiltration (NF) and (4) RO. In all four processes, a mixture of different components is brought to the surface of a semipermeable membrane; under the driving force gradient, some components permeate the membrane, whereas others are more or less retained. Thus, a feed solution is separated into a filtrate that is depleted of particles or molecules, and a retentate in which these components are concentrated.

As we move from MF through UF to NF and RO, the size (molecular weight) of the particles or molecules that are separated diminishes and, consequently, the pore size of the membrane becomes smaller. This implies that the resistance of the membranes to mass transfer increases and the applied pressure (which is the driving force) has to be increased to achieve the same flux.

The performance of a pressure-driven membrane separation process is determined by the filtration rate (membrane flux) and membrane separation properties. The separation capability of a membrane with respect to a component can be expressed in terms of *membrane rejection R_i* as follows:

$$R_i = \left(1 - \frac{C_i^p}{C_i^f}\right) \leq 1 \qquad (1.5)$$

where R_i is the rejection of the membrane for a given component i at a defined hydrostatic pressure, C is the concentration of component i, superscripts f and p refer to the feed and permeate or filtrate solutions, respectively.

For a simple filtration system similar to the one reported in Figure 1.8, the recovery ratio Δ is the ratio of the filtrate V^p to the feed volume V^f:

$$\Delta = \frac{V^p}{V^f} \qquad (1.6)$$

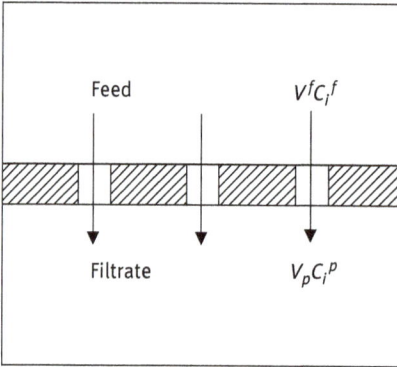

Figure 1.8: General filtration system.

Since in a practical application the concentration in the retentate r is limited by factors such as osmotic pressure and viscosity, the solvent of the feed solution cannot be completely recovered as filtrate. The recovery ratio has a value between 0 and 1.

When the system reported in Figure 1.8 is characterized by complete mixing, the membrane flux is identical over the entire membrane area, and the relation between the concentration of feed f, filtrate p and retentate r is given by the following mass balance:

$$V^f C_i^f = V^r C_i^r + V^p C_i^p \qquad (1.7)$$

Rearranging previous equations, the concentration of a component in the retentate and filtrate at a given recovery rate is obtained from the mass balance by introducing recovery ratio and rejection coefficient as follows:

$$C_i^r = C_i^f (1 - \Delta)^{-R} \qquad (1.8)$$

$$C_i^p = C_i^f (1 - R)(1 - \Delta)^{-R} \tag{1.9}$$

The filtrate concentration expressed in eq. (1.9) is the concentration corresponding to a given recovery rate, that is, during an infinitely small time interval.

1.5 Solute losses in membrane filtration processes

For membranes that are not strictly semipermeable, some solutes will permeate through the membranes with the filtrate. This may affect the quality of the filtrate, or lead to product losses.

The *fractional solute loss* δ of a component i is usually expressed by the amount of solute lost with the filtrate divided by the total amount of solute in the feed solution:

$$\delta = \frac{V^p \overline{C}_i^p}{V^f C_i^f} = 1 - (1 - \Delta)^{1-R} \tag{1.10}$$

Equation (1.10) shows that the solute lost can be significant even with high rejection membranes.

1.6 Microfiltration

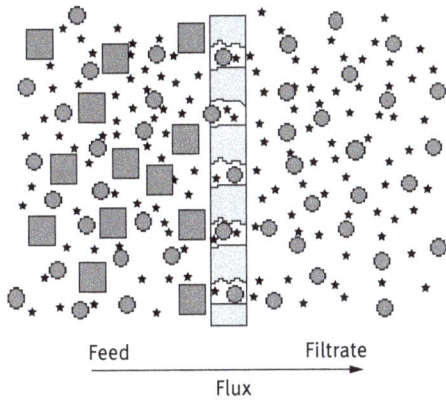

Feed   Filtrate

Flux

Figure 1.9: Schematic representation illustrating the principle of microfiltration. Adapted from Strathmann et al. [1].

The term MF is used when particles with a diameter of 0.1–10 µm are separated from a solvent or other low molecular components. The separation mechanism is based on a sieving effect, and particles are separated according to their dimensions. The membranes used for MF have pores of 0.1–10 micron in diameter. The hydrostatic pressure differences used are in the range of 0.05–0.2 MPa. Since only large particles are separated by the membrane (Figure 1.9), the diffusion of particles and the osmotic pressure difference between the feed and the filtrate solution are negligibly low and the mass flux through a MF membrane is given by the following equation:

$$J_v = \sum_i J_i \bar{V}_i \cong L_v \frac{\Delta p}{\Delta z} \qquad (1.11)$$

where J_v is the volumetric flux across the membrane, $\bar{V}$ is the partial molar volume, L_v is the hydrodynamic permeability of the membrane, Δz is membrane thickness, Δp is the pressure difference between the feed and filtrate solution and i refers to the component in the solution.

The mass transport in MF membranes takes place by viscous flow through the pores. If the membrane consists of straight capillaries, hydrodynamic permeability can be expressed in terms of membrane pore size, porosity and solution viscosity according to the following Hagen–Poiseuille's law:

$$J_v = \frac{\varepsilon r^2}{8\eta\tau} \frac{\Delta p}{\Delta z} \qquad (1.12)$$

where ε is the membrane porosity, r is the pore radius, η is the viscosity and τ is the tortuosity factor.

Tortuosity (Figure 1.10) is defined as the ratio of the actual pore length L to the thickness l of the membrane:

$$\tau = \frac{L}{l} \geq 1 \qquad (1.13)$$

Tortuosity is always ≥ 1 because pore length is, in general, longer than the cross-section of the membrane.

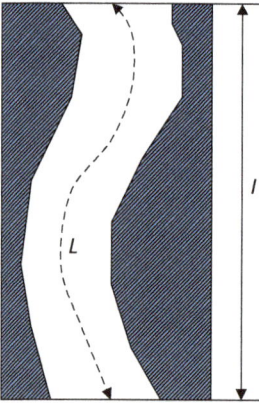

Figure 1.10: Schematic diagram of membrane tortuosity.

Porosity ε (Figure 1.11) is always ≤ 1 because it is defined as the total area of pores to the total area:

$$\varepsilon = \frac{\text{Total area of pores}}{\text{Total area}} \leq 1 \qquad (1.14)$$

Figure 1.11: Schematic representation of membrane porosity.

When a nodular structure exists, the following Kozeny–Carmen equation can be employed:

$$J = \frac{\varepsilon^3}{K\eta S^2}\frac{\Delta p}{\Delta x} \qquad (1.15)$$

where K is a constant that depends on the geometry of the pore, S is the superficial area of the spherical particles per unit volume and ε is the porosity. From eqs. (1.14) and (1.15), it appears that, in order to optimize MF, it is essential that the porosity should be as high as possible and a pore size distribution should be as narrow as possible. MF membrane can be prepared from a large number of different materials, based on organic materials (polymers) or inorganic materials (ceramics, metals and glasses).

Two modes of process operation exist: (1) dead-end (in which the feed flow is perpendicular to the membrane surface so that the retained particles accumulate and form a cake layer at the membrane surface) and (2) cross-flow filtration (in which the feed flow is along the membrane surface so that part of the retained solutes accumulate). Moreover, hydrophobic MF membranes were observed to be more prone to fouling than hydrophilic MF membranes, especially in the case of proteins, hydrophilic neutral and colloidal components of the natural organic matter (NOM).

1.7 Ultrafiltration

In UF, the components to be retained by the membrane are macromolecules or submicron particles. Generally, hydrostatic pressures of 0.1–0.5 MPa are used. UF membranes are asymmetric (Figure 1.12) having the smallest pores on the surface facing the feed solution, with pores in the skin layer having a diameter of 2–10 nm (significantly smaller than those of a MF membrane).

Since UF membranes also retain some relatively low molecular weight solutes, osmotic pressure differences between the feed and the filtrate can be significant and diffusive fluxes of the solutes across the membranes are no longer negligibly low.

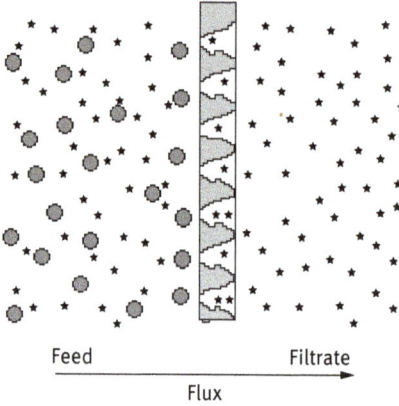

Feed   Filtrate

Flux

Figure 1.12: Schematic diagram illustrating the principle of ultrafiltration.

The flux of individual components in UF is given by the sum of the fluxes of the individual components and can be expressed as a function of the chemical potential gradient and hydrostatic pressure gradient:

$$J_v = \sum_i J_i \overline{V}_i = \sum_i \overline{V}_i L_i \frac{d\mu_i}{dz} + L_v \frac{dp}{dz} = \sum_i \overline{V}_i L_i \frac{d}{dz}\left(\overline{V}_i p + RT \ln a_i\right) + L_v \frac{dp}{dz} \qquad (1.16)$$

Here J is the flux, L is a phenomenological coefficient referring to interactions of the permeating components with the membrane matrix, $\overline{V}_i$ is the partial molar volume, μ is the chemical potential, p is the hydrostatic pressure, z is a directional coordinate, a is the activity and the subscripts v and i refer to volume flow and individual components, respectively.

The first term in eq. (1.16) describes the diffusive fluxes of all components in the pores of the membrane and the second term is the volume flow.

In UF the total volume flux, that is, the filtration rate in a dilute solution can be expressed to a first approximation by the flux of the solvent, that is, $J_v \cong J_w$, and the activity of the solvent in the solution a_w can be expressed by an osmotic pressure. Assuming a linear relation for the pressure and activity gradients across the membrane, integration of eq. (1.16) gives the flux through an UF membrane as a function of pressure difference between feed and permeate solution, the hydrodynamic permeability for the viscous flow, the osmotic pressure difference between feed and permeate solution and the phenomenological coefficient determining the diffusive flow of water through the membrane pores as follows:

$$J_v \cong J_w = \overline{V}_w^2 L_w \frac{\Delta p - \Delta \pi}{\Delta z} + L_v \frac{\Delta p}{\Delta z} \tag{1.17}$$

where J_v and J_w are total volume and solvent fluxes, respectively, Δp and $\Delta \pi$ are the hydrostatic and the osmotic pressure gradients across the membrane, L_v is the hydrodynamic permeability and L_w is the diffusive permeability of the solvent, Δz is the thickness of the selective barrier of the membrane.

In most practical applications of UF, the first term of eq. (1.17) can be neglected since $L_w \ll L_v$ and the filtration rate in UF can be described as follows:

$$J_v = L_v \frac{\Delta p}{\Delta z} \tag{1.18}$$

1.8 Nanofiltration

The separation properties of NF membrane (Figure 1.13) are determined, in general, by two distinct properties: 1) the pore size of the membranes, which corresponds to a molecular weight cutoff value of about 400 (±100) Da, and 2) the surface charge that can be positive or negative and affects the permeability of charged components such as salt ions.

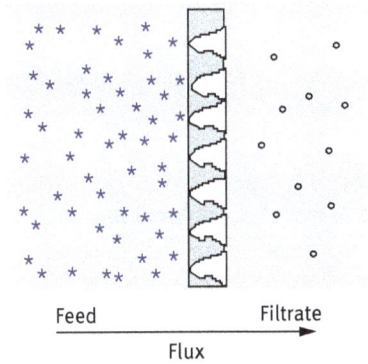

Feed   Filtrate

Flux

Figure 1.13: Schematic diagram illustrating the principle of nanofiltration. Adapted from Strathmann et al. [1].

The transport of individual component i-th (J_i) can be described as follows:

$$J_i = L_i \frac{d\mu_i}{dz} = L_i \frac{d}{dz} \left(\overline{V}_i p + RT \ln a_i \right) + L_v{}^m C_i \frac{dp}{dz}$$

where L_i is phenomenological coefficients referring to diffusion of the permeating components within the membrane matrix, L_v is the hydraulic permeability of the membrane, V is the partial molar volume, μ is the chemical potential, p is the hydrostatic pressure and a is the activity.

Therefore, the total volumetric flux (J_v) through the membrane is as follows:

$$J_v = \sum_i J_i \bar{V}_i = \sum_i \bar{V}_i L_i \frac{d}{dz}\left(\bar{V}_i p + RT \ln{}^m a_i\right) + L_v \frac{dp}{dz}$$

Expressing the phenomenological coefficient L_w by

$$L_w = \frac{D_w C_w}{RT}$$

and integration of the total volumetric flux (J_v) over the pore length and expressing the water activity gradient in the pores by the osmotic pressure difference between the feed and the filtrate leads to the following equation:

$$J_v \cong \frac{\bar{V}_w^2 L_w}{\Delta z}(\Delta p + \Delta \pi) + L_v \frac{\Delta p}{\Delta z} = \left(\frac{D_w \bar{V}_w}{RT} + L_v\right)\frac{\Delta p}{\Delta z} + \frac{D_w \bar{V}_w}{RT}\frac{\Delta \pi}{\Delta z}$$

For NF membranes with pore sizes in the range of ca. 1 nm the term $\frac{D_w \bar{V}_w}{RT}$ is of the same order of magnitude or larger than L_v. Thus, unlike in UF the effect of the osmotic pressure on the solvent flux cannot be neglected in NF.

If it is assumed that the solutions treated in NF are relatively dilute and that to a first approximation the activities of the individual components can be replaced by their concentrations, the transport of the individual J_i can be written as follows:

$$J_i = {}^m D_i \frac{d}{dz}\left(\frac{\bar{V}_i {}^m C_i}{RT} dp + d^m C_i\right) + L_v {}^m C_i \frac{dp}{dz}$$

where ${}^m D_i = L_i \frac{RT}{{}^m C_i}$

In NF, a partition coefficient correlates the concentration at the membrane interface ${}^m C_i$ to the concentration in the solution ${}^s C_i$ as follows:

$$ {}^m C_i = k_i {}^s C_i $$

However, if the membrane carries positive or negative electric charges at the surface, the partition coefficient for ionic components such as salt ions is not only determined by size exclusion but also by the so-called Donnan exclusion that postulates that ions carrying the same charge as the membrane, that is, the so-called coions, will be excluded from the membrane.

The number of electrical charges carried by all ions of an electrolyte under the driving force of an electrical potential gradient through a certain area A is given by the following equation:

$$J_e = \sum_i z_i u_i v_i C \ e \ N_A \ \Delta\varphi = \sum_i z_i F J_i$$

where J_e is the flux of electrical charges and J_i that of the individual ions; z, u and n are the charge number, the ion mobility and the stoichiometric coefficient, respectively; C is the concentration of the electrolyte; e the charge of an electron, N_A the Avogadro number, j is an electrical potential and the subscript i refers to the ions in the solution and F is the Faraday constant.

The flux of electrical charges represents an electrical current, which is a solution of a single electrolyte according to Ohms's law given by the following equation:

$$I = \frac{U}{R} = \sum_i z_i F J_i A = \frac{\sum_i z_i u_i v_i\, C\, F\, A\, \Delta\varphi}{l}$$

where I is the current, U is the applied voltage, R is the resistance, A is the area through which the current passes, Dj is the voltage difference between two points and l is the distance between the two points.

If the solution contains charged components, that is, ions, and the membrane is permeable for at least one ionic component, the membrane will be in equilibrium with the adjacent solution if the electrochemical potential of all ions in the membrane and the solution are equal. For each ion in equilibrium:

$$\tilde{\mu}_i^m = \tilde{\mu}_i^s = \mu_i^m + z_i F \varphi^m = \mu_i^s + z_i F \varphi^s$$

The electrochemical potential of an ion is composed of two additive terms, the first is the chemical potential and the second is the electrical potential multiplied by the Faraday constant and the valence of the ion as follows:

$$\varphi^m - \varphi^s = \frac{1}{z_i F}\left[RT \ln \frac{a_i^s}{a_i^m} +_i (p^s - p^m) \right] = \varphi_{Don}$$

where j is the electrical potential, a is the ion activity, V is the partial molar volume, z is the valence, F is the Faraday constant, p is the pressure, T is the absolute temperature and R is the gas constant; the subscript i refers to individual components and the superscripts m and s refer to the membrane and the electrolyte solution, respectively.

Introducing the osmotic pressure into the previous equation gives the Donnan potential as a function of the ion and the water activities in the membrane and the solution as follows:

$$\varphi_{Don} = \frac{1}{z_i F}\left[RT \ln \frac{a_i^s}{a_i^m} - \overline{V}_i\, \Delta\pi \right]$$

The Donnan equilibrium describes the electrochemical equilibrium of an ion in an ion-exchange membrane (IEM) and an adjacent solution and can be calculated for a single electrolyte as follows:

$$\left(\frac{a_a^s}{a_a^m}\right)^{\frac{1}{z_a}}\left(\frac{a_c^m}{a_c^s}\right)^{\frac{1}{z_c}} = e^{-\frac{\Delta\pi\,\overline{V}_s}{RTz_c\overline{V}_c}}$$

where z is the valence, n is the stoichiometric coefficient of the electrolyte, F is the Faraday constant, R is the gas constant, T is the absolute temperature, V is the partial molar volume and a is the activity; the subscripts a and c refer to anion and cation, respectively and the superscripts s and m refer to the solution and the membrane, respectively.

The Donnan potential between an ion-exchange membrane and a dilute electrolyte solution is given to a first approximation by the following equation:

$$\varphi_{Don} = \sum_i \frac{1}{z_i F}\left[RT \ln \frac{{}^sC_i}{{}^mC_i}\right]$$

where φ_{Don} is the Donnan potential, ${}^sC_i$ and ${}^mC_i$ are the concentrations of an ion in the solution and the membrane, respectively.

The exclusion of coions in a dilute solution of a single monovalent electrolyte is given to a first approximation by the following equation:

$$^mC_{co} = \frac{{}^sC_s^{\,2}}{C_{fix}}$$

where ${}^mC_{co}$, ${}^sCs$ and C_{fix} are the coion concentration in the membrane, the electrolyte concentration in the solution and the fixed-ion concentration of the membrane, respectively.

As a result of the Donnan exclusion, the partition coefficient for a component between a NF membrane carrying positive or negative fixed charges and an electrolyte solution depends on two parameters: one is the size exclusion (k_{size}) and the other is the Donnan exclusion (k_{Don}). It is as follows:

$$k_i = k_{size}k_{Don}$$

The total value of the partition coefficient is always $0 \le k \le 1$.

Therefore, the flux of individual components through a NF membrane containing fixed positive or negative charges is given by the following equation:

$$J_i = {}^mD_i\left[\left(\frac{\overline{V}_i k_{size}k_{Don}{}^sC_i}{RT}\frac{dp}{dz} + k_{size}k_{Don}\frac{d^sC_i}{dz}\right) + \frac{z_i F k_{size}k_{Don}{}^sC_i}{RT}\frac{d\varphi_{Don}}{dz}\right] + L_v k_{size}k_{Don}{}^sC_i\frac{dp}{dz}$$

Since concentrations are different at feed and permeate side of NF membrane, an electrical potential difference across the membrane is established which affects the transport of charged components through the membrane.

The consequence of the additional driving force of the Donnan potential difference between two solutions separated by a NF membrane is that components with the same electrical charge such as mono- and divalent cations or anions can be separated when their diffusivity in the membrane is different.

1.9 Reverse osmosis

The principle of the RO is to force a solvent through the molecular structure of a membrane while trapping impurities and salts. To obtain this, the applied pressure Δp (between the concentrated side and the dilute side) must be higher than the osmotic pressure $\Delta \pi$.

As reported by Jonsson and Macedonio in [2], several models on RO transport mechanisms have been developed to describe solute and solvent fluxes through RO membranes. The general purpose of a membrane mass transfer model is to relate the fluxes to the operating conditions.

The solution–diffusion model assumes that the (i) membrane surface layer is homogeneous and nonporous and (ii) both solute and solvent dissolve in the surface layer and then they diffuse across it independently. Water and solute fluxes are proportional to their chemical potential gradient. The latter is expressed as the pressure and concentration difference across the membrane for the solvent, whereas it is assumed to be equal to the solute concentration difference across the membrane for the solute. Thus, the flux of the component i can be described as follows:

$$J_i = L_i d\mu_i = L_i(\overline{V}_i dp + RTd \ln a_i) = \frac{D_i{}^m C_i}{RT}(\overline{V}_i dp + RTd \ln a_i)$$

where J is the flux, L is phenomenological coefficient, μ is the chemical potential, p is the hydrostatic pressure, a is the activity, $\overline{V}$ is the partial molar volume, D is the diffusion coefficient and $^m C_i$ is the concentration in the membrane.

It is assumed that for the solvent ($i=w$), $J_v \cong J_w$ and $C_w \overline{V}_w = 1$ and for the solute ($i=s$), $d\mu_s \cong RTd \ln a_s$ and $\pi \cong \frac{RT \ln a_w}{\overline{V}_w}$.

Integrating over the cross-section of the membrane and expressing the activity of the solvent by the osmotic pressure and the activity of the solute by its concentration the fluxes can be written as follows:
- For the solvent $J_v = L_w \overline{V}_w^2(\Delta p - \Delta \pi)$
- for the solute $J_s = D_s{}^m C_s d\ln{}^m C_s = D_s d^m C_s$

By introducing the partition coefficient k_i: $^m C_i = k_i C_i$, the fluxes can be written as follows:

$$J_v = \frac{k_w D_w C_w \overline{V}_w^2}{RT}(\Delta p - \Delta \pi) = \frac{k_w D_w \overline{V}_w}{RT}(\Delta p - \Delta \pi)$$

$$J_s = k_s D_s\left(C_s^f - C_s^p\right)$$

RO membranes have an asymmetric structure, with a thin dense top layer (thickness ≤ 1 μm) supported by a porous sublayer (thickness in the range of 50–150 μm). The selectively permeable layer is reduced to a very fine skin in order to limit the resistance to the transfer related to the thickness of the layer. The porous thicker sublayer has much larger pores that intends to provide the membrane with satisfactory mechanical properties without significantly impeding the water flow.

On the basis of the internal structure, there are two main types of asymmetric membranes for NF/RO: (1) asymmetric homogeneous membranes and (2) composite membranes.

– In asymmetric homogeneous membranes, both top layer and sublayer consist of the same material. Cellulose esters (especially cellulose diacetate and triacetate) were the first commercially used materials. Unfortunately, these materials have poor chemical stability and tend to hydrolyze over time depending on temperature and pH operating conditions. They are also subjected to biological degradation. Other materials that are frequently used for RO membranes are aromatic polyamides, polybenzimidazoles, polybenzimidazolones, polymidehydrazide and polyimides [3]. Polyimides can be used over a wider pH range. The main drawback of polyamides (or of polymers with an amide group –NH–CO in general) is their susceptibility against free chlorine Cl_2 that causes degradation of the amide group.
– Composite membranes are made by assembling two distinct parts composed of different polymeric materials. Composite membranes can combine various materials and provide optimum properties depending on their use.

Despite major earlier breakthroughs such as the Loeb–Sourirajan asymmetric membrane (1960s), fully cross-linked aromatic thin film composite (TFC) membrane (1970s to 1980s) and controlling morphological changes by monitoring polymerization reactions (1990s), the evolutionary improvement of a commercial RO membrane has been rather slow during the first decade of this century. One motivation for this is that the development of thin-film composite membranes with selectivity higher than the existing RO commercial membrane modules (between 99.40% and 99.80%) is difficult. This is a direct consequence of the separation mechanism of thin-film composite membranes, where increasing selectivity to allow higher removal of ions will substantially reduce the membrane permeability and will increase energy consumption. Developing RO membranes with higher selectivity without sacrificing water permeability will necessitate a major paradigm shift, as it will require membranes that do not follow the solution–diffusion mechanism. Various nanostructured RO membranes have been proposed that offer attractive characteristics and that could possibly bring revolutionary advancements (example e.g., mixed matrix membranes, biomimetic RO membranes, carbon nanotubes and ceramic/inorganic membranes).

Ceramic/inorganic membranes
The interest in ceramic membranes is due, in particular, to their robustness. Ceramic membranes are mostly made from alumina, silica, titania, zirconia or any mixture of these materials. Because of the high manufacturing cost, their use is currently limited to applications where polymeric membranes cannot be used (i.e., high operating temperatures, radioactive/heavily contaminated feeds and highly reactive environments [4]).

Mixed matrix membranes (MMM)

The concept of MMM with the combination of organic and inorganic material is not new; however, it started in 1980 in the field of GS. The main objective of MMM is to combine the benefits offered by each material, that is, the high packing density, good perms-electivity and long operational experience of polymeric membranes, coupled with the superior chemical, biological and thermal stability of inorganic membranes [5].

Carbon nanotubes (CNTs)

CNTs have been studied extensively, especially in the past 20 years, owing to their broad range of applications [6]. Experimental results by Holt et al. [7] showed that the flow rate of water through CNTs is three orders of magnitude higher than that predicted from no-slip hydrodynamic flow by the Hagen–Poiseuille equation. When the pore size is less than 20Å, the permeability is higher than that of conventional polycarbonate membranes [8]. The main drawback is that CNTs must be aligned and a proper alignment of CNTs in a membrane skin layer is very difficult to be achieved.

Biomimetic RO membranes

Aquaporins (AQPs) are water conducting channels found in biological membranes and have a unique hourglass architecture with a "pore opening" of 2.8Å; the narrow pore prevents the passage of large molecules [9]. Membranes incorporating bacterial AQP Z proteins have been reported to show at least an order of magnitude improvement in permeability compared to commercially available TFC RO membranes [10]. Many practical issues, such as identification of appropriate support materials, understanding of the resistance to membrane fouling and even identification of an appropriate range of operating conditions must be carried out to develop this membrane for practical use.

1.10 Limiting factors: concentration polarization and fouling phenomena

Real membrane processes are limited by concentration polarization and fouling. These phenomena strongly reduce the performance of membrane operations because they decrease mass flux and/or separation performance, that is, salt rejection. Consequently, their control is one of the major problem in the design of membrane systems.

1.10.1 Concentration polarization

When in a mass separation procedure, a molecular mixture is brought to a membrane surface, some components will permeate the membrane under a given driving force while others are retained. This leads to an accumulation of retained material and to a depletion of permeating components in the boundary layers adjacent to the membrane surface. Thus, a concentration gradient between the solution at the membrane surface and the bulk is established that leads to a back transport of the material accumulated at the membrane surface by diffusion. This phenomenon is referred to as concentration polarization.

A typical concentration profile is shown in Figure 1.14.

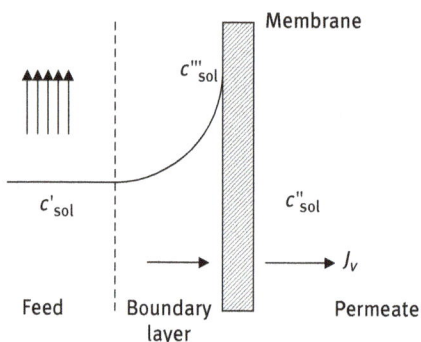

Figure 1.14: Concentration profile. Adapted from Strathmann et al. [1].

Concentration polarization can be minimized by hydrodynamic means such as the feed flow velocity and the membrane module design. The causes and the consequences of concentration polarization as well as necessary means to control it depend on the feed water composition and on the membrane process.

Concentration polarization complicates the modeling of membrane systems because experimental calculation of the wall concentration is difficult. For high feed flow rates, it has often been assumed that the wall concentration is equal to the bulk concentration because of high mixing, which is however seldom the case. At low flow rates, this assumption is certainly no more applicable and can cause substantial errors. To estimate the extent of concentration polarization, the following film theory is the most well-used technique [11, 12]:

$$\frac{c'''_{sol} - c''_{sol}}{c'_{sol} - c''_{sol}} = \exp\left(\frac{J_v}{k}\right)$$

In the above-mentioned equation, k denotes the mass transfer coefficient that can be estimated using a Sherwood correlation such as the following derived by Gekas and Hallstrom [13].

$$Sh = 0.023 Re^{0.8} Sc^{0.33} \text{ for turbulent flow}$$

$$Sh = 1.86 \cdot (Re \cdot Sc \cdot d_h/L)^{0.33} \text{ for laminar flow.}$$

Under conditions of precipitation layer formation at the membrane surface, the membrane flux can only be increased by a decrease of the boundary layer thickness or bulk solution concentration. The experimental results show that in the case of precipitation of solute at the membrane surface, the hydrodynamic resistance for membrane flux is not only a function of the membrane properties but also it is strongly affected by the gel or cake layer, which is generally formed by the retained solutes at the membrane surface.

A simple approach to describe the membrane flux in case of a gel or a cake layer formation is to assume that the osmotic pressure of the feed solution in MF and UF can be neglected, and expressing the flux in terms of the resistances of the membrane and the layer in series according to the following relation:

$$J_v = A \frac{1}{R_m + r_1 \Delta z_1} \Delta p$$

where J_v is the membrane flux, R_m is the hydrodynamic resistance of the membrane, r_1 is the specific resistance of the layer, D_{z1} is the thickness of the layer, D_p the hydrostatic pressure driving force and A is the membrane area.

As above-mentioned described, under conditions of precipitation of retained components at the membrane surface, the membrane flux can be increased by decreasing the thickness of the boundary layer, or decreasing the concentration of the bulk solution as can be seen from the following equation that describes the mass transport in the laminar boundary layer in a filtration device with turbulent bulk flow:

$$\frac{C_s^w}{C_s^b} = \exp \frac{J_v Z_b}{D_s}$$

where, for simplicity, it is assumed that the dissolved components are completely retained by the membrane, that is, $R=1$. Moreover, C_s^b and C_s^w are the solute concentrations in the bulk solution and at the membrane surface, which is identical to the gel layer concentration C_s^g, and which is constant for a given temperature and pressure.

1.10.2 Fouling

Membrane fouling is a consequence of adsorption or deposition of constituents of feed solution at the membrane surface and also within the membrane structure.

The transition between concentration polarization and fouling can be expressed by the concept of "critical flux." When operating below the critical flux, a linear correlation between flux and transmembrane pressure can be observed. Above this, further increase in transmembrane pressures leads to deposits of additional layers on the membrane surface, until a point where the deposit fully compensates the increase in pressure drop. At this stage, the limiting flux is reached, which represents the maximum stationary permeation flux. After operating above the critical flux value, decreasing the transmembrane pressure will not lead to the previous flux behavior. Critical flux depends on numerous parameters such as properties of solutions to be treated and hydrodynamics conditions.

The means of preventing or at least controlling membrane fouling effects are as heterogeneous as different materials and mechanisms that causes fouling. The main procedures to avoid or control fouling involves the following:

–   Pretreatment of the feed solution
–   Membrane surface modifications
–   Hydrodynamic optimization of the membrane module
–   Membrane cleaning with the proper chemical agents

1.11 Operation modes in filtration processes

Three different operation modes can be utilized in filtration: (1) batch process, (2) continuous process and (3) feed and bleed process.

In a batch process, under a hydrostatic pressure certain components permeate through the membrane and are collected as filtrate. The components retained by the membrane are concentrated. When a certain concentration in the retentate is achieved, the process is terminated.

On the contrary, in a continuous process the solution is continuously fed into the filtration device. The retained components are concentrated in the process through a device that leaves at the end of the process path as the retentate.

In a feed and bleed process, part of the retentate is recycled in the device inlet and mixed with the feed solution. This operation mode is used when the recovery rate achieved in one process path is not satisfactory and a higher recovery rate needs to be obtained. If the concentration of partly retained components in the filtrate exceeds the desired maximum value in the given process then part of the filtrate may be recycled to the feed inlet.

Moreover, membrane processes can be carried out by a single-pass configuration or double-pass arrangements.

In a single-pass configuration, one or more membrane modules are installed in parallel to provide the filtrate with the required characteristics.

In a double-pass operation, each stage is fed by the rejection of the previous stage. This arrangement is shown in Figure 1.15.

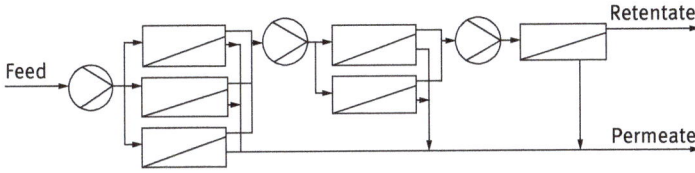

Figure 1.15: Double-pass process.

1.12 Case study: membrane operations in seawater and brackish water desalination

Desalination is a technology that converts saline water into clean water. It offers one of the most important solutions to water shortage problem. There has been a rapid growth in the installation of seawater desalination facilities in the past decade for augmenting water supply in water-stressed countries. Notable examples are the large-scale seawater reverse osmosis (SWRO) desalination plants recently constructed in Israel (such as the Sorek SWRO desalination plant), the United States (such as the Carlsbad Desalination SWRO Plant in San Diego County), Oman (such as the Al Ghubrah plant or the Barka independent water and power project (IWPP) expansion – both SWRO), United Arab Emirates (e.g., the Al Fujairah IWPP expansion). According to the International Desalination Association [14], in the first half of 2016, the amount of global contracted capacity in desalination plants was 95.59 million m^3/d and the global online capacity was 88.56 million m^3/d, a total increase of 2.1 million m^3/d in new desalination capacity over 2015. In terms of technology, the general trend is the adoption of membrane over thermal technologies, and this became more acute from 2000 to 2016. The large diffusion of RO desalination plants was, in particular, due to their lower capital costs from less-expensive construction materials, their versatility in feedwater and application and stabilization in the price of produced desalted water (Table 1.3).

As above described, membrane fouling is one of the major problems of RO. It can never fully be prevented but it can be reduced and controlled through an adequate pretreatment of the feed solution. Pressure-driven membrane operations (such as MF and UF) are the new trends in designing pretreatment systems because they can handle a large variation in raw water quality and still produce water for the RO unit that is of better quality than water produced by conventional technology. Membrane pretreatment systems are also more compact and have lower operating costs than conventional processes.

Microfiltration use include, among other things, sterilization, clarification and the treatment of oily wastewaters (because of the high oil removal efficiency, low energy cost and compact design of MF compared with traditional treatments such as mechanical separation, filtration and chemical de-emulsification). In general, MF is utilized for the elimination of particulates with particle sizes in the submicrometer range [16].

Table 1.3: Main advantages and drawbacks of the most widespread thermal and nonthermal desalination systems. From Brunetti et al. [15] (Open Access Article).

Desalination processes	Advantages	Drawbacks
Multistage flash distillation (MSF)	– Ease of the process – No risk of reduced heat transfer by scaling since heat exchange with the saline water does not occur through heat transfer surfaces – MSF is also insensitive to the initial feed concentrations and to the presence of suspended particles – Desalted water contains about 50 ppm of total dissolved salts.	The most important disadvantages of MSF are that the top brine temperature is limited to about 110° C with the risk of scaling, thus limiting the performance ratio at about 11. This results in a much higher energy consumption, which makes MSF a more expensive technique than MED. Precipitation can be reduced by applying acid or anti-scalants.
Multieffect distillation (MED)	– Reliable design, technological maturity, high quality of distillate produced, good operating records and high unit capacity are the main merits of MED technology. – Cogeneration desalination plants that include MED combined with thermal or absorption heat pumps, and waste-heat steam generators or gas turbines show promising performance.	The main problems are related to corrosion and scaling of oversaturated compounds (such as $CaSO_4$); the gain output ratio is generally very high.
Reverse osmosis	– High recovery factor, low energy consumption, low investment and total water cost are the main advantages of RO technology. – Relatively small footprint and modularity enabling easy adaptation of process scale.	Some of the major problems in RO applications are fouling and concentration polarization phenomena. Concentration polarization can be minimized by hydrodynamic means (such as an appropriate feed flow velocity, an adequate membrane module design and spacer as also turbulence promoters). Fouling can be reduced and controlled through proper pretreatment of the feed solution.

Table 1.3 (continued)

Desalination processes	Advantages	Drawbacks
Electrodialysis	ED is competitive for brackish waters with up to 3000 ppm of salt, whereas it is rarely used for seawater desalination. For water with low salt concentrations, ED/EDR is considered to be the most advantageous technique.	ED process is noneconomical for waters with high salt concentrations.
Membrane distillation (MD)	The main MD advantages are as follows: – Theoretically complete rejection of nonvolatile components – Low operating temperature with respect to the distillation separation, with consequent possibility to utilize low-grade waste heat streams and/or alternative energy sources (solar, wind or geothermal) – Low operating pressure, lower equipment costs and increased process safety – Robust membranes – High system compactness – Less membrane fouling – Extremely low sensitivity to concentration polarization phenomenon.	Temperature polarization, membrane wetting, development of proper membranes and modules for MD application are the main drawbacks of MD.

Membranes for UF have been developed and proven for many years in a wide range of applications, such as highly polluted municipal and industrial wastewaters. In recent years, UF has also been considered in seawater desalination installations, especially when treating surface seawater and for retrofit upgrades to existing conventional RO pretreatment systems. Nanofiltration is a type of pressure-driven membrane operation that has properties in between those of UF and RO. NF membranes have relatively high charge and are typically characterized by lower rejection of monovalent ions than that of RO membranes; however, maintaining high rejection of divalent ions. NF membranes have been employed in pretreatment unit operations in both thermal and membrane seawater desalination processes, for softening brackish and seawaters as well as in membrane-mediated wastewater reclamation and other industrial separations. RO is usually used to separate dissolved salts and ions. Its applications range from the production of ultrapure water for semiconductor and pharmaceutical use to the desalination of seawater for drinking water production and the purification of industrial wastewater.

ED has been in commercial use for desalination of brackish water for the past three decades, particularly for small- and medium-scale processes. The ED process is not economical for waters with high salt concentrations but is competitive for brackish waters with up to 3000 ppm salt. Later, capacitive deionization (CDI) and membrane capacity deionization (MCDI) generated a lot of interest. A CDI cycle consists of two steps: the first being an ion electrosorption, or charging, step to purify water, where ions are immobilized in porous carbon electrode pairs. In the second step, ions are released, that is, are desorbed from the electrodes, and thus the electrodes are regenerated [17]. CDI is suitable for desalination of brackish water. One of the most promising recent developments in CDI is to include IEMs in front of the electrodes, called MCDI. IEMs can be placed in front of both electrodes, or just in front of one. The inclusion of IEMs in the cell design significantly improves the desalination performance of the CDI process.

Recently, the thermally driven membrane process called MD has gained popularity because of some unique benefits associated with the process, such as the possibility to concentrate the seawater till its saturation point without any significant flux decline, to utilize low-grade waste heat streams and/or alternative energy sources (solar, wind or geothermal) and to theoretically reject 100% nonvolatile components [18]. Owing to these attractive benefits, MD might become one of the most interesting desalination techniques. It can overcome not only the limits of thermal systems (such as distillation) but also those of membrane systems (such as RO). Concentration polarization does not affect the driving force of the process significantly and therefore high recovery factors and high concentrations can be achieved in the operation, compared with the RO process. All the other properties of membrane systems (easy scale-up, easy remote control and automation, no chemicals, low environmental impact, high productivity/size ratio, high productivity/weight ratio, high simplicity in operation and flexibility) are also present.

Membrane Crystallization (MCr) is conceived as an alternative technology for producing crystals and pure water from supersatured solutions; the use of the MD technique in the concentration of a solution by solvent removal in the vapor phase is utilized in this application. Owing to their low energy requirement, MD/MCr coupled with solar energy, geothermal energy or waste heat can achieve cost and energy efficiency.

If as single units MC operations are more efficient than corresponding traditional unit operations (i.e., water recovery from conventional thermal desalination processes is not more than 40%), their combination into existing water treatment processes generates important synergistic effects and improves the overall process efficiency. For example, water recoveries as high as 76.6%–88.9% were achieved in a brackish water desalination system constituted by pretreatment/RO/wind evaporation/MCr, whereas less than 0.75%–0.27% of the raw brackish water fed to the plant was discharged to the environment [19].

References

[1]   Strathmann, H, L. Giorno, and E. Drioli. "An Introduction to Membrane Science and Technology," Roma: CNR, 2006. ISBN 88-8080-063-9.

[2]   G. Jonsson, F. Macedonio. Fundamentals in Reverse Osmosis, Chapter 2.01 in: Comprehensive Membrane Science and Engineering – Vol. 2. Editor: Enrico Drioli and Lidietta Giorno. Elsevier B.V, 2010.

[3]   Marcel Mulder. Basic Principles of Membrane Technology, 1996 Kluwer Academic Publishers, London.

[4]   C.A.M, Siskens. Applications of Ceramic Membranes in Liquid Filtration, Chapter 13 in: Membrane Science and Technology, A.J. Burggraaf, L. Cot (Eds.). Elsevier, 1996, pp.619–639.

[5]   A.F. Ismail, P.S. Goh, S.M. Sanip, M. Aziz. Transport and separation properties of carbon nanotube-mixed matrix membrane. Sep. Purif. Technol. 2012.

[6]   LeDuc Y, Michau M, Gilles A, Gence V, Legrand Y-M, et al. Imidazole-quartet water and proton dipolar channels. Angew. Chem. Int. Ed. 2011; 50:11366–72.

[7]   Peter C, Hummer G. Ion transport through membrane-spanning nanopores studied by molecular dynamics simulations and continuum electrostatics calculations. Biophys. J. 2005;89:2222–34.

[8]   B.-H. Jeong, E.M.V. Hoek, Y. Yan, A. Subramani, X. Huang, G. Hurwitz, A.K. Ghosh, A. Jawor. Interfacial polymerization of thin film nanocomposites: A new concept for reverse osmosis membranes. J. Membr. Sci. (2007); 294: 1–7.

[9]   F. Macedonio, E. Drioli, A. A. Gusev, A. Bardow, R. Semiat, M. Kurihara. Efficient technologies for worldwide clean water supply. Chemical Engineering and Processing: Process Intensification, 51 (2012) 2–17.

[10] M. Kumar, M. Grzelakowski, J. Zilles, M. Clark, W. Meier. Highly permeable polymeric membranes based on the incorporation of the functional water channel protein aquaporin Z. PNAS. (2007); 104:20719–20724.

[11] Bhattacharyya, D. and Williams, M. E. Reverse Osmosis. Chapter VI in: Membrane Handbook, W. S. Winston Ho and K. K. Sirkar (Eds.). Springer. 1992.

[12] Fritzmann, C. Löwenberg, J. Wintgens, T. Melin, T. State-of-the-art of reverse osmosis desalination. Desalination. 2007; 216: pp 1–76.

[13] Gekas, V. and Hallstrom, B. Mass transfer in the membrane concentration polarization layer under turbulent cross flow. J. Memb. Sci. 1987; 30: pp 153.

[14] IDA Desalination Yearbook 2016–2017. Published by Media Analytics Ltd., United Kingdom. ISBN: 978-I-907467-49-3.

[15] Brunetti, A., Macedonio, F., Barbieri, G., & Drioli, E. Membrane engineering for environmental protection and sustainable industrial growth: Options for water and gas treatment. Environ. Eng. Res. (2015);20(4): 307–328.

[16] Drioli, Enrico, and Francesca Macedonio. "Membrane engineering for water engineering." Ind. Eng. Chem. Res. (2012); 51(30): 10051–10056.

[17] Porada, S., Zhao, R., Van Der Wal, A., Presser, V., & Biesheuvel, P. M. Review on the science and technology of water desalination by capacitive deionization. Prog. Mater. Sci. (2013); 58(8): 1388–1442. http://dx.doi.org/10.1016/j.pmatsci.2013.03.005. Article published under the terms of Creative Commons Attribution-NonCommercial-No Derivatives License (CC BY NC ND).

[18] Drioli E, Aamer A, and Macedonio F. Membrane distillation: Recent developments and perspectives. Desalination. 2015; 356: 56–84.

[19] Macedonio, F., Katzir, L., Geisma, N., Simone, S., Drioli, E., & Gilron, J. Wind-Aided Intensified eVaporation (WAIV) and Membrane Crystallizer (MCr) integrated brackish water desalination process: Advantages and drawbacks. Desalination. (2011); 273(1):127–135.

Juntae Jung, Enrico Drioli

2 Pervaporation

2.1 Introduction

Pervaporation (PV) is one of the most efficient, commercialized membrane process for liquid/liquid separation. Unlike solid–liquid separation, a porous membrane is no longer needed, and nonporous membranes are required for the separation. The separation occurs from the difference in transport of species through the solution and diffusion mechanism. Hence, the intrinsic properties play a crucial role in the PV process. It is often compared with the membrane distillation process due to similarity in the process; however, the underlying concept and separation behavior is completely dissimilar. In the first part of this chapter, an overview of membrane transport mechanism in PV (solution-diffusion mechanism) will be covered from thermodynamic and kinetic perspectives. In the following part, the membrane properties required for PV and applications will be described.

2.2 Definition of PV process

2.2.1 The common membrane transport mechanism

PV is a membrane process where the permeation of certain components through a membrane from a pure liquid or a liquid feed mixture into a vapor phase is combined with the evaporation. Generally, the feed side consists of a flow of liquid mixture, whereas the vacuum or low vapor pressure gas phase environment is set at the permeate side. The feed and permeate are separated by the membrane, and the permeates are removed in the form of vapor due to a low vapor pressure in the permeated side. The PV process seems similar to membrane distillation in the sense that the liquid phase turns into a gas phase; however, the separation of components in PV is determined not only by the difference in vapor pressure but also by the permeation rate through the dense membrane. The membranes and transport mechanism of a component (through the dense membranes) used in PV are the same as used in gas separation. The driving force for the transport is the chemical potential gradient of the permeating components in the membrane. The chemical potential gradient in the membrane can be related to partial

Juntae Jung, Department of Energy Engineering, Hanyang University, Seoul, South Korea
Enrico Drioli, National Research Council of Italy, Institute for Membrane Technology (ITM-CNR), University of Calabria, Rende, Italy

https://doi.org/10.1515/9783110281392-002

vapor pressures in the liquid and vapor phases. The mass transport in PV can be expressed as follows:

$$J_i = -D_i k_i \frac{X_i^p \varphi p^p - X_i^f \gamma_i^f p_i^f}{\Delta z} \qquad (2.1)$$

where D_i is the diffusion coefficient of component i in the membrane, k_i is the sorption coefficient of component i into the matrix, X is the molar fraction, d is the fugacity, p is the pressure, γ is the activity coefficient and the subscriptions f and p refer to saturation pressure feed and permeate, respectively.

The product of diffusion and distribution coefficient can be presented as a permeability coefficient (eq. 2.2), which is commonly used to evaluate the PV performance. Both membrane selectivity and the separation factor are based on the permeability coefficient and indicate the PV performance, however, and focus of each parameter is rather dissimilar. The membrane selectivity for the two components incorporates the ratio of the permeability coefficients or permeance of two components (eq. 2.3), while the separation factor is determined by the weight fraction of components in the permeate and feed (eq. 2.4):

$$D_i k_i = P_i \qquad (2.2)$$

$$S_{j,k} = \frac{P_j}{P_k} = \frac{P_{j/l}}{P_{k/l}} \qquad (2.3)$$

$$\alpha_{j/k} = \frac{y_j/y_k}{x_j/x_k} \qquad (2.4)$$

$$\alpha_{j/k} = \frac{x_k}{x_j} \times \frac{J_j}{J_k} = \frac{x_k}{x_j} \times \frac{D_i k_i \frac{X_i^p \varphi p^p - X_i^f \gamma_i^f p_i^f}{\Delta z}}{D_j k_j \frac{X_j^p \varphi p^p - X_j^f \gamma_j^f p_j^f}{\Delta z}} = \frac{x_k}{x_j} \times S_{j,k} \times \frac{X_i^f \varphi p^f}{X_j^f \varphi p^f} \qquad (2.5)$$

where P refers to the permeability of the membrane, S refers to the membrane selectivity, α refers to the separation factor, y and x refer to the weight fraction of components in the permeate and feed and j and k refer to the components of the mixture.

As presented earlier, the permeability coefficient mainly attests to the separation performance in the membrane, whereas the flux incorporates the activity coefficient and saturated vapor pressure that is determined by the process parameter. Likewise, assuming permeate vapor pressure is negligible, one can derive the relation between the selectivity and separation parameter (2.5) by combining eqs. (2.1) and (2.4). These relations suggest that using permeance and selectivity in lieu of flux and separation factor well clarifies the attributes of the membranes in the separation process. On the other hand, flux and separation factor well reflect the effect of processing conditions on separation process, such as temperature.

The mass transport in a PV membrane can be described by the same mathematical relations used in the gas transport, when it is assumed that the chemical potential in the membrane on the feed side of the membrane is expressed by the molar fraction and the activity coefficient in the liquid phase. Contrary to gas separation, the separation factor achieved in PV is a function not only of the membrane properties but also of the vapor pressure of the different components in the mixture. In this regard, assuming constant thermodynamic factors, selectivity is used here to elucidate the transport of a component in the PV process. The total selectivity is the sum of two terms, that is, the evaporation selectivity that is a function of vapor pressure of the components and the permeation selectivity that is a function of the membrane properties:

$$S_{j,k}^{\text{total}} = S_{j,k}^{\text{perm}} \times S_{j,k}^{\text{evap}} \tag{2.6}$$

where $S_{j,k}^{\text{total}}$, $S_{j,k}^{\text{perm}}$ and $S_{j,k}^{\text{evap}}$ are the total, the permeation and the evaporation selectivity of a membrane for the components j and k.

The evaporation selectivity is a thermodynamic parameter and the permeation selectivity is determined by the membrane properties. The permeability and the evaporation selectivity can be both positive or both negative, or one might be positive and the other negative. It means the fraction of transported vapor can vary depending on the membrane property, even in the same evaporation environment. For instance, Figure 2.1 shows the permeate composition as a function of the feed solution composition of a

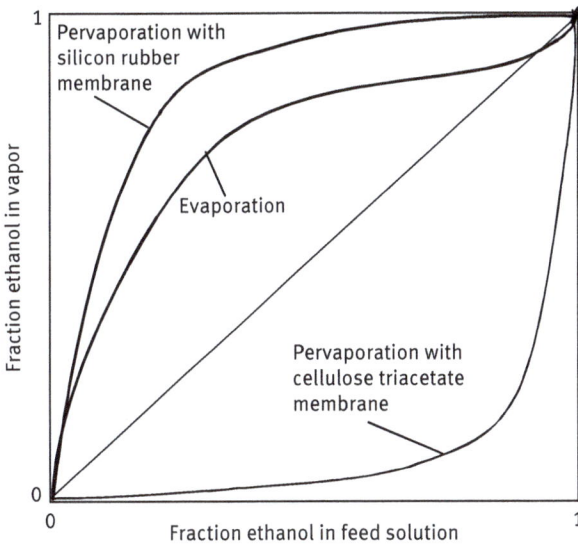

Figure 2.1: A schematic diagram of transported ethanol in PV with two different membranes. From Reprinted with permission from An introduction to membrane science and technology / CNR. Institute on Membrane Technology (ITM); Heiner Strathmann, Lidietta Giorno, Enrico Drioli. Roma: CNR, 2006. ISBN 88-8080-063-9. Pag. 122 [1].

water–ethanol mixture for a PV process using a silicon rubber membrane and a cellulose triacetate membrane. It can be observed that the silicon rubber membrane has a preferred ethanol permselectivity, whereas the cellulose triacetate membrane has a preferred water permeability, that is, negative ethanol selectivity.

As in other membrane processes, it is also desirable that the minor component in the mixture permeates the membrane in the PV process. Therefore, it is crucial to select optimum component/membrane combination considering the membrane transport; in this case, the silicon rubber membrane rather than cellulose triacetate is more appropriate to treat ethanol in low ethanol-concentrated feed stream, concentrating ethanol at the permeate stream.

2.3 Membranes in pervaporation

As described earlier, similar to the gas separation process, each component is transported by solution-diffusion mechanism, which means the selectivity arises from the nonporous, dense layer of the membranes rather than size-sieving effect by macrosize pores. For providing low mass transport resistance, the asymmetric membranes containing dense layer with a porous substrate are preferred over symmetric membranes. As shown in Figure 2.2, it was investigated the effect of selective layer thickness on the permeation flux and selectivity in the PV process using the two kinds of membranes – a symmetric dense membrane with a selective layer thickness around 140 μm and an asymmetric dense membrane with selective layer thickness around 10 μm. The flux from the thick selective layer thickness exhibited one-order difference compared to asymmetric membrane. However, no difference was observed in terms of selectivity, which proves that the flux is a function of thickness and the selectivity is the ratio of permeability of the membranes as discussed in the previous section.

For this reason, membrane manufacturers provide end users with an asymmetric or a composite membrane. The asymmetric and composite membranes have the same characteristics, where both have the thin selective layer for PV; however, the substrate for the membrane is different. The asymmetric membrane is composed of only single material throughout the membrane, whereas the composite membrane contains polymeric fibers or metal wires as the porous substrate. The asymmetric membrane has an advantage that the manufacturing method of the membrane is simple; however, the dimensional stability of an asymmetric membrane in an organic solvent is rather low due to inevitable swelling by the organic solvent. On the other hand, manufacturing composite membrane usually includes a series of processes for substrate fabrication and coating for the selective layer. In addition to membrane morphology, membrane materials can be classified into two classes depending on the target component preferentially permeated. If the target component is water, the membrane should separate water from an aqueous feed solution by selectively permeating water through the membrane. Accordingly, hydrophilic materials such as crosslinked poly(vinyl

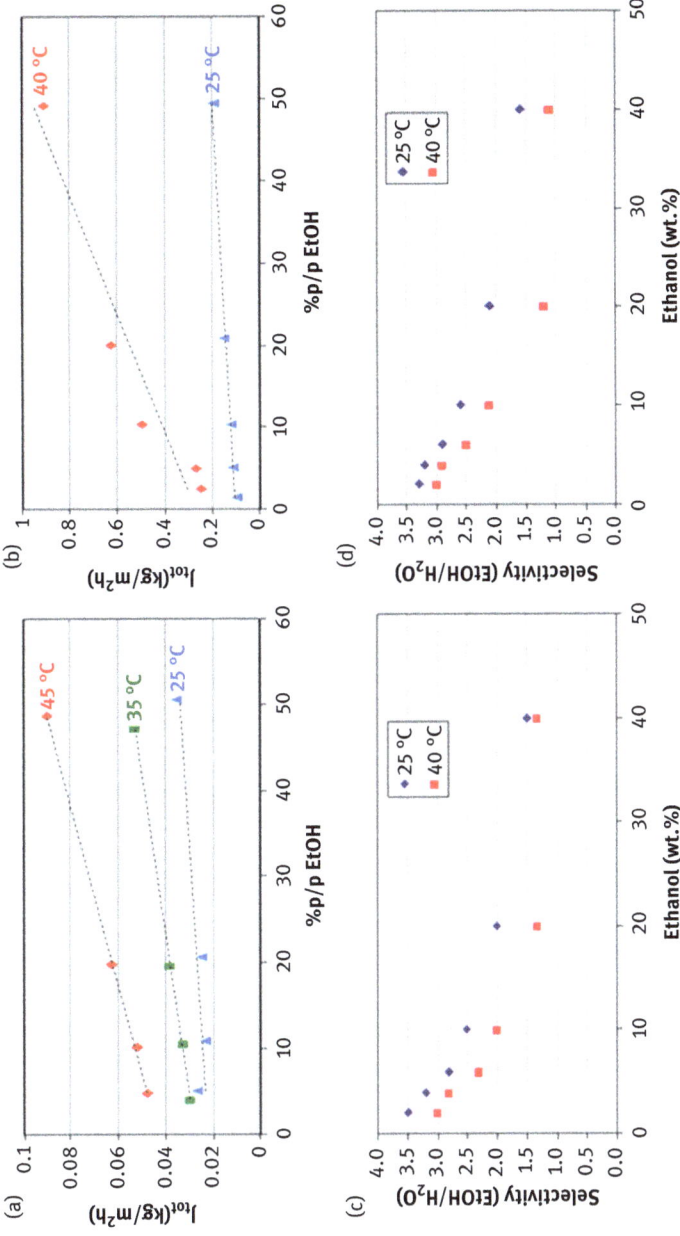

Figure 2.2: Effect of membrane morphology on permeation flux and selectivity in PV process: (a) and (c) These refer to the permeation flux and selectivity from a symmetric dense membrane (thickness around 140 μm), respectively; (b) and (d) These refer to the permeation flux and selectivity from an asymmetric dense membrane (thickness around 10 μm), respectively.

alcohol), sulfonated poly(sulfone), polyamide and poly(acrylonitrile) are commonly applied to the membranes. On the other hand, if the target component is an organic compound, organophilic materials such as poly(sulfone), poly(ether sulfone), poly (propylene), poly(carbonate), poly(vinylidene fluoride), poly(dimethylsiloxane) and styrene-butadien-*co*-styrene [2] are usually employed for the membrane preparation. In addition to the organic material, an inorganic material such as zeolite group mineral – NaY zeolite, mordenite and ferrierite, is commonly used for PV as well. The inorganic materials attain very high fluxes and selectivity as well as good thermal and chemical resistance; however, the reduction of cost remains as a challenge.

2.4 Industrial applications of pervaporation

PV is a combined separation process, and the separation kinetic is highly dependent on the thermodynamic parameters such as composition, temperature and chemical affinity between the components and the membranes. In terms of energy consumption, it is more desirable to separate or remove the minor component in the feed mixture. Therefore, the material of the membrane in the PV membrane process should be selected by taking into consideration the minor component in the feed mixture. For instance, if the minor compound in the mixture is water, hydrophilic PV using a hydrophilic membrane should be applied. Likewise, when the mixture contains a small amount of organic compound in an aqueous mixture, the hydrophobic PV using a hydrophobic membrane is more desired. The selectivity from the membrane material causes a large difference in the separation factor, particularly when compared to the distillation separation performance. For example, the separation factor of isopropanol/ water mixture (90/10 wt.%) is around 2; however, a PV process can provide the separation factor around 2,000–1,000 [3, 4]. For this reason, it is accepted that the latent heat is only consumed by the minor component in the mixture in the PV process; hence, the PV process has more advantages in terms of energy efficiency. Furthermore, this superior separation factor becomes attractive when the separation by distillation is not favorable in the cases such as removing very small amount of a minor component and separation of azeotropes. Owing to these advantages, the PV process is being actively applied in the industrial process, and a classification in the PV process and corresponding industrial applications are presented in Figure 2.3 and Table 2.1.

2.4.1 Hydrophilic pervaporation

The main industrial applications of PV is organic solvent dehydration. The most common example remains the production of anhydrous ethanol for the pharmaceutical industry. More than 50 plants have been installed for the hydration of ethanol by PV, mainly commercialized by Sulzer Chemtech formerly known as GFT (polymeric

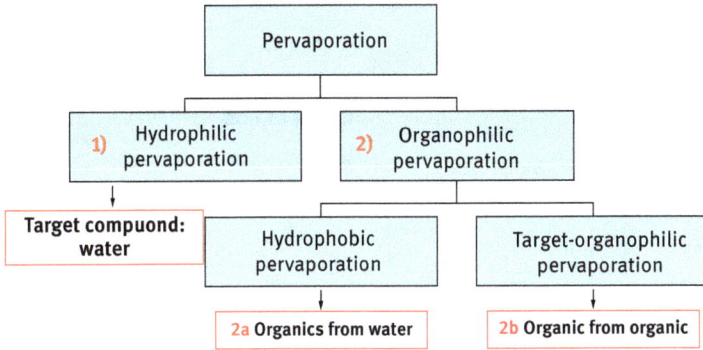

Figure 2.3: A schematic diagram of applications using the PV process.

Table 2.1: Examples of industrial applications in the PV process.

Membrane	Composition	Application
	A: Hydrophilic PV	
PERVAP 2210 (Sulzer)	PVA (poly(vinyl alcohol)) lightly crosslinked/PAN (polyacrylonitrile)	Final dehydration of alcohols
PERVAP 2510	PVA specially crosslinked/PAN	Dehydration of isopropanol and ethanol
	2a: Hydrophobic PV	
PERVAP1070 (Sulzer)	Zeolite filled PDMS (polydimethylsiloxane)/PAN support	Wastewater phenol removal [5]
PEBA (GKSS)	PEBA (Poly (ether-block-amide))/porous support	Extraction of phenols from aqueous effluents
	2b: Target organophilic PV	
PERVAP2256 1 (Sulzer)	Confidential	Separation (metOH/MTBE)
PERVAP 2256 2	Confidential	Separation ethOH/ETBE

PVA membrane). The PV process is efficient in terms of separation factor; however, the overall production rate can be retarded by the presence of the membrane when compared to the distillation process. Therefore, the PV process is not very advantageous if the entire separation is carried out solely by PV. In fact, the practical separation is mostly carried out by distillation, and PV is applied to break the azeotrope composition. Therefore, the closer the composition of azeotrope to the pure component, more advantages PV have. In this respect, the production of anhydrous ethanol for the pharmaceutical industry is a favorable application for PV because ethanol exhibits an azeotrope with water at 95% and a 99.5% of ethanol. Most of these plants are made by an integration of distillation and PV, as illustrated in Figure 2.4.

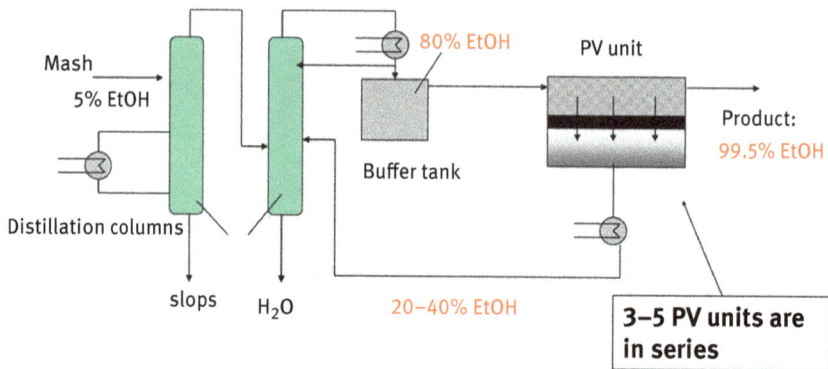

Figure 2.4: A schematic diagram of an integrated distillation/PV plant for ethanol recovery from fermentation.

As discussed earlier, PV is advantageous when the separation should be of a high-purity grade. Sulzer Chemtech applies the PV unit to feed containing 80/20 wt.% EtOH/water mixture. Since the minor (target) component is water, the hydrophilic membranes made of crosslinked poly(vinyl alcohol) were applied. To maximize the permeation flux, the membrane has a membrane thickness of about 250 μm; the thickness of the separation layer (selective layer) is less than 5 μm. The water permeation rate is about 600 g/m^{2}h, and the selectivity is about 200. In the case of SepraTek, the feed is 95/5 wt.% IPA/water mixture and the hollow fiber composite membrane has a selective layer thickness of less than 5 μm. Therefore, the hollow fiber membrane system allows water permeation rate of about 900 g/m^{2}h, with a separation factor of 850–1,200 depending on the feed temperature. The details of the PV unit and the used membranes are illustrated in Figure 2.5.

In addition to organic membranes, a commercial zeolite membrane was also developed and commercialized as well. Zeolites structures such as FAU, MFI and LTA are commonly used in PV process. An example is the hydrophilic tubular zeolite NaA (LTA structure) membranes developed by Mitsui Engineering & Shipbuilding. NaA zeolite crystals were synthesized hydrothermally on the surface of a porous tubular support. The membrane was highly selective for permeating water preferentially with a high permeation flux. The first large-scale PV plant produced 530 L/h of solvents (EtOH, IPA, MtOH, etc.) at less than 0.2 wt.% of water from 90/10 solvent/water mixture at 120 °C, with a permeation flux of 1–4 (kg/m^{2}h) and a separation factor as high as 10,000 [6]. The plant was equipped with 16 modules, each of which consists of 125 pieces of NaA zeolite membrane tubes.

SepraTek™ Inc. (http://www.sepratek.com/home/en/) and Jiangsu Nine Heaven Hi-Tech Co. Ltd (http://www.9t-tech.com/en/) are two companies manufacturing PV membranes. The novel technologies of SepraTek™ make it possible to fabricate a hollow fiber form of PV composite membranes, utilizing the significant benefits of

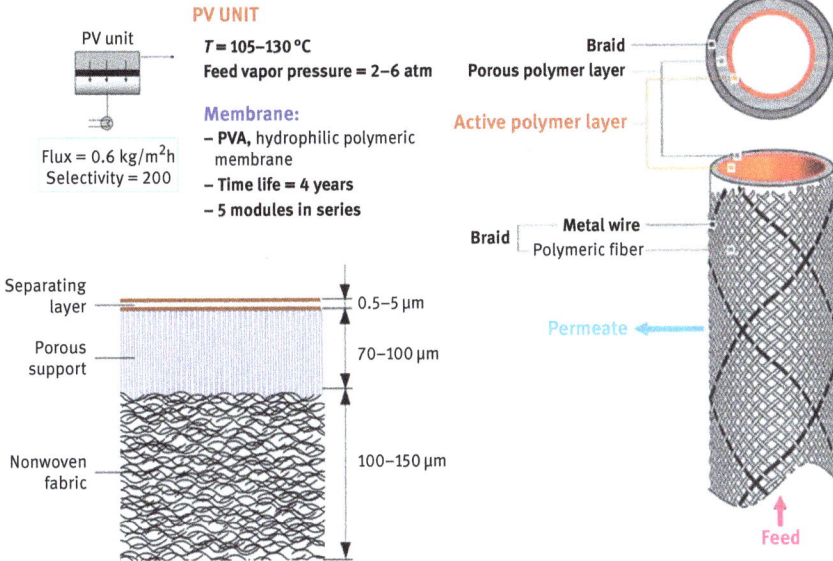

Figure 2.5: A schematic diagram of the PV plant unit and morphology of the membrane used in the PV processes (SepraTek catalog). Reprinted with permission.

hollow fiber configuration: high membrane packing density, low fabrication cost, flexibility in membrane module design and so on. SepraTek™ hollow fiber composite membranes have a braid fabric structure of reinforcement, a porous layer and an active layer. The braid fabrics are fabricated by braiding polymeric fibers and metal wires together, so SepraTek™ hollow fiber composite membranes reinforced by the braid possess an excellent pressure resistance as well as a good dimensional stability in organic solvents even at a high temperature.

Jiangsu Nine Heaven Hi-Tech Co. Ltd has world-leading NaA zeolite PV membrane. It owns more than 40 patents on membrane manufacturing and applications. The company first industrialized the technology of PV dehydration using zeolite membranes in China by 2009, and over 100 industrial plants were built ever since. The company is building a world-oriented R&D platform for PV membranes and an international technology exchange center, making it a global leader and the locomotive of the development of the PV industry. The pore size of NaA zeolite membrane is 0.42 nm, which is larger than a water molecule (~2.9 A) and less than most of organic molecular diameters. Owing to the molecular sieving properties and strong hydrophilicity of NaA zeolite membranes, water could be preferentially separated from organic solvents. Compared to organic membranes, the zeolite membranes have several advantages including higher permeation flux, higher separation factor and better thermal/chemical stability.

Figure 2.6: First large-scale PV plant made with zeolite membranes. From Morigami et al. [6]. Reprinted with permission.

2.4.2 Hydrophobic pervaporation

Volatile organic compounds (VOCs) are organic chemicals whose vapor pressure is high at room temperature. Due to this attribute, VOCs should be treated well to prevent human health and the environment from potential risks. For this reason, VOCs are usually removed from wastestreams by stripping (air stripping) or adsorption on adsorbents (i.e., activated carbon). However, removal of VOCs by air stripping does not enable to recover VOCs but includes direct discharge to the atmosphere; in addition, VOCs recovery via adsorption or adsorbents does not guarantee high energy efficiency. PV can be considered as an alternative to achieve the removal and recovery of VOCs in the 50–150's ppm range and up concentrating by a factor of 10–7,000 times or more, permitting recovery in a concentrated form for recycle and reuse or disposal.

Since VOCs are the major contaminants in industrial and ground water, needs for removing or recycling VOCs with energy efficient and economical technologies become higher [7, 8]. In particular, attention for the treatment of the chlorinated hydrocarbon that can cause negative impact on the ozone layer and birth defect in organisms is getting higher. Table 2.2 summarizes the examples of PV process for removing 1,1,1-trichloroethane. PDMS, Silicone, SBS and SBS/ceramic composite

Table 2.2: The summary of PV process for removing 1,1,1-trichloroethane.

Membrane	Type of VOC	Conc. (ppm)	Permeation flux (g/m^{2}h)	Selectivity (VOC/H$_2$O)	Ref.
PDMS comm	1,1,1-Trichloroethane	250	16	1,000	[9]
PDMS	1,1,1-Trichloroethane	160	16.5	2,300	[10]
Silicone/AC	1,1,1-Trichloroethane	300	19	2,700	[7]
SBS/PTFE	1,1,1-Trichloroethane	450	26	3,000	[5]
SBS/ceramic	1,1,1-Trichloroethane	45–50	5–9	>10,000	[8]
SBS	1,1,1-Trichloroethane	130	18	~5,000	[2]

membranes were chosen for the membrane material. Unlike carbon adsorption and air stripping is only economical when the feed VOC concentration is below 100 ppm, PV can be efficient at a wider VOC concentration range.

In addition to chlorinated hydrocarbon, PV can be applied to process or concentrate the aroma compounds that are very sensitive to heat treatment. During conventional juice concentration, both physical and chemical losses of aroma compounds occur to a great extent, as a result of evaporation and chemical alteration. This leads to a decreased quality of the final product. An example is shown in Figure 2.7, where the flow sheet of an integrated membrane process for kiwifruit juice production is shown. It was demonstrated that, by introducing PV to the integrated process, the aroma compounds could be removed from the kiwifruit juice and fed back later to the concentrated juice, resulting in an improved organoleptic quality [11, 12].

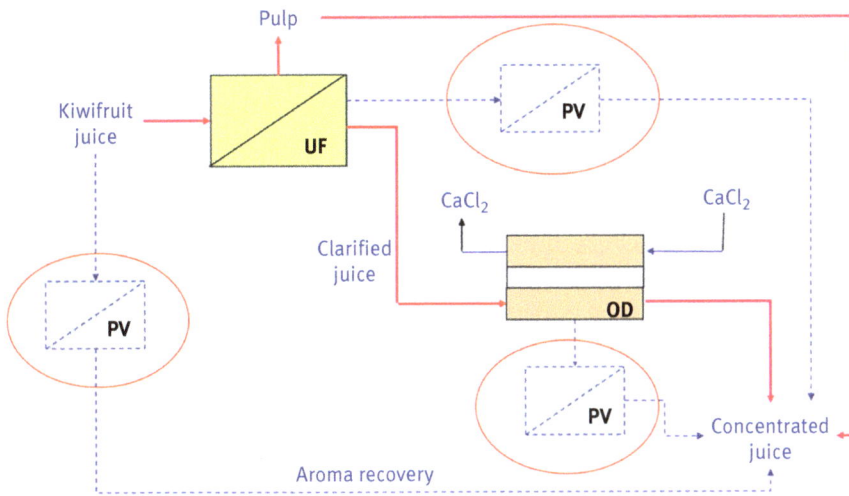

Figure 2.7: A schematic diagram of integrated membrane process in the production of kiwifruit juice. From Cassano et al. [11]. Reprinted with permission.

The aroma complex of kiwifruit is a highly volatile fraction of principally esters, alcohols, aldehydes and ketones. Flavor profile of samples coming from ultrafiltration, osmotic distillation and PV treatment was determined by gas chromatography–mass spectrometry. As expected, the PV process provided the best aroma selectivity.

2.4.3 Target organophilic pervaporation

The last application field of PV is the organophilic PV. The feed for this application only comprise organic/organic chemical mixture. As always, distillation is the biggest competitive technology in this organic/organic separation. However, PV can be more attractive at organic/organic separation where multiple azeotropes exist and close boiling point among the chemicals is common. Nevertheless, the commercial use of PV for organic/organic separation is rather subdued, which represents one of the remaining challenges for the development of PV technology in the industry. The main problem regarding the development of commercial systems for organic/organic separations is lack of membranes and modules able to withstand long-term exposure to organic compounds at elevated temperatures required for PV; in addition, the existing membranes exhibit only moderate selectivity because the difference in sorption between different organic molecules is very low.

In spite of these limitations, applications of PV for organic/organic separation have been studied to replace chemicals for the protection of the environment [13]. Particularly, it is applied to the separation of methanol from an isobutene/MTBE (methyl tert-butyl ether) mixture and purification of ETBE (ethyl tert-butyl ether). MTBE and ETBE can be used as a high-octane fuel additive, which is comparably nontoxic and nonpolluting in contrast to the formerly used lead alkyl additives. In fact, as shown in Figure 2.8, the series of separation process include the synthesis of methyl ester using PV to extract methanol in a continuous way and it is commercialized by Sulzer. (It ensures a production of 8.5 t per day in the fine chemical industry.)

At the same time, studies on the development of organic solvent-resistant membrane were carried out to reduce environmental impact. In general, the membranes with high chemical resistance cause higher environmental impact upon disposal. One of the example is the use of PEEK-WC, which is an amorphous modified poly (ether ether ketone) with cardo group. PEEK-WC has been found useful as a membrane material due to its good stability, biocompatibility and permeability. PEEKWC/PVP blend membranes have been successfully employed in PV separation of EtOH/CH_x and MetOH/MTBE azeotropic mixture [14]. PEEKWC is inherently selective to EtOH/MetOH; it can be applied to MTBE purification. However, it was not free from the low biodegradability issue. On the other hand, the use of polylactic acid or polylactide (PLA) that is a thermoplastic, aliphatic polyester derived from renewable resources, such as corn starch (in the US) or sugarcanes (rest of world), is rather irrelevant to the issue. The use of PLA, which has been known for more than a

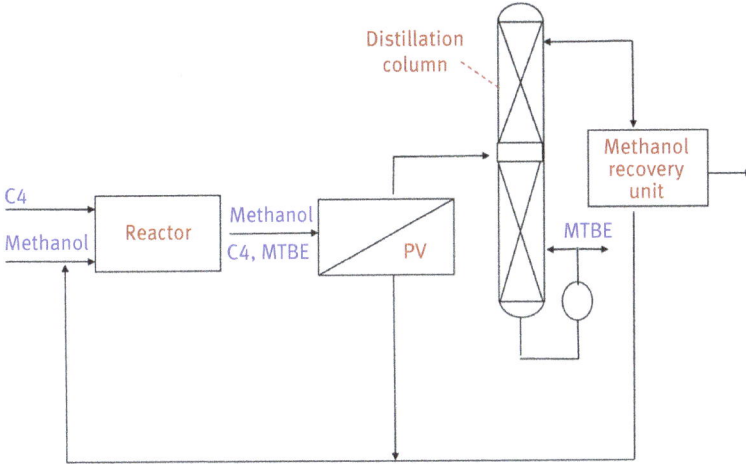

Figure 2.8: A hybrid system of synthesis of methyl ester using PV, commercialized by Sulzer.

century, is of commercial interest in recent years, in light of its biodegradability. Zereshki et al. [15] have shown the possibility of preparing organic solvent-resistant membrane using PLA in PV application. They have successfully demonstrated the use of PLA/PVP blend membranes in the separation of $EtOH/CH_x$ azeotropic mixture using PV. It has been found to be resistant to alcohols (EtOH, MeOH, etc.) and aliphatic and cyclic hydrocarbons (ETBE, MTBE, cyclohexane, etc.).

In addition, due to the advantages in cost reduction of PV, PV is applied to applications such as the purification of dimethyl carbonate that forms an azeotropic mixture containing almost 70 wt.% of methanol, aromatic/aliphatic mixtures in refining crude oils, and removal of butanol from the acetone–butanol–ethanol fermentation process. For instance, purification of dimethyl carbonate that forms an azeotropic mixture containing almost 70 wt.% of methanol is treated with PV. PV coupled with distillation enables to simply break the azeotrope and inject the corresponding mixture on a lower distillation plate, resulting in 60% cost reduction in the capital cost of each process.

2.5 Conclusion

Energy consumption is a major factor in the separation process. The PV process has the advantage that only latent heat is required for the evaporation of the permeating component. However, as discussed earlier, the economic application of PV is highly dependent upon the efficiencies of the membranes developed for PV applications. In fact, distillation and air stripping are more economic for a certain separation, which allows PV to be used in only a small scale. Although PV has a discriminative merit that it can easily break the azeotropes and extract minor component in the mixture, systematical approaches over increasing energy efficiency by changing the nature of the membranes or optimize the module design are still required for the increase of PV in the market share.

References

[1] Strathmann, H., Giorno, L., & Drioli, E. An introduction to membrane science and technology. Roma: CNR, 2006. ISBN 88-8080-063-9.
[2] Sikdar, S. K., Burckle, J. O., Dutta, B. K., Figoli, A., & Drioli, E. Method for fabrication of elastomeric asymmetric membranes from hydrophobic polymers, Google Patents, 2013.
[3] Huang, R., Pal, R., & Moon, G. Characteristics of sodium alginate membranes for the pervaporation dehydration of ethanol–water and isopropanol–water mixtures. J. Membr. Sci. 1999; 160:101–113.
[4] Huang, R., & Rhim, J. Separation characteristics of pervaporation membrane separation processes, Pervaporation Membrane Separation Processes, Elsevier, Amsterdam, 1991,111–180.
[5] Dutta, B. K., & Sikdar, S. K. Separation of volatile organic compounds from aqueous solutions by pervaporation using S– B– S block copolymer membranes. Environ. Sci. Technol. 1999; 33:1709–1716.
[6] Morigami, Y., Kondo, M., Abe, J., Kita, H., & Okamoto, K. The first large-scale pervaporation plant using tubular-type module with zeolite NaA membrane. Sep. Purif. Technol. 2001; 25:251–260.
[7] Ji, W., Sikdar, S. K., & Hwang, S.-T. Modeling of multicomponent pervaporation for removal of volatile organic compounds from water. J. Membr. Sci. 1994; 93:1–19.
[8] Ganapathi-Desai, S., & Sikdar, S. K. A polymer-ceramic composite membrane for recovering volatile organic compounds from wastewaters by pervaporation. Clean Products and Processes. 2000; 2:140–148.
[9] Visvanathan, C., Basu, B., & Mora, J. C. Separation of volatile organic compounds by pervaporation for a binary compound combination: trichloroethylene and 1, 1, 1-trichloroethane. Ind. Eng. Chem. Res. 1995; 34:3956–3962.
[10] Ji, W., Sikdar, S. K., & Hwang, S.-T. Sorption, diffusion and permeation of 1, 1, 1-trichloroethane through adsorbent-filled polymeric membranes. J. Membr. sci. 1995; 103:243–255.
[11] Cassano, A., Figoli, A., Tagarelli, A., Sindona, G., & Drioli, E. Integrated membrane process for the production of highly nutritional kiwifruit juice, Desalination. 2006; 189:21–30.
[12] Figoli, A., Tagarelli, A., Cavaliere, B., Voci, C., Sindona, G., Sikdar, S., et al. Evaluation of pervaporation process of kiwifruit juice by SPME-GC/Ion Trap Mass Spectrometry. Desalination. 2010; 250:1113–1117.
[13] Hömmerich, U., & Rautenbach, R. Design and optimization of combined pervaporation/ distillation processes for the production of MTBE. J. Membr. Sci. 1998; 146:53–64.
[14] Zereshki, S., Figoli, A., Madaeni, S., Simone, S., Esmailinezhad, M., & Drioli, E. Pervaporation separation of MeOH/MTBE mixtures with modified PEEK membrane: Effect of operating conditions. J. Membr. Sci. 2011; 371:1–9.
[15] Zereshki, S., Figoli, A., Madaeni, S., Simone, S., & Drioli, E. Pervaporation separation of methanol/methyl tert-butyl ether with poly (lactic acid) membranes. J. Appl. Polym. Sci. 2010; 118:1364–1371.

Enrica Fontananova, Enrico Drioli

3 Ion-exchange membranes

3.1 Fundamentals of ion-exchange membranes

Ion-exchange membranes (IEMs) are charge-selective membranes characterized by the presence of ionizable groups linked by covalent bonds with the polymeric membrane matrix. Ions having the same charge of these ionizable fixed groups (or coions) are theoretically impermeable through the IEM by electrostatic repulsion (Donnan or Gibbs–Donnan effect), while ions with the opposite charge (or counterions) can be transported through the IEM under the effect of an electrochemical potential [1].

IEMs can be classified as cation-exchange membranes (CEMs), containing negatively charged fixed groups (e.g., $-SO_3^{3-}$, $-COO^-$, $-PO_3^{2-}$, $-AsO_3^{2-}$, $-SeO_3^-$), anion-exchange membranes (AEMs), characterized by positively charged fixed groups (e.g., $-NR_3^+$, $-NHR_2^+$, $-NH_2R^+$, $-PR_3^+$, $-PS_2^+$) and bipolar membranes, composed of laminated anion- and cation-exchange layers (Figure 3.1).

In general, the permselectivity of a membrane is defined as the ratio of the flux of specific components to the total mass flux through the membrane under the effect of a given driving force. In the case of an IEM, the permselectivity is related to the selective transport of electric charges by the counterions (Figure 3.2).

Considering the IEM's structure, it is possible to categorize the IEM as homogeneous and heterogeneous membranes. In a homogeneous IEM, the fixed charged groups are evenly distributed over the entire membrane polymer matrix (Figure 3.1(a) and (b)). Heterogeneous IEMs are instead composed of finely powdered ion-exchange material and a binder polymer [2] defining distinct conductive and nonconductive domains (Figure 3.3). The transport of the ions occurs only through the conductive material.

IEMs are successfully used in several electromembrane processes coupling mass transport with an electrical current transport through ion permselective membranes.

Electromembrane process plays an important role in addressing environmental and energy issues in three main areas:
- the de-ionization of salt solutions (e.g., by electrodialysis [ED]),
- the electrochemical synthesis of inorganic and organic compounds (e.g., chlorine–alkaline process),
- the conversion of chemical into electrical energy (e.g., by fuel cells and reverse electrodialysis [RED]).

Moreover, it is important to mention that IEMs have been contributing successfully in the past and continue to contribute today in the development of membrane engineering,

Enrica Fontananova, Enrico Drioli, National Research Council of Italy, Institute for Membrane Technology (ITM-CNR), University of Calabria, Rende, Italy

https://doi.org/10.1515/9783110281392-003

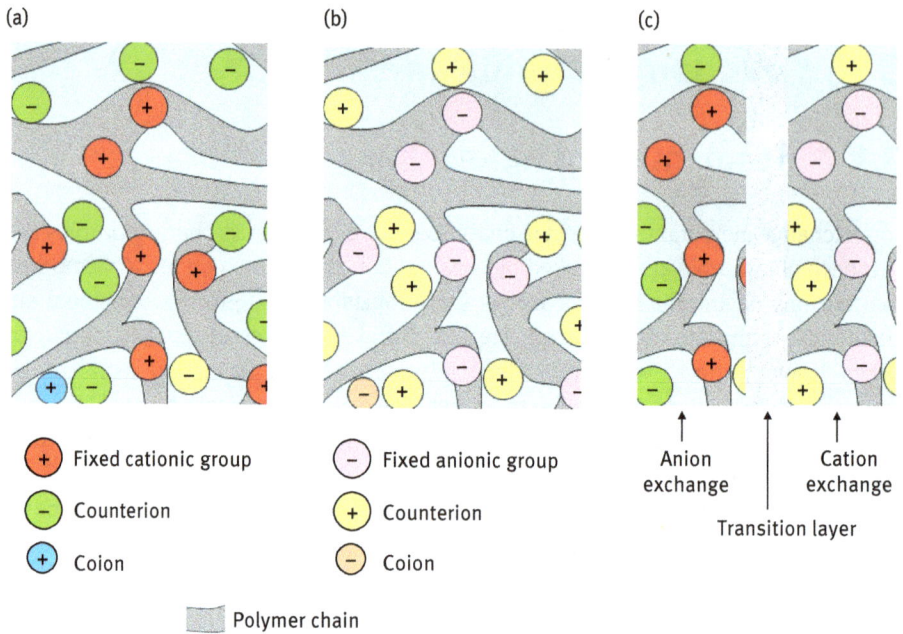

Figure 3.1: Schematization of (a) an AEM; (b) a CEM and (c) a bipolar membrane.

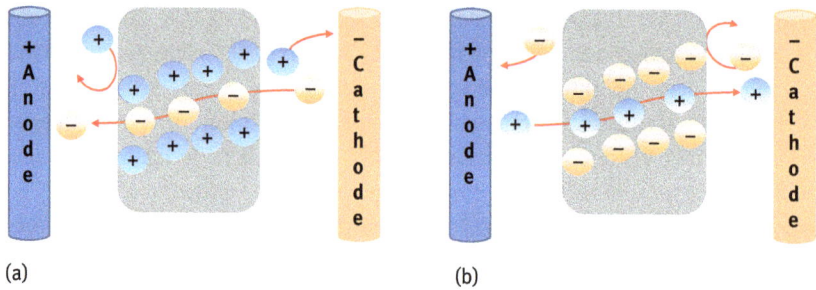

Figure 3.2: Representation of the selective transport of ions through (a) an AEM and (b) a CEM under the effect of an electrochemical potential.

being used in various industrial applications such as chlor-soda processes, blue energy production (with reverse electrodialysis), electrochemical specific sensors, and so on. Details about reverse electrodialysis are discussed in Chapter 9.

The main demanding requirements for IEM applications in electromembrane processes are as follows:

- high permselectivity,
- high ionic conductivity (low ionic resistance),
- good mechanical, chemical and thermal stability.

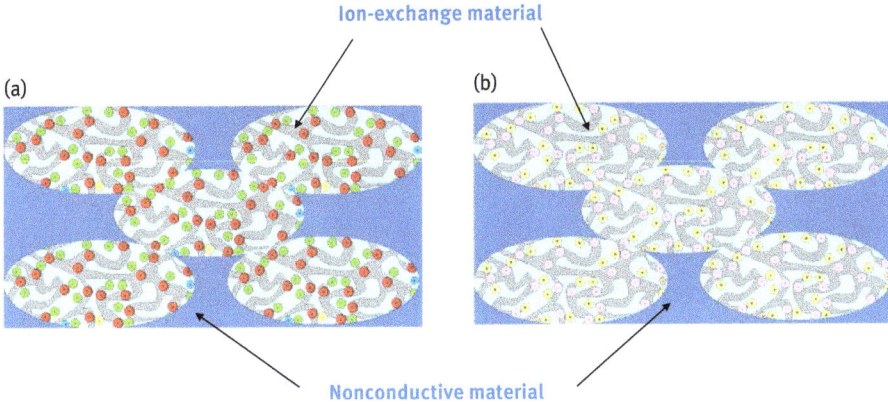

Figure 3.3: Structure of heterogeneous: (a) AEM and (b) CEM.

These properties are mainly determined by the type, concentration and distribution of the fixed ion-exchange moieties (which determine the ion-exchange capacity of the membrane), as well as by the composition of the polymer chain (which influences the microphase separation in the membrane structure). A trade-off relationship between the permselectivity and conductivity is commonly observed for IEMs: membranes with higher permselectivity are generally less conductive [3].

Homogeneous IEMs have usually higher permselectivity and conductivity in comparison with heterogeneous IEMs [1]. However, heterogeneous membranes are usually produced at lower costs and are characterized by higher mechanical strength than the homogeneous ones.

An efficient strategy used to improve the mechanical properties of homogeneous IEMs is to reinforce by a nonwoven backing material (Figure 3.4).

Ion-exchange polymers can be prepared by polymerization of monomers that contain a moiety that either is, or can be made, an anionic or a cationic exchanger. Alternatively, the exchanger moieties are introduced into a dissolved polymer by a chemical reaction.

Figure 3.5 shows some examples of typical synthesis of ion-exchange polymers [4]:
(a) polymerization of styrene and divinylbenzene and its subsequent sulfonation;
(b) introduction of a quaternary amine group into polystyrene by a chloromethylation procedure, followed by an amination with a tertiary amine;
(c) sulfonation of dissolved polyethersulfone;
(d) halomethylation of dissolved polyethersulfone and subsequent reaction with a tertiary amine.

Homogeneous IEMs are frequently prepared by solubilization of an ion-exchange polymer in an appropriate solvent or solvent's mixture, solution casting and

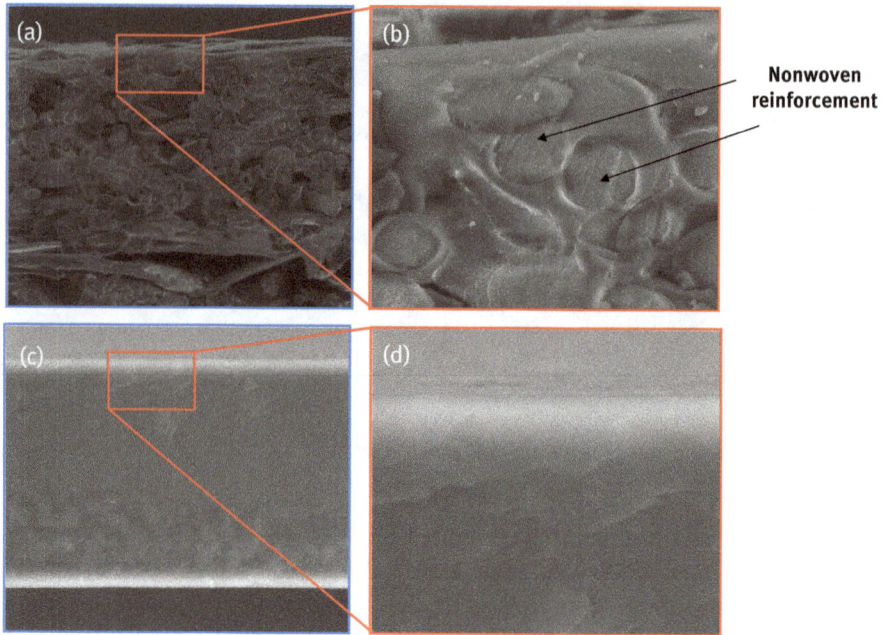

Figure 3.4: Scanning electron images of the cross section of homogenous membranes: (a) and (b) reinforced with nonwoven backing material; and (c) (d) not reinforced. Note that (b) and (d) are particulars of (a) and (c), respectively, taken at higher magnification.

solvent evaporation method. Another method used to produce this type of membranes is the grafting of ion-exchange groups on a preformed film. Melting and extrusion of ion-exchange resins is the third method used to prepare homogeneous IEMs.

Heterogeneous IEMs are usually prepared by mixing finely powdered organic and/or inorganic ion-exchanger materials and heating with a thermoplastic polymer (i.e., polymers that get soften while heating and can be remolded in different shapes) such as poly(vinyl chloride), polyethylene, polypropylene or other engineered plastics. Finally, the mixture is extruded under conditions of high pressure and temperature.

An alternative method for the preparation of heterogeneous IEMs is the dispersion of ion-exchange particles in a solution of a film-forming binder polymer, casting dispersion and solvent evaporation. The content of the ion-exchange material in the final membrane is usually >60% [4].

It is worth to note that at the present state of knowledge, the AEM's technology is less mature in comparison to the CEM's one. More research work is needed to improve AEM's alkaline stability and performance.

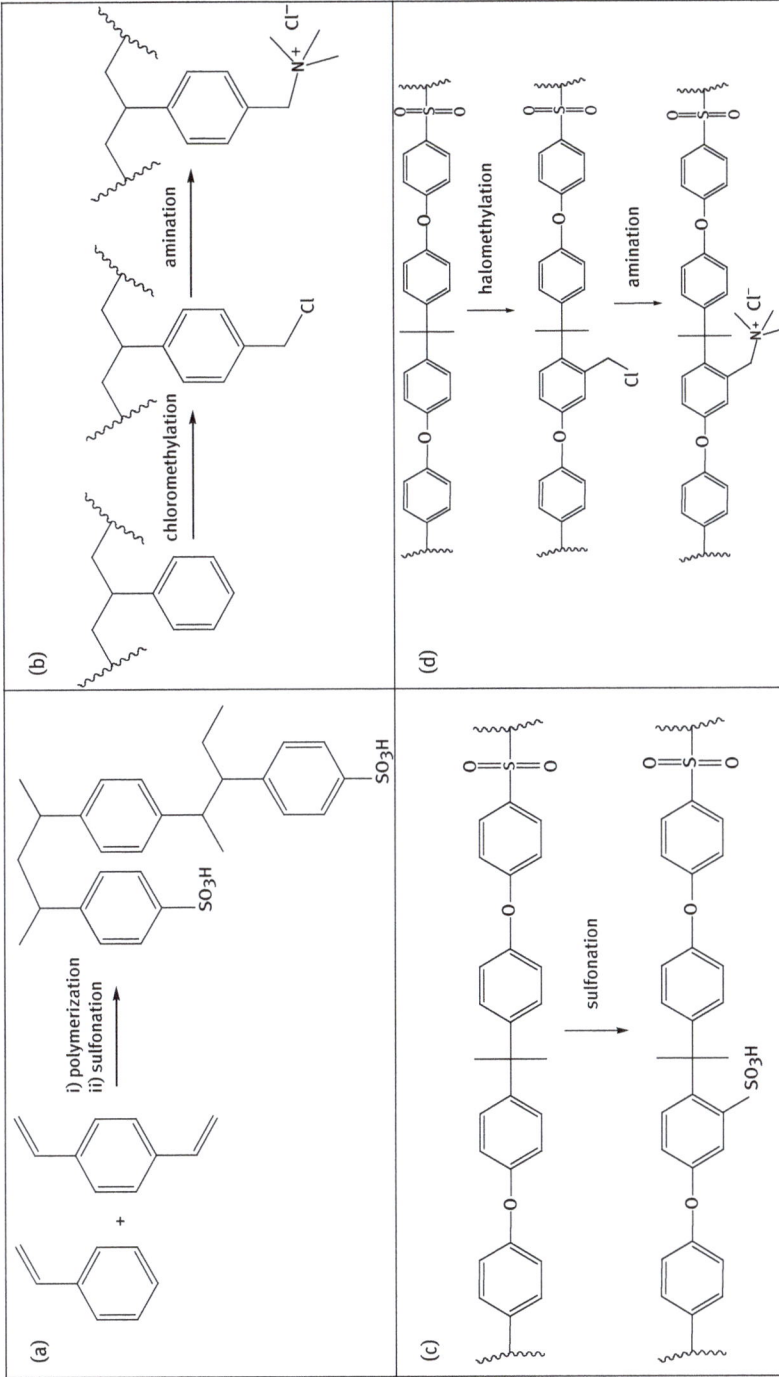

Figure 3.5: Reaction schemes of typical synthesis of ion-exchange polymers: (a)–(c) cation-exchange and (d) anion-exchange polymers (adapted from [4]).

3.2 Transport of ions in membranes

The transport of a component i having charge z_i through an IEM is due to a gradient in electrochemical potential (i.e., the sum of chemical potential and electrical potential), which at a constant temperature can be expressed as follows [4, 5]:

$$\nabla \tilde{\mu}_i = \nabla \mu_i + \nabla \varphi = \overline{V}_i \nabla p + RT \nabla \ln a_i + z_i F \nabla \varphi \tag{3.1}$$

where $\tilde{\mu}_i$ is the electrochemical potential of the component i; μ_i is the chemical potential; φ is the electrical potential; V_i the partial molar volume; a_i the activity; p is the hydrostatic pressure; z_i is the charge; F is the Faraday constant (96,485 C/mol); R is the gas constant (8.314 J/molK) and T is the temperature. Considering the one-dimensional case (x coordinate), the electrochemical potential can be written as follows:

$$\frac{d\tilde{\mu}_i}{dx} = \frac{d\mu_i}{dx} + \frac{d\phi}{dx} = \overline{V}_i \frac{dp}{dx} + RT \frac{d \ln a_i}{dx} + z_i F \frac{d\phi}{dx} \tag{3.2}$$

The flux (J_i) of the component i at a constant temperature can be expressed as a function of the driving force by a phenomenological equation [4–6]:

$$J_i = -L_i \frac{d\tilde{\mu}_i}{dx} = -L_i \left(\overline{V}_i \frac{dp}{dx} + RT \frac{d \ln a_i}{dx} + z_i F \frac{d\varphi}{dx} \right) \tag{3.3}$$

where J_i is a phenomenological coefficient related to the working driving force.
   The activity a_i is given as follows:

$$a_i = C_i \gamma_i \tag{3.4}$$

where γ_i is the activity coefficient and C_i the concentration.
   The derivative of the activity (x coordinate) is the following:

$$\frac{d \ln a_i}{dx} = \frac{d \ln C_i + d \ln \gamma_i}{dx} = \frac{d \ln C_i}{dx} \left(1 + \frac{d \ln \gamma_i}{d \ln C_i} \right) = \frac{1}{C_i} \left(1 + \frac{d \ln \gamma_i}{d \ln C_i} \right) \frac{dC_i}{dx} \tag{3.5}$$

Assuming that the activity coefficient is 1 (assumption valid for diluted solution), eq. (3.5) becomes

$$\frac{d \ln a_i}{dx} = \frac{d \ln C_i}{dx} = \frac{1}{C_i} \frac{dC_i}{dx} \tag{3.6}$$

Consequently, eq. (3.3) can be rewritten as follows:

$$J_i = -L_i \frac{d\tilde{\mu}_i}{dx} = -L_i \left(\overline{V}_i \frac{dp}{dx} + \frac{RT}{C_i} \frac{dC_i}{dx} + z_i F \frac{d\varphi}{dx} \right) \tag{3.7}$$

It is important to note that the three terms in the brackets (eq. 3.7) refer to convective, diffusive and migration transport mechanism, respectively.

If the pressure is constant, the convective contribution can be eliminated and the previous equation becomes:

$$J_i = -L_i \frac{d\tilde{\mu}_i}{dx} = -L_i \left(\frac{RT}{C_i} \frac{dC_i}{dx} + z_i F \frac{d\varphi}{dx} \right) \tag{3.8}$$

Expressing the phenomenological coefficient as a function of the diffusion coefficient D_i:

$$L_i = \frac{D_i C_i}{RT} \tag{3.9}$$

The flux of the component i can be expressed by an equation identical to the Nernst–Planck flux equation:

$$J_i = -D_i \left(\frac{dC_i}{dx} + \frac{z_i F C_i}{RT} \frac{d\phi}{dx} \right) \tag{3.10}$$

3.3 Membrane permselectivity

The permselectivity (Ψ) of an IEM quantifies the capacity of the membrane to retain coions while counterions are transported, and it is defined as a function of ions transport number in solution and in membrane phase. The permselectivity IEM results from the exclusion of coions from the membrane phase (Donnan exclusion).

In an electrolyte solution, the electrical current is carried by both, cations and anions; however, in an IEM the current is carried prevalently by the counterions (coions are ideally excluded).

The transport number of the ith ion (t_i) is defined as the fraction of the current carried by a given ion with respect to the overall current:

$$t_i = \frac{z_i J_i}{\sum\limits_i z_i J_i} \tag{3.11}$$

where J_i is the flux of ion and z_i is the valence. The flux J_i can be calculated by eq. (3.10).

In an ideally permselective AEM, the current is transported only by anions, and the transport number of anionic species is 1; in a strictly permselective CEM, the current is transported only by cations, and the transport number of cationic species is 1.

Obviously, the sum of the transport numbers of all ions in an electrolyte is equal to 1.

The permselectivity (Ψ^{cem}) of an AEM is defined by the following relations:

$$\psi^{aem} = \frac{t^{aem} - t_a}{t_c} \tag{3.12}$$

and the permselectivity of a CEM (Ψ^{cem}) is given by

$$\psi^{cem} = \frac{t^{cem} - t_c}{t_a} \tag{3.13}$$

where the superscripts aem and cem refer to AEM and CEM, respectively; the subscripts a and c refer to anion and cation, respectively.

Considering an IEM separating two electrolytic solutions, these solutions are in equilibrium (Donnan equilibrium) when their electrochemical potential are equal:

$$\frac{d\tilde{\mu}_i}{dx} = 0 = \bar{V}_i \frac{dp}{dx} + RT \frac{d \ln a_i}{dx} + z_i F \frac{d\varphi}{dx} \tag{3.14}$$

Integrating and rearranging the previous equation, it is possible to calculate the Donnan potential (φ_{Don}), defined as the difference of potential between the membrane and the solution phase:

$$\varphi_{Don} = \varphi^m - \varphi^s = \frac{1}{z_i F}\left(RT \ln \frac{a_i^s}{a_i^m} + \bar{V}_i(p^s - p^m) \right) \tag{3.15}$$

where φ^m and φ^s are the electrical potential in membrane and solution phase, respectively; a_i is the activity of the ith ion; z_i and $\bar{V}_i$ are the charge and partial molar volume; p is the pressure and the superscripts m or s indicate the membrane and solution phase, respectively.

Introducing in eq. (3.15) the difference of osmotic pressure ($\Delta\pi$), the expression of the Donnan potential for both anion and cation is given as follows:

$$\varphi_{Don} = \varphi^m - \varphi^s = \frac{1}{z_i F}\left(RT \ln \frac{a_i^s}{a_i^m} + \bar{V}_i \Delta\pi \right) \tag{3.16}$$

As a consequence, the relation between the anion and cation distribution is given as follows:

$$\frac{1}{z_a F}\left(RT \ln \frac{a_a^s}{a_a^m} + \bar{V}_a \Delta\pi \right) = \frac{1}{z_c F}\left(RT \ln \frac{a_c^s}{a_c^m} + \bar{V}_c \Delta\pi \right) \tag{3.17}$$

If one mole of electrolyte dissociates in υ_a moles of anions (with charge z_a) and υ_c moles of cations (with charge z_c), it is possible to write, for the electroneutrality principle and the volume balance, the following equations:

$$z_a \upsilon_a = -z_c \upsilon_c \tag{3.18}$$

$$\overline{V_e} = \upsilon_a \overline{V_a} + \upsilon_c \overline{V_c} \tag{3.19}$$

where the subscripts a, c and e refer to anion, cation and electrolyte, respectively.

Introducing the two last equations in eq. (3.17) and rearranging, the Donnan equilibrium between solution and membrane phase is described by the following equation [4–6]:

$$\left(\frac{a_a^s}{a_a^m}\right)^{\frac{1}{z_a}}\left(\frac{a_c^s}{a_c^m}\right)^{\frac{1}{z_c}} = e^{-\frac{\Delta\pi\bar{V}_s}{RT v_c z_c}} \tag{3.20}$$

Introducing the concentration and activity coefficient of the coions and counterions and rearranging, the equation becomes

$$\left(\frac{C_{co}^s}{C_{co}^m}\right)^{\frac{1}{z_{co}}}\left(\frac{C_{cou}^m}{C_{cou}^s}\right)^{\frac{1}{z_{cou}}} = \left(\frac{\gamma_{co}^m}{\gamma_{co}^s}\right)^{\frac{1}{z_{co}}}\left(\frac{\gamma_{cou}^s}{\gamma_{cou}^m}\right)^{\frac{1}{z_{cou}}} e^{-\frac{\Delta\pi\bar{V}_s}{RT v_c z_c}} \tag{3.21}$$

where C and γ are the concentration and the activity coefficient, respectively, and the subscripts co and cou refer to coion, and counterion, respectively.

For a monovalent electrolyte, the coion concentration in an IEM can be approximated under some assumption valid in numerous practical applications [4]:

$$C_{co}^m = \frac{C_s^2}{C_{fix}} \tag{3.22}$$

The last equation indicates that the coion concentration in the membrane (C_{co}^m) increases (i.e., the permselectivity decreases) with the increase in the external solution concentration (C_s). An opposite effect has the fixed ion concentration in the membrane (C_{fix}) because increasing Cfix the permeselectivity increases.

The membrane permselectivity can be calculated from the measured potential (ΔV_{exp}) across the membrane separating the two solutions of different concentrations and its theoretical value (ΔV_{the}) is given as follows [1]:

$$\psi = \frac{\Delta V_{exp}}{\Delta V_{the}} \star 100 \tag{3.23}$$

The theoretical membrane potential can be calculated by the Nernst equation:

$$\Delta V_{the} = \sum_i \frac{RT}{z_i F} \ln \frac{\gamma_{\pm i}^c c_i^c}{\gamma_{\pm i}^d c_i^d} \tag{3.24}$$

where R is the universal gas constant (8.314472 J/Kmol), T is the absolute temperature, z is the electrochemical valence of the ith salt, F is the Faraday constant (96,485 C/mol) and $\gamma_{i}\pm$ and C_i are the medium activity coefficient and molality of the ith salt in the solution, respectively. The summation is extended to all the ions in solution.

The superscript d and c indicate the diluted and concentrated solution, respectively.

The value obtained by this method is indicated as "apparent permselectivity" to distinguish it from the value of permselectivity obtained by measuring experimentally the increase or decrease in concentration of the various ions in the concentrated or the diluted compartment, respectively, in an ED-type experimental set-up.

3.4 ED, polarization phenomena and limiting current

ED is an electrically driven membrane separation process that is able to separate, concentrate and purify electrolyte solutions. The selective transport of ions through IEMs results in concentration of these ions at one side of the membrane and dilution at the other side.

In an ED stack, alternating AEM and CEM form diluate and concentrate compartments (Figure 3.6).

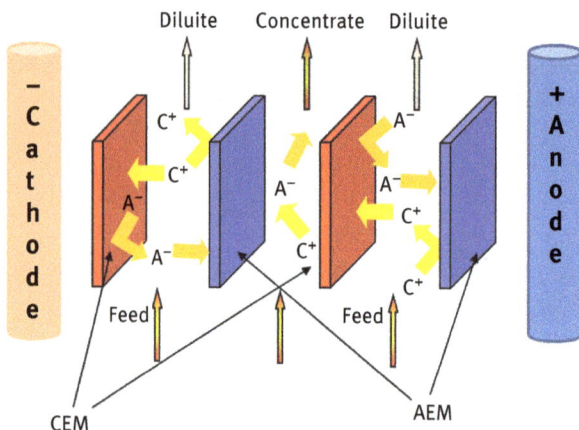

Figure 3.6: The scheme of an ED stack.

Mass transport in ED is described by phenomenological equations [1, 4, 5]. Under the assumptions of ideal solutions, no pressure gradients, no kinetic coupling of fluxes and expressing the phenomenological coefficient by the diffusion coefficient, the ionic flux J_i is given by the following equation:

$$J_i = -D_i \left(\frac{dc_i}{dx} - c_i \sum_{i=1}^{n} \frac{T_i}{c_i} \frac{dc_i}{dx} \right) + T_i \frac{I}{z_i F} \tag{3.25}$$

where D_i is the diffusion coefficient, c_i the concentration, T_i the transport number, z_i the valence of ith ion, x the axial coordinate, I the current and F the Faraday constant.

The current I is related to the flux of the ions:

$$I = F \sum_{i=1}^{n} z_i J_i \tag{3.26}$$

Concentration polarization at the IEM surface facing the dilute-containing compartment induces the existence of a limiting current density. Concentration polarization is a result of the differences in the transport number of ions in the solution and in the membrane phase. In an IEM, the current is transported principally by counterions, while both anions and cations transport the current in solution.

If the limiting current density is overcome, the electric resistance in the dilute compartment increases and water dissociation may occur at the membrane interface with relevant consequences in current utilization. Moreover, water splitting can lead to pH changes in the solutions.

Considering the mass balance between the flux through the membrane and the fluxes in the boundary layers, it is possible to write the following equation:

$$J_i^m = J_i^{\text{diff}} + J_i^{\text{mig}} \tag{3.27}$$

where J_i^m is the total flux of a counterion through an IEM; J_i^{diff} and J_i^{mig} are the diffusive and migration fluxes in the boundary layers, respectively.

The total flux of the counterion through the membrane is given as follows:

$$J_i^m = T_i^m \frac{i}{|z_i| F} \tag{3.28}$$

where T_i^m is the transport number of the counterion in the membrane.

The transport of the counterions through the boundary layer at the dilute side of the cell is a combination of migration and diffusion:

$$J_i^{\text{diff}} + J_i^{\text{mig}} = -D \frac{dC_i^d}{dz} + T_i^s \frac{i}{|z_i| F} \tag{3.29}$$

Considering eq. (3.26), the current through an IEM can be expressed as follows:

$$i = \frac{F D_i}{z_i \left(T_i^m - T_i \right)} \frac{dC_i^d}{dz} \tag{3.30}$$

Substituting the diffusion coefficient and the concentration of the individual ions and integrating over the thickness of the boundary layer, the current through the cell results:

$$i = \frac{F D_s}{z_c v_c \left(T_c^m - T_c \right)} \left(\frac{\Delta C_s^d}{\Delta z} \right) = \frac{F D_s}{z_c v_c \left(T_c^m - T_c \right)} \left(\frac{{}^b C_s^d - {}^m C_s^d}{\Delta z} \right) = \frac{F D_s}{z_a v_a \left(T_a^m - T_a \right)} \left(\frac{{}^b C_s^d - {}^m C_s^d}{\Delta z} \right) \tag{3.31}$$

where T^m and T are the transport numbers in the membrane and in the solution phase; D_s is the diffusion coefficient of the salt in the solution; ${}^m C_s^d$ and ${}^b C_s^d$ are the salt concentration in the solution at the membrane surface and in the well mixed bulk solution, respectively; Δz is the thickness of the of the boundary layer, z is the valence of the salt, v is its stoichiometric coefficient and the subscripts c and a refer to cation and anion.

The limiting or maximum current density $(i = i_{\text{lim}})$ is reached when the salt concentration at the membrane interface with the dilute compartment becomes zero:

$$i_{\text{lim}} = \frac{FD_s}{z_c v_c \left(T_c^m - T_c\right)} \left(\frac{^b C_s^d}{\Delta z}\right) \tag{3.32}$$

After the limiting current, the electrical resistance of the dilute compartment increases drastically.

3.5 Effect of the membrane's microstructure and contacting environment on IEM's transport properties

The ion conductivity and the permselectivity of an IEM at a given temperature strongly depend not only on membrane's composition and microstructure, but also on the liquid or gaseous phase contacting the membrane. The activity of water in the surrounding environment indeed influences the water volume fraction in the membrane phase and, as a consequence, the membrane microstructure [7, 8].

It is well known that the proton conductivity of CEM used for polymer electrolyte membrane fuel cells (PEMFC) increases with the increase of relative humidity [9, 10]. In order to understand this effect, it is necessary to consider that in the presence of water the ionic fixed groups of IEMs tend to interconnect by hydrogen bonds forming swelled hydrophilic domains. This network of water hydration molecules and ionizable fixed groups linked by hydrogen bonds represent the pathway for the ions (e.g., protons in PEMFC) transport. Protons transport in membranes can occur by a vehicular or a Grotthus mechanism [11]. In the vehicular mechanism, the protons are linked to water (the vehicle) forming H_3O^+ and $H_5O_2^+$ ions, which diffuse under a gradient of electrochemical potential. In the Grotthus mechanism, the vehicles' molecules are the fixed charged groups. They are stationary and the transport occurs by the intermolecular structural reorganization of hydrogen bonds with the concomitant reorientation of the vehicle molecules by the so-called protons hopping.

Currently, the most commonly used CEM for PEMFC applications is an expensive perfluorosulfonic acid (PFSA) membrane known as Nafion produced by Du Point (Figure 3.7(a)). Nafion is characterized by high proton conductivity and elevated chemical, thermal and mechanical stability. Nafion belongs to a group of PFSA membranes made from long-side-chain (LSC) polymers. A second group of PFSA is composed of short-side-chain (SSC) polymers. In addition to Nafion, other examples of LSC membranes are Aciplex (Asahi Chemical), Flemion (Asahi Glass) and Gore-Select (Gore and Associates) [10]. The SSC ionomers were initially produced in the 1980s by the Dow company and commercialized under the trade name Dow Ionomer. More recently, Solvay Solexis produced Hyflon Ion ionomer, now called Aquivion (Figure 3.7(b)) [12]. The SSC ionomers are characterized by shorter pendent group carrying the ionizable

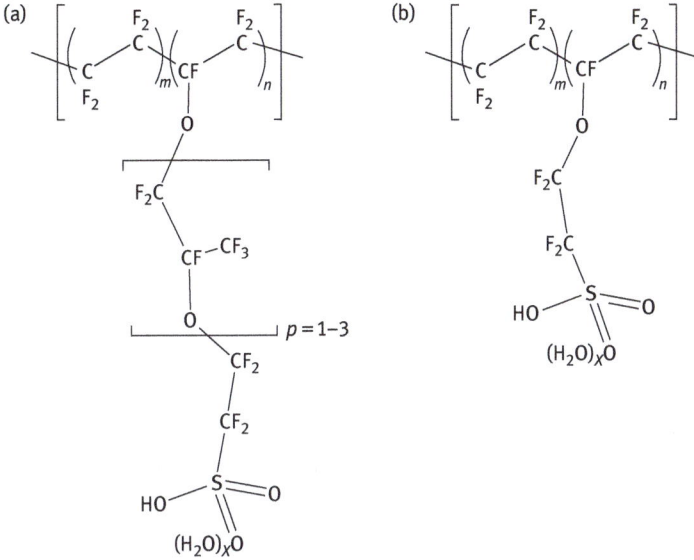

Figure 3.7: Chemical formula of (a) Nafion and (b) Hyflon ion (or Aquivion).

functionality, higher crystallinity and higher glass transition temperature (T_g) than LSC ionomers at the given equivalent weight [13].

It is important to highlight that Nafion and, more in general, PFSA-based membranes have high costs [14], which limit PEMFC implementation at large scale. In addition, PFSA membranes are characterized by an elevated cross-over of the fuel (H_2 in H_2-PEMFC and methanol in direct-methanol PEMFC, respectively), which reduces the productivity of the system. Moreover, the high water permeability of these membranes favors membrane dehydration at medium-high temperature, with a consequent decrease of conductivity. Another potential consequence of the high water permeability of PFSA is the cathode flooding with resulting restricted oxygen transport through the pores of the gas diffusion electrode [14]. Alternative membranes widely investigated to potentially replace PFSA-type are partially fluorinated, nonfluorinated hydrocarbon, nonfluorinated aromatic and acid–base blend membranes [15–19].

An interesting case of nonfluorinated polymers investigated as possible IEM material is the group of the sulfonated aromatic polymers (SAP), such as the sulfonated poly(etheretherketone) (SPEEK) [17, 20–22].

The SPEEK polymer is synthesized by sulfonation of the semicrystalline thermoplastic polymer polyetheretherketone (PEEK, Figure 3.8(a)). PEEK is not soluble in common organic solvents under mild conditions and therefore it is not suitable for the preparation of membranes by solution casting technique [23].

After sulfonation, the resulting SPEEK polymer is amorphous and is soluble in several organic solvents under mild conditions. However, the sulfonation of PEEK is typically a heterogeneous reaction because of the limited solubility of the starting

(a)   (b)

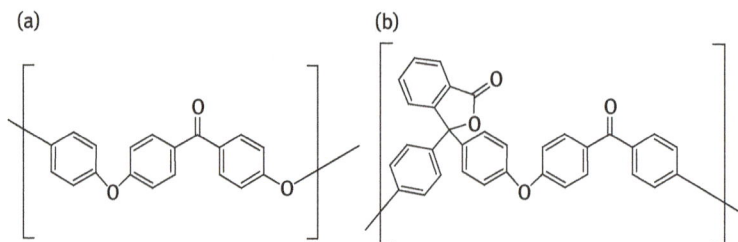

Figure 3.8: Chemical formula of the (a) PEEK and (b) PEEK-WC.

polymer. As a consequence, the resulting SPEEK product has a wide distribution of the degree of sulfonation (DS) [20].

On the other hand, a modified PEEK, known as PEEK-WC (Figure 3.8(b)), is amorphous, thanks to the presence of a lactonic group, called Cardo group (WC in the polymer name in fact means with Cardo), which hinders the crystalline organization of the polymeric chains. PEEK-WC is soluble in organic solvents with medium polarity (e.g., chloroform, dimethylsulfoxide, dimethylacetamide, dimethylformamide) and it can be used as a membrane material by casting technique [24, 25].

The SPEEK-WC ionomer can be obtained when PEEK reacts with chlorosulfonic acid at the room temperature [26]. The DS is a key parameter for determining the ion-exchange capacity and the proton conductivity of the system, and it can be tailored for controlling the reaction conditions.

SPEEK-WC membranes have an higher resistance to water and methanol vapor transport than Nafion, as well as a lower permeability to H_2 and O_2 [27]. Consequently, minor crossover and electrode flooding problems are expected to occur with SPEEK-WC membranes compared with Nafion in PEMFC applications. The higher mass transport resistance of the SPEEK-WC membranes is a result of lower diffusion coefficients, determined by the higher stiffness of the SPEEK-WC with respect to Nafion [27].

The lower water and methanol permeability compared to PFSA membranes is a general characteristic of the SAP due to the peculiar differences in their microstructures [10]. PFSA membranes present high hydrophobic perfluorinated regions combined with hydrophilic sulfonic groups. When the membrane is humidified, the hydrophilic portions of the polymer aggregate, forming hydrophilic nanodomains. In the case of Nafion membranes, the formation of hydrophilic pockets of few nanometers in diameter, composed of sulfonic acid groups, counterions and absorbed water, is reported [28]. These pockets are connected by a ramified network of hydrophilic channels, which are approximatively 1 nm in diameter, separated from the hydrophobic part of the polymer formed by the fluorinated backbone of the polymer responsible of the morphological stability of the system.

The transport of hydrophilic species, such as proton, water and methanol, occurs through these hydrophilic regions. The dimension of the hydrophilic pockets in Nafion with an equivalent weight 1,200 shrinks from 4 to 2.44 nm when the relative humidity is reduced from 100% to 34% [29]. This also means that the mass transport resistance of Nafion and, in general, of an IEM depends on the level of humidification.

In the case of SAP membranes, the hydrophilic nanochannels are more narrow, more branched and less separated from the hydrophobic regions in comparison with Nafion (and, in general, with PFSA membranes) because of the lower hydrophobicity of the polymer backbone [10, 30]. Moreover, the sulfonic groups of SAP membranes are characterized by a lower acidity degree and the membranes have usually a lower proton conductivity than in the case of the PFSA [10]. Also the water and methanol diffusion coefficient of this family of polymers are typically lower [22, 31].

Concerning the effect of the contacting environment on IEM's transport properties, additional considerations are necessary when the membrane is in contact with liquid electrolyte solutions like in the case of RED [32]. The ionic strength and composition of the external liquid solution may indeed impact in a relevant way on membrane microstructure and mobile ionic species uptake. It is important to note that it is more appropriate to consider the electrolyte solution ionic strength to relate the effect different solutions instead the simple molar or molal concentration of the same. The ionic strength (I) indeed reflects the effect of charges and interionic interactions on electrolyte activities and it is defined as follow:

$$I = \frac{1}{2} \sum_{i} m_i \cdot z_i^2 \qquad (3.33)$$

where m_i is the molality or molarity and z_i the charge of the ith ion.

Increasing the solution ionic strength, the water content in an IEM decreases for osmotic effect, the hydrophilic channels of the membrane shrinks and the ionic transport resistance increases [7, 8].

The water uptake reduction determines the increase of the concentration of the fixed charged groups in membrane and, consequently, ion migration through in membrane phase results more hindered because of the stronger interactions with the fixed charged groups that can form isolated ionic domains not sufficiently interconnected. In particular, an IEM with a lower fixed charged group concentration results more sensitivity to the increase of the external solution ionic strength [7]. Moreover, in the case of multi-components electrolyte solutions, the ionic concentration profiles inside the membrane and at its interfaces are influenced by counterion exchange phenomena, as well as by the coion diffusion occurring for nonideally permselective membranes.

Homogeneous reinforced nonfluorinated AEM and CEM, produced by Fujifilm Manufacturing Europe B.V. and indicated as AEM-80045 and CEM-80050, were applied for salinity-gradient power harvesting by RED using seawater and brine as diluted and concentrated feed solutions, respectively [33]. The effect of multivalent ions on membrane electrochemical properties was deeply investigated, focusing on the most common multivalent ions in seawater [8]. A critical role of the Mg^{2+}, the third most abundant ion in seawater after sodium and chloride, on the ionic conductivity of CEM-800050 was observed. The sulfonic groups (i.e., the fixed groups of the CEM-80050) have an higher affinity for Mg^{2+} than for Na^+ [4, 34], but the bivalent Mg^{2+} ion also has lower mobility than the monovalent Na^+ caused by the higher hydrated radius of the first one [35]. In addition, the multivalent ions can also alter the membrane microstructure forming bridges between different fixed charged groups, reducing the effective dimension of the hydrophilic nanochannels of the IEM contributing to the reduction of membrane ionic conductivity.

On the contrary, the conductivity of the AEM-80045 was not influenced in relevant way by the presence of this bivalent cation. AEM indeed tend to reject Mg^{2+}, and in general cations, by Donnan effect, thanks to quaternary ammonium groups (i.e., positively charged groups) covalently linked on its polymer chains.

Membrane permselectivity is also influenced by the solution ionic strength and compositions. In general, the permselectivity tends to decrease with the increase of the external solutions ionic strength because of the increase in the difference of ions activity between the electrolyte solution and the membrane. The result is a diffusive transport of the coions [4].

The permselectivity of AEM-80045 and CEM-80050 was tested in mixed electrolyte solutions containing Mg^{2+}.

The permselectivities of both membranes measured in the presence of Mg^{+2} ions decreased, but in more relevant way for the CEM-80050. This decrease of permselectivities was for the CEM due to the shielding effect of the bivalent cation on the fixed charged groups, which reduces the capacity of the membrane to exclude coions [36].

The presence of Mg^{2+} also influences the permselectivity of the AEM-80045. This effect was related to a decrease in dielectric exclusion of the noncharged portion of the AEM800 [37] induced by the Mg^{2+} ion having a higher charge density than that of Na^+. Consequently, Mg^{2+} was able to induce a partial polarization of the polymer with an increase of coion uptake in the AEM-80045 [8].

Considering that the properties of IEMs strongly depend on chemical, physical and structural environment of ion nanochannels, the development of a new nanostructured IEM, in which the building blocks and/or their periodic alignment are controlled at the nanoscale, has been receiving a growing attention in the last few years [38]. Nanostructured IEM include nanostructured polymeric IEM prepared from ionomers block copolymers containing hydrophilic and hydrophobic polymer moieties which forms self-organized nanochannels [39], and nanostructured mixed matrix IEM combining in a synergic way a polymeric component with a nanostructured filler material [19].

3.6 Membranes and interfaces characterization by impedance spectroscopy

The knowledge of the membrane electrical and dielectric properties is one of the major concerns of membranologists working with electromembranes processes. At the interfaces between a solid ionic conductor and a liquid electrolyte solution, physical and electrical properties change suddenly because of a nonhomogeneous distribution of charged species (i.e., polarization phenomena by accumulation or depletion of ionic species near an interface), which reduces the overall electrical conductivity of the system (Figure 3.9) [40–42].

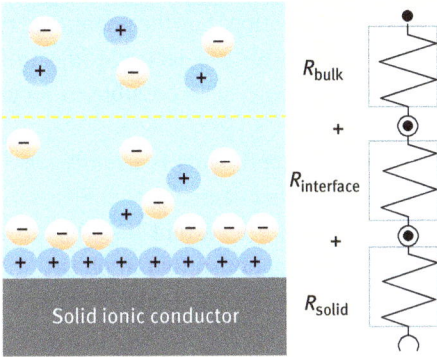

Figure 3.9: Schematization of polarization phenomena occurring at the interface between a solid ionic conductor (the case of a system with positive surface charge has been represented) and an electrolyte solution.

It is worth noting that the charge polarization phenomenon is not limited to electro-membrane processes, but concentration polarization and fouling phenomena in pressure-driven membrane process (e.g., reverse osmosis, nanofiltration, ultrafiltration and microfiltration) might induce relevant changes in electrical properties of the membrane and boundary layers [43].

As a consequence, the relevance of the electrical and dielectrical properties characterization of the of the membranes and interfaces is not limited to IEMs, but it spreads over all types of the membrane used in liquid separations.

The classical approach used to measure membrane electrical resistance is the direct current (DC) method. In DC measurements, in agreement with Ohm's law, the electrical resistance (R [Ω]) is given by the slope of the current (I [A]) versus the voltage drop (U [V]) curve [4]:

$$R = \frac{U}{I} \tag{3.34}$$

The DC method is simple, but it is not possible to distinguish the membrane from the interface resistance. On the contrary, in electrochemical impedance spectroscopy (EIS), an alternate current over a frequency range is used. In this way, it is possible to separate the contribution to the total resistance of phenomena proceeding at different rates.

In EIS experiments, a sinusoidal electrical stimulus (current or voltage) is applied over a frequency range to a pair of electrodes and the response of the system under investigation is observed (current or voltage) by the same, or different, electrodes. In the first case, the configuration is indicated as two-probes (or two electrodes) configuration. When two additional electrodes are used to collect the response of the system, the configuration is indicated as a four-probes configuration [44].

A two-probes configuration is usually applied when the membrane is pressed between two solid conductive electrodes, like in the case of the membrane electrode assembly for PEMFC [27, 45–47].

The four-electrodes configuration is the most appropriate one for the study of the ion transport through a membrane separating two liquid electrolyte solutions. This configuration, with respect to the two-electrodes configuration, has the advantage to eliminate from the impedance spectra the contribution of the electrode injecting stimulus/electrolyte charge transfer resistance. In this way, it is possible to focus the analyses on the membrane and its interfaces [48–54].

The sinusoidal electrical stimulus applied in EIS experiments is usually small enough to comply with the assumptions of a pseudolinear segment of the current versus voltage curve. In linear or pseudolinear systems, the current response to a sinusoidal potential is a sinusoid with the same frequency but shifted in phase.

Voltage ($U_{(\omega)}$ [V]) and current ($I_{(\omega)}$ [A]) are functions of the circular velocity or circular frequency $\omega(s^{-1})$ [4, 40]:

$$U_{(\omega)} = U_o \sin \omega t \tag{3.35}$$

$$I_{(\omega)} = I_o \sin\ (\omega t + \varphi) \tag{3.36}$$

$$\omega = 2\pi v \tag{3.37}$$

where t (s) is the time, φ (°) is the phase shift between voltage and current (Figure 3.10), and the subscript ° refers to the amplitude of voltage and current in phase, $v(s^{-1})$ is the frequency.

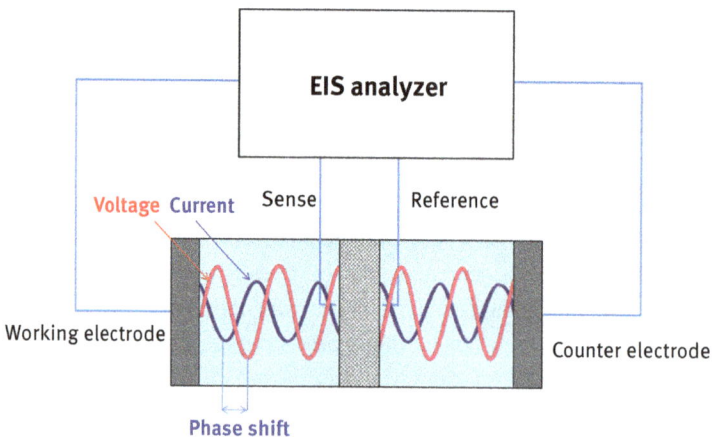

Figure 3.10: Typical set up of EIS experiments carried out with the four-electrodes configuration.

By using Eulero's form:

$$e^{j\varphi} = \cos\varphi + j\sin\varphi \tag{3.38}$$

the impedance $Z_{(\omega)}$ (Ω) can be expressed in a form similar to Ohm's law:

$$Z_{(\omega)} = \frac{U_0 e^{j\omega t}}{I_0 e^{j(\omega t + \varphi)}} = \frac{U_0}{I_0} e^{j\varphi} = |Z|\cos\varphi + j\,|Z|\sin\varphi \tag{3.39}$$

where j is the imaginary number ($j = \sqrt{-1}$) and $|Z|$ is the impedance module.

Equation (3.30) shows that the impedance is composed of two parts: a real part (Z') and an imaginary part (Z''):

$$Z' = |Z|\cos\varphi \tag{3.40}$$

$$Z'' = |Z|\sin\varphi \tag{3.41}$$

The real part of the impedance is called resistance; the imaginary part is the reactance. The reciprocal of the impedance is the admittance $Y_{(\omega)}$ (S):

$$Y_{(\omega)} = \frac{1}{Z_{(\omega)}} = G + j\omega C \tag{3.42}$$

where G (S) is the conductance and C (F) is the capacitance.

The conductance measures of the ability of the system to conduct electric charges, while the capacitance quantify the capacity to store them.

The conductance and capacitance are related to the impedance magnitude and the phase angle as follows [4, 40]:

$$G = \frac{1}{|Z|}\cos\varphi \tag{3.43}$$

$$C = \frac{1}{\omega|Z|}\sin\varphi \tag{3.44}$$

Physically meaningful properties of the electrochemical system under investigation are obtained by equivalent circuit modeling of the impedance data (Table 3.1).

EIS data can be represented in the form of Nyquist or Bode plots. In the Nyquist plot, the imaginary part ($-Z''$) versus the real part of the impedance (Z') is showed. In the Bode plot, the dependence module of the impedance (Z) or the phase shift ($-$Phase) from the frequency is reported.

The circuit showed in Figure 3.11(a) was used to fit experimental EIS data of homogeneous-reinforced AEM and CEM membranes tested in various electrolyte solutions [7, 8]. This circuit includes three different type of resistances: membrane and solution resistance (R_{m+s}), electrical double layer (EDL) resistance (R_{edl}) and the diffusion boundary layer resistance (R_{dbl}). The formation of the EDL is due to

Table 3.1: Examples of equivalent circuit elements and their relationship with the impedance.

Element	Relationship with impedance
Resistor (*R*)	$Z_R = R$
Capacitor (*C*)	$Z_C = \frac{1}{j\omega C}$
Inductor (*L*)	$Z_L = j\omega L$
Constant phase element (CPE or *Q*)	$Z_{CPE} = \frac{1}{Y_0 (j\omega)^n}$
Warburg impedance (*W*)	$Z_W = \frac{1}{Y_0 \sqrt{j\omega}}$

the presence of a net charge on the membrane surface that influences the ions distribution at the membrane/solution interface, resulting in an increase in the concentration of counterions. The region over which this influence is being felt is called EDL [42, 55, 56] schematically divided in two layers: an inner layer, called *Stern layer* (SL), formed by ions strongly bound by electrostatic interactions with the membrane surface, and an outer layer, called *diffuse layer* (DL), constituted by loosely bond ions. The *Stern layer* has a thickness in the order of Angstroms (i.e., one or two radius of solvated ions away from the surface); the thickness of the *diffuse layer* is in the order of nanometers.

Moreover, charges polarization phenomena occur at the interface membrane/solution because of the different mobility and flux between the coion and the counterion as well as because of the difference in transport number of the ions between the IEM and the solution phase [52, 53]. In an IEM, the electrical current is transported almost exclusively by the counterions having a transport number in membrane close to one. The coions are instead ideally excluded within the IEM and the transport number tends to zero. On the contrary, in the liquid electrolyte solution, the current is transported by both co- and counterions (in the case of symmetric salts, the transport number is 0.5 for both). Consequently, an additional polarization layer, called diffusion boundary layer (DBL), having a thickness in the order of several hundred of micrometers, is formed.

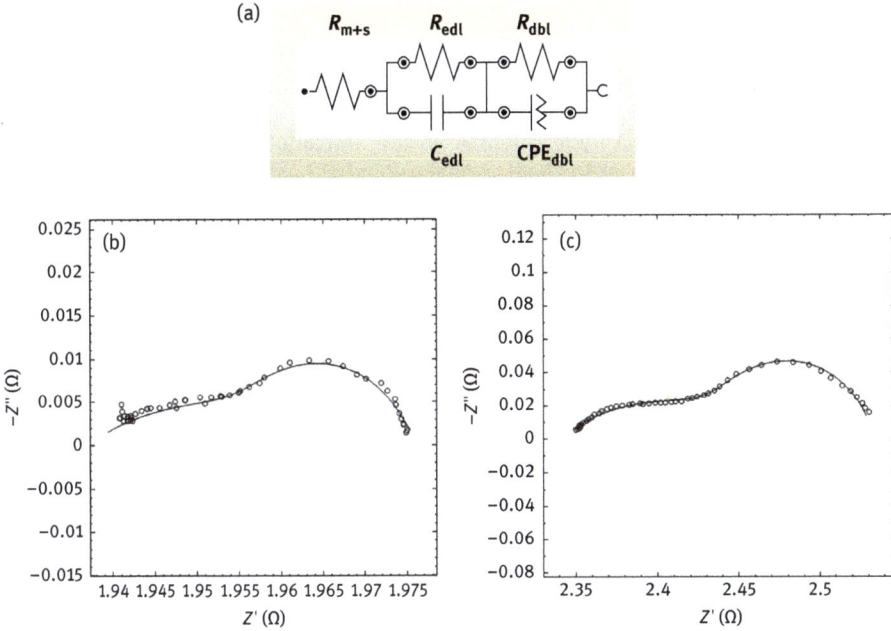

Figure 3.11: (a) Equivalent circuits used to fit the EIS spectra of (b) AEM and (c) CEM in contact with NaCl 0.5 M. The resistor is indicated as R; the capacitor as C; the constant phase element (a nonideal capacitor) as CPE; the Warburg impedance as W. The subscript "m+s" refers to membrane plus solution, the subscript "edl" to the electrical double layer; the subscript "dbl" diffusion boundary layer.

The Nyquist plots provide a visual verification of the good fitting of the experimental data obtained with AEM and CEM tested in 0.5 M NaCl with the equivalent circuit (Figure 3.11(b) and (c)). Multiple parameters were determined from a single experiment (membrane and interface resistance and capacitance) with an high accuracy and sensitivity [7, 8].

Impedance spectroscopy might be applied not only for membrane characterization, but also for the in situ nondestructive study and monitoring of fouling phenomena [43, 50, 57, 58].

EIS test can be in fact carried out under real operative conditions in noninvasive and not destructive way, giving information in real time and the measurements can be easily automatized. The main limits of the EIS are associated with the possible ambiguities in the data interpretation. In many cases, several equivalent circuits can successfully fit the experimental data, but the selection of the correct one needs to be based on the physical fundaments of the system and process under investigation. In some other cases, the ordinary ideal circuit elements, representing ideal lumped-constant properties, are inadequate to represent the electrical behavior of complex systems and distributed elements need to be introduced in the circuit (e.g., Warburg impedance).

References

[1]   Strathmann H. Ion exchange membrane separation processes, Elsevier; 2004.

[2]   Mitsuru Higa. Heterogeneous Ion-Exchange Membranes, Springer-Verlag Berlin Heidelberg,
      E. Drioli, L. Giorno (eds.), Encyclopedia of Membranes, DOI 10.1007/978-3-642-40872-4_278-1,
      2014.

[3]   Geise G M, Hickner M A, Logan B. Ionic resistance and permselectivity tradeoffs in anion
      exchange membranes. ACS Appl. Mater. Interfaces 2013; 5(20): 10294–10301.

[4]   Strathmann, H., Giorno, L., Drioli, E. An introduction to membrane science and technology,
      Consiglio Nazionale delle Ricerche (CNR), Rome (Italy), 2006.

[5]   Strathmann, H. Electromembrane processes: basic aspects and application, in: Enrico Drioli
      and Lidietta Giorno (eds.), Comprehensive Membrane Science and Engineering, volume 2,
      391–429, Elsevier, Oxford, 2010.

[6]   Kedem, O., Katchalsky, A. A physical interpretation of the phenomenological coefficients of
      membrane permeability. J. Gen. Physiol. 1961; 45: 143–179.

[7]   Fontananova E, Zhang W, Nicotera I, Simari C, van Baak W, Di Profio G, Curcio E, Drioli E. Probing
      membrane and interface properties in concentrated electrolyte solutions. J. Membr. Sci. 2014;
      459: 177–189.

[8]   Fontananova E, Messana D, Tufa RA, Nicotera I, Kosma V, Curcio E, van Baak W, Drioli E, Di
      Profio G. Effect of solution concentration and composition on the electrochemical properties of
      ion exchange membranes for energy conversion. J. Power Sources. 2017; 340: 282–293.

[9]   Marechal M, Souquet J-L, Guindet J, Sanchez J-Y. Solvation of sulphonic acid groups in
      Nafion® membranes from accurate conductivity measurements. Electrochem. Commun.
      2007; 9: 1023–1028.

[10] Kreuer KD. On the development of proton conducting polymer membranes for hydrogen and
      methanol fuel cells. J. Membr. Sci. 2001; 185: 29–39.

[11] Kreuer KD. On the complexity of proton conduction phenomena. Solid State Ionics. 2000;
      136–137:149–160.

[12] Arcella V, Troglia C, Ghielmi A. Hyflon ion membranes for fuel cells. Ind. Eng. Chem. Res. 2005;
      44: 7645–7651.

[13] Ghielmi A, Vaccarono P, Troglia C, Arcella C. Proton exchange membranes based on the short-
      side-chain perfluorinated ionomer. J. Power Sources. 2005; 145: 108–115.

[14] Neburchilov V, Martin J, Wang H, Zhang J. A review of polymer electrolyte membranes for direct
      methanol fuel cells. J. Power Sources. 2007; 169: 221–238.

[15] Roziere J, Jones D J. Non-fluorinated polymer materials for proton exchange membrane fuel
      cells. Annu. Rev. Mater. Res. 2003; 33: 503–555.

[16] Hickner M A, Ghassemi H, Kim Y S, Einsla B R, McGrath J E. Alternative polymer systems for
      proton exchange membranes (PEMs). Chem. Rev. 2004; 104: 4587–4612.

[17] Smitha S, Sridhar S, Khan A. Solid polymer electrolyte membranes for fuel cell applications—a
      review. J. Membrane Sci. 2005; 259: 10–26.

[18] Tripathi B P, Shahi V K. Organic–inorganic nanocomposite polymer electrolyte membranes for
      fuel cell applications. Prog. Polym. Sci. 2011; 36: 945–979.

[19] Thiam H S, Daud W R W, Kamarudin S K, Mohammad A B, Kadhum A A H, Loh K S, Majlan
      E H. Overview on nanostructured membrane in fuel cell applications. Int. J. hydrogen Energy.
      2011; 36: 3187–3205.

[20] Mikhailenko S D, Wang K, Kaliaguine S, Xing P, Robertson G P, Guiver M D. Proton conducting
      membranes based on cross-linked sulfonated poly(ether ether ketone) (SPEEK). J. Membrane
      Sci. 2004; 233: 93–99.

[21] Zaidi SMJ, Mikhailenko SD, Robertson GP, Guiver MD, Kaliaguine S. Proton conducting composite membranes from polyether ether ketone and heteropolyacids for fuel cell applications. J. Membr. Sci. 2000; 173: 17–34.

[22] Li L, Zhang J, Wang Y. Sulfonated poly(ether ether ketone) membranes for direct methanol fuel cell. J. Membr. Sci. 2003; 226: 159–167.

[23] Beck H N, Solubility characteristics of poly(ether ether ketone) and poly(phenylene sulphide). J. Appl. Polym. Sci. 1992; 45: 36–40.

[24] Fontananova E, Basile A, Cassano A, Drioli E. Preparation of polymeric membranes entrapping β-cyclodextrins and their molecular recognition of naringin. J. Inclusion Phenom. Macrocyclic Chem. 2003; 47: 33–37.

[25] Tasselli F, Jansen JC, Drioli E. PEEKWC ultrafiltration hollow fibre membranes: Preparation, morphology and transport properties, J. Appl. Polym. Sci. 2004.

[26] Trotta F, Drioli E, Moraglio G, Baima Poma E. Sulfonation of polyetheretherketone by chlorosulfuric Acid. J. Appl. Polym. Sci. 1998; 70: 477–482.

[27] Fontananova E, Trotta F, Jansen JC, Drioli E. Preparation and characterization of new non-fluorinated polymeric and composite membranes for PEMFCs. J. Membr. Sci. 2010; 348: 326–336.

[28] Mauritz KA, Moore RB. State of understanding of Nafion. Chem. Rev. 2004; 104: 4535–4585.

[29] Hsu, W. Y. and Gierke, T. D. Elastic theory for ionic clustering in perfluorinated ionomers. Macromolecules. 1982; 15: 101–105.

[30] Xue S, Yin G. Methanol permeability in sulfonated poly(ethretherketone) membranes: A comparison with Nafion membranes. Europ. Pol. J. 2006; 42: 776–785.

[31] Yang B, Manthiram A. J. Comparison of the small angle X-ray scattering study of sulfonated poly (etheretherketone) and Nafion membranes for direct methanol fuel cells. Power Sources. 2006; 153: 29–35.

[32] Logan B E, Elimelech M. Membrane-based processes for sustainable power generation using water. Nature. 2012; 488: 313–319.

[33] Tufa RA, Curcio E, Brauns E, Van Baak W, Fontananova E, Di Profio G. Membrane Distillation and Reverse Electrodialysis for Near-Zero Liquid Discharge and low energy seawater desalination. J. Membr. Sci. 2015; 496: 325–333.

[34] Ersöz M, Güğül İ H, Çimen A, Leylek B, Yildiz S. The sorption of metals on polysulfone cation exchange membranes. Turkish J. Chem. 2001; 25: 39–48.

[35] Saracco G. Transport properties of monovalent-ion-permselective membranes. Chem. Eng. Sci. 1997; 52: 3019–3031.

[36] Jörissen J, Breiter S M, Funk C. Ion transport in anion exchange membranes in presence of multivalent anions like sulfate or phosphate. J. Membr. Sci. 2003; 213: 247–261.

[37] Yaroshchuk A. E. Dielectric exclusion of ions from membranes. Adv.Colloid Interface Sci. 2000; 85: 193–230.

[38] He G, Li Z, Zhao J, Wang S, Wu H, Guiver M D, Jiang Z. Nanostructured ion-exchange membranes for fuel cells: Recent advances and perspectives. Adv. Mater. 2015; 27: 5280–5295.

[39] Ran J, Wu L, He Y, Yang Z, Wang Y, Jiang C, Ge L, Bakangura E, Xu T. Ion exchange membranes: New developments and applications. J. Membr. Sci. 2017; 522: 267–291.

[40] Barsoukov, E.; Macdonald, J. R. Impedance Spectroscopy. Theory, Experiment, and. Applications, Second Edition. John Wiley & Sons, New Jersey, 2005.

[41] Islam N, Bulla N A, Islam, S. Electrical double layer at the peritoneal membrane/electrolyte interface. J. Membr. Sci. 2006; 282: 89–95.

[42] Sang S, Wu Q, Huang K. A discussion on ion conductivity at cation exchange membrane/ solution interface. Colloids Surf. A. 2008; 320: 43–48.

[43] Kavanagh J M, Hussain S, Chilcott T C, Coster H G L. Fouling of reverse osmosis membranes using electrical impedance spectroscopy: Measurements and simulations. Desalination. 2009; 236: 187–193.

[44] Fontananova E, Di Profio G, Giorno L, Drioli E. 2017 Membranes and Interfaces Characterization by Impedance Spectroscopy. In: Enrico Drioli, Lidietta Giorno, Enrica Fontananova (Editors), Comprehensive Membrane Science and Engineering, 2nd Edition, Elsevier B.V. Volume 2: 393–410.

[45] Alberti G, Casciola M, Massinelli L, Bauer B. Polymeric proton conducting membranes for medium temperature fuel cells (110–160°C). J. Membr. Sci. 2001; 185: 73–81.

[46] K.-V. Peinemann, S. Pereira Nunes. Membranes for energy conversion, Wiley–VCH, Chichester, 2008.

[47] Fontananova E, Cucunato V, Curcio E, Trotta F, Biasizzo M, Drioli E, Barbieri G. Influence of the preparation conditions on the properties of polymeric and hybrid cation exchange membranes. Electrochim. Acta. 2012, 66: 164–172.

[48] Bockris J O'M, Diniz F B. Aspects of electron transfer at a conducting membrane-solution interface. Electrochim. Acta. 1989; 34: 561–575.

[49] Park J S, Chilcott T C, Coster H G L, Moon S H. Characterization of BSA-fouling of ion-exchange membrane systems using a subtraction technique for lumped data. J. Membr. Sci. 2005; 246: 137–144.

[50] Antony A, Chilcott T, Coster H, Leslie G. In situ structural and functional characterization of reverse osmosis membranes using impedance spectroscopy. J. Membr. Sci. 2013; 425–426: 89–97.

[51] Park J-S, Choi J-., Woo J-J, Moon S.-H. An electrical impedance spectroscopic (EIS) study on transport characteristics of ion-exchange membrane systems. J. Colloid Interface Sci. 2006; 300: 655–662.

[52] Długołecki P, Ogonowski P, Metz S J, Saakes M, Nijmeijer K, Wessling M. On the resistances of membrane, diffusion boundary layer and double layer in ion exchange membrane transport. J. Membr. Sci. 2010; 349: 369–379.

[53] Xu Y, Wang M, Ma Z, Gao C. Electrochemical impedance spectroscopy analysis of sulfonated polyethersulfone nanofiltration membrane. Desalination. 2011; 271: 29–33.

[54] Wang Y, Wang A, Zhang X, Xu T. The concentration, resistance, and potential distribution across a cation exchange membrane in 1: 2(Na_2SO_4) type aqueous solution. Desalination. 2012; 284: 106–115.

[55] Hunter R J. Zeta potential in colloid science, Academic Press, London, 1981.

[56] Manzanares J A, Murpby W D, Maffè S, Reiss H. Numerical simulation of the nonequilibrium diffuse double layer in ion-exchange membranes. J. Phys. Chem. 1993; 97: 8524–8530.

[57] Gao y, Li W, Lay W C L, Coster H G L, Fane A G, Tang C Y. Characterization of forward osmosis membranes by electrochemical impedance spectroscopy. Desalination. 2013; 312: 45–51.

[58] Cen J, Kavanagh J, Coster H, Barton G. Fouling of reverse osmosis membranes by cane molasses fermentation wastewater: Detection by electrical impedance spectroscopy techniques. Desalin. Water Treat. 2013; 51(4–6): 969–975.

Adele Brunetti, Enrico Drioli, Giuseppe Barbieri

4 Membrane gas separation

Executive summary

Membrane-based gas separation (GS) systems are widely accepted today and in some cases, these systems are used as an operating unit for generation, separation and purification of gases in gas, chemical, petroleum and allied industries. There are many field applications, membrane materials and module solutions that are available worldwide for various fields of interest. However, the growth of large-scale industrial applications is still far from all the real potentialities that the membrane GS offers. Combining the investigations of new materials with improved properties, a key role for the widespread use of this technology is represented by the development of a new knowledge for better utilization of unit operations that are already available on the market in integrated membrane systems and combining various membrane operations in the same industrial process. On this hurdle, the membrane engineering plays a crucial role.

This chapter describes the main aspects related to membrane GS. After a brief introduction to the main transport mechanisms that are involved in the gas permeation and main variables that are usually used for describing it, the chapter provides an overview of the membrane materials and membrane modules now available at the market level or are under study at laboratory scale.

Afterward, a description of the main GS processes involving membranes is presented together with a comparison with traditional technologies currently in use.

4.1 Introduction

In the past few years, membrane engineering has been growing significantly so that membrane operations are the dominant technology in various areas today, for example, in seawater desalination, waste-water and gas treatment and reuse, energy, petrochemistry, artificial organs, food juice treatment and so on. The intrinsic properties of membranes such as molecular separations, possibility of coupling reaction and separation in the same unit and so on contribute to prove membrane engineering as a powerful tool to pursue the process intensification strategy, which is currently the best answer to sustainable industrial growth. In various areas, membrane technology has become competitive also in the GS field with the traditional operations. Since 1970s

Adele Brunetti, Enrico Drioli, Giuseppe Barbieri, National Research Council of Italy, Institute for Membrane Technology (ITM-CNR), University of Calabria, Rende, Italy

https://doi.org/10.1515/9783110281392-004

when commercial-scale GS membrane systems were used for the first time for the separation of hydrogen from petroleum refineries and in the H_2/CO ratio adjustments of the synthesis gas, membrane-based GS systems have made tremendous progress and gained wider acceptance in a variety of applications.

Today, membrane technology for GS is a well-consolidated technique with its applications in various sectors, such as separation of air, H_2 from industrial gases refinery, natural gas dehumidification and separation and recovery of CO_2 from biogas and natural gas.

The significant positive results achieved in GS membrane systems are, however, still far from concretizing all the potentialities of this technology. Currently, problems related to pretreatment of the streams, membrane lifetime and their selectivity and permeability, still slow down the growth of large-scale industrial applications. Combining the investigation of new polymeric, inorganic and hybrid materials, the design and optimization of new membrane plant solutions, also integrated with the traditional operations, will lead significant innovation toward the large-scale diffusion of the membranes for GS. The development of new knowledge for the better utilization of these unit operations in integrated membrane systems, combining various membrane operations in the industrial process will be the key role of membrane engineering in the field of a sustainable industrial growth.

4.2 Fundamentals of GS

In GS, a feed stream enters a membrane module. Under a pressure difference acting as the driving force, some components permeate through the membrane, whereas others are retained in the retentate (Figure 4.1).

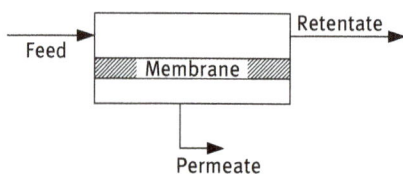

Figure 4.1: Schematic diagram of a membrane module for gas separation.

The transport mechanisms that operate through the membranes strictly depend on the structure, nature and morphology of the membranes used. Basically, two widely used categories can be distinguished within the membranes that are suitable for GS: (1) nonporous and (2) porous. In the case of porous membranes, Knudsen diffusion and molecular sieving are the transport mechanisms involved.

Figure 4.2 shows two membrane pores of different diameter with a number of gas molecules. In the pore with the larger diameter, the gas molecules have more

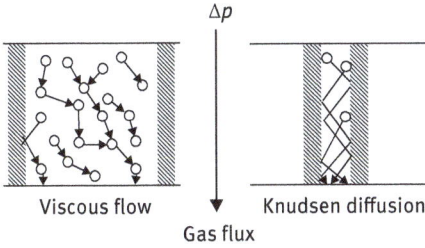

Figure 4.2: Schematic drawing of gas permeation through membrane pores.

interaction with each other, contrast to the narrow pores where the gas molecules have more interactions with the pore wall than with each other. In the case of the larger pores, the energy loss during transportation is mainly owed to the interaction of the molecules with each other. It is expressed by the solution viscosity and the flux that is referred to as the viscous flow. In the pores with smaller diameter, the energy loss during the transportation of gas molecule is because of their interaction with the pore wall. The process is similar to diffusion in a homogeneous phase and is therefore referred to as Knudsen diffusion. It can be considered as flow in narrow pores, that is, pores with a diameter that is smaller than the mean free path length of the diffusing gas molecules. The mean free path length is defined as the average distance a gas molecule travels before it will collide with another gas molecule

$$\lambda = \frac{kT}{\pi d_{gas} P \sqrt{2}} \tag{4.1}$$

where λ is the mean free path length of a gas molecule, k is the Boltzmann constant, d_{gas} is the diameter of the gas molecule and P is the hydrostatic pressure.

The flux in Knudsen diffusion can be described by the following relation:

$$J_i = \frac{\pi n r^2 D_i^k \Delta P}{RT \tau \Delta z} \tag{4.2}$$

where J is the flux through the membrane, n is the number of pores in the membrane per unit of surface area, ΔP is the pressure difference across the membrane, Δz is the thickness of the membrane, τ is the tortuosity and D^k is the Knudsen diffusion.

The Knudsen diffusion is given as follows:

$$D_i^k = 0.66 \, r \sqrt{\frac{8RT}{\pi M_i}} \tag{4.3}$$

It is inversely proportional to the square root of the molecular weight of the gas molecule considered. Thus, the separation of two gases based on Knudsen diffusion is given by the ratio of the square root of the molecular weights

$$\alpha_{j,k} \propto \sqrt{\frac{M_k}{M_j}} \qquad (4.4)$$

Solution-diffusion transport mechanism

The most basic discovery in the field of GS, made in the mid-nineteenth century, was the recognition of the fact that mass transfer of gases and liquids through polymeric films or membranes, liquid films and inorganic septa can proceed even in the absence of opened pores, that is, the idea behind nonporous, permeable membranes was introduced [1].

The basic observations were made by Thomas Graham and John Mitchell. Afterward, an important contribution to the quantitative investigation of the process of gas permeation through membranes was made by von Wroblewski (1879) who proposed the equation for the gas permeation rate or flux J (mol m^{-2} s^{-1}):

$$J_i = Pe_i \frac{P^{\text{Feed}}x_i^{\text{Feed}} - P^{\text{Permeate}}x_i^{\text{Permeate}}}{\Delta z} \qquad (4.5)$$

where J_i, that is, the flux of a component through the membrane, is a function of its molar fractions in the feed and permeate streams, its permeability Pe_i, membrane thickness Δz and the pressures of feed and permeate.

Considering a steady-state isothermal flux through a homogeneous film with thickness λ that separates two gas phases, according to Fick's first law:

$$J_i = -D\frac{dC}{dx} \qquad (4.6)$$

where C is the concentration, x is the coordinate through the film and diffusion D in the first approximation does not depend on C and x. This equation predicts a linear concentration profile within the membrane (Crank, 1975). It can easily be integrated, but boundary conditions $C(x = 1)$ and $C(x = 0)$ are usually unknown (in contrast to pressures P_1 and P_2). If sorption equilibrium condition is assumed, the concentration in the membrane can be correlated with the pressure in gas phase. The simplest case of sorption isotherms is Henry's law:

$$C = S\,P \qquad (4.7)$$

where S is the solubility coefficient. Substituting C with P in eq. (4.6) gives the following equation:

$$J_i = SD\frac{P^{\text{Feed}}x_i^{\text{Feed}} - P^{\text{Permeate}}x_i^{\text{Permeate}}}{\Delta z} \qquad (4.8)$$

By comparing eqs. (4.5) and (4.8), it results in the following:

$$Pe = SD \qquad (4.9)$$

Pe includes two independent terms that characterize a gas–polymer system. The thermodynamic term S determines the affinity of gas membrane and, hence, the driving force within the membrane; the kinetic term D characterizes the response of the system to the superimposed driving force. The Barrer in the honor of Richard Barrer is a widely accepted unit of permeability, which is by definition equals 10^{-10} cm^3(STP) cm/(cm^2 s cmHg) or mol/(m s Pa) in SI units.

The quantity that characterizes the gas permeation rate of a membrane, that is, permeance ($Pe/\Delta z$), is expressed in SI units as cm^3(STP)/(cm^2 s cmHg) or m^3(STP)/m^2 h bar or mol/m^2 s Pa. Permeance is also often expressed in gas permeation units (GPU), where 1 GPU $= 10^{-6}$ cm^3(STP)/(cm^2 s cmHg).

Another key characteristic of membranes is their selectivity in GS. There are several definitions of this property. The ideal selectivity is defined as follows:

$$\alpha_{AB} = \frac{Pe_A}{Pe_B} = \frac{Permeance_A}{Permeance_B} = \frac{Flux_A}{Flux_B}\bigg|_{\Delta P} \qquad (4.10)$$

where Pe_A and Pe_B are the permeability of gases A and B measured as single gases. Commonly, a "fast" gas is taken as A, that is, $\alpha_{AB} > 1$.

Selectivity is a useful parameter for characterizing the capability of membrane separation. For designing a membrane plant, however, the separation factor is more useful. For a binary mixture, the separation factor is defined by the following equation:

$$Separation\ factor_{AB} = SF_{AB} = \frac{x_A^{Permeate} \Big/ x_B^{Permeate}}{x_A^{Feed} \Big/ x_B^{Feed}} \qquad (4.11)$$

4.3 Materials and membranes in gas separation

The criteria for selecting membranes for a given application are strongly related to several factors such as durability, mechanical integrity at the operating conditions, productivity and separation efficiency and so on [2]. Different aspects must be deeply investigated for assuring a good membrane separation process:

1. Membrane transport properties (flux, permeability, selectivity and permeance)
2. Membrane material and structure
3. Module
4. Process design

Polymeric membranes

These are specifically used for GS; they are generally asymmetric and are based on a solution-diffusion transport mechanism. These membranes, made as flat sheet or hollow fibers, have a thin, dense skin layer on a microporous support that provides mechanical strength to them. Currently, however, only eight or nine polymer materials are used for preparing at least 90% of the membranes in use.

Polymeric membranes are a good solution for application in large-scale separations owing to the low cost, ease of processing and high packing density. However, they cannot withstand high temperatures and aggressive chemical environments. Many polymers can be swollen or plasticized when exposed to hydrocarbons or CO_2 at a high partial pressure; their separation capabilities can be dramatically reduced, or the membranes are irreparably damaged. Therefore, pretreatment selection and condensate handling are critical decision factors for the proper operation of GS membrane modules [3].

The key factors to qualify the performance of a specific membrane material for GS applications are permeability and selectivity. It is generally recognized that there is a trade-off limitation for polymeric materials between these two parameters: as selectivity increases, permeability decreases and vice versa. Robeson first proposed this diagram in 1991 [4] and then an update in 2008 [5], it virtually summarized all the existing membrane materials, giving an indication of the best performance achievable by polymeric membrane materials for specific separation (Figure 4.3).

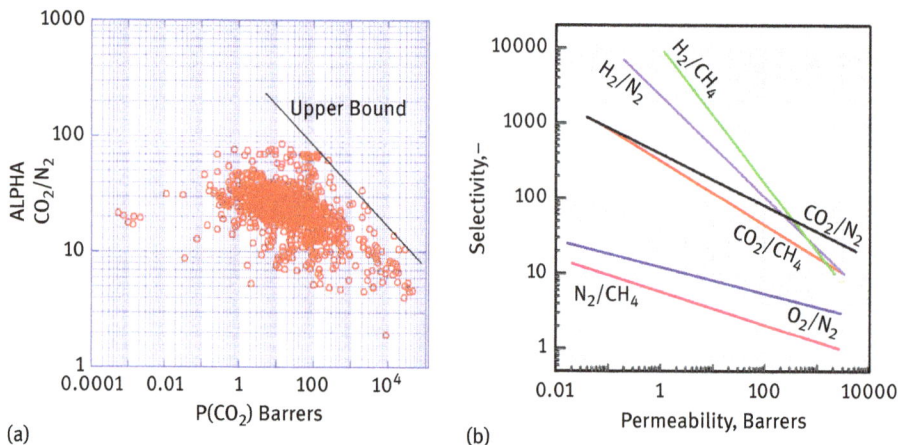

Figure 4.3: (a) Robeson plot example (Published from Robeson [5] with permission from Elsevier); (b) trade-off of Robeson's plot for different GS applications (Published from Brunetti et al. [24] with permission from Wiley).

Owing to the trade-off limitation, there is a strong requirement to develop high-performance polymer membranes with superior thermal, chemical, mechanical and long-term stabilities for GS. Thus, recently developed polymer materials are suggested to be excellent candidates. Based on solution-diffusion mechanism, glassy polymers represent superior diffusivity selectivities as well as extraordinary diffusivities, whereas rubbery polymers have displayed high solubility selectivities. Polyether membranes, which contain ether linkages possessing strong chemical interaction with carbon dioxide, have very promising permeabilities as well as selectivities as an efficient CO_2 separation membrane [6]. So far, *poly(ethylene oxide) (PEO)* with ether oxygen was reported to be the most useful groups to provide excellent CO_2 separation and permeation properties [7]. However, these polymers possess low mechanical and thermal resistances and are strongly affected by plasticization. To compensate these weak properties, maintaining the advantages of *poly (ethylene oxide), block copolymers* are prepared to have a hard block composed of glassy polymer unit as well as a soft block of ethylene oxide unit [8]. CO_2 permeabilities usually range from 50 to 200 Barrers (at most, 650 Barrer), whereas CO_2/N_2 selectivities are more than 40–50; this is enough to enrich CO_2 concentration over 90% at a single stage from flue gas. On the other hand, performances are significantly affected by the operating temperature that is mainly because of the solubility of CO_2 and morphologies of the multi-block copolymers.

Although conventional glassy polymers are known to be dense without the presence of micropores except free volume elements resulting from chain rotation in the matrix, *polymers with intrinsic microporosity (PIMs)* is a kind of microporous polymer containing microporous cavities owing to the mixed conformation of highly stiff ladder-like domain and flexible benzodioxan structure. Their rigid structures can retain an efficient size of free volume elements for permeation of small gas molecules, which allow high diffusivities. PIMs have a large number of free volume elements by which they have high surface areas ($500–1000 \ m^2 \ g^{-1}$) with microcavity diameters in the range of 0.6–0.8 nm. A *thermally rearranged polymer* is a novel microporous material prepared by chain rearrangement of polyimide at elevated temperature that results in the evolution of microcavities with an intermediate cavity size, a narrow cavity size distribution and a shape reminiscent of bottlenecks connecting adjacent chambers [9]. The free volume elements in thermally rearranged polymers are believed to be three-dimensional networks of intermolecular microcavities, which are accessible for small gas molecules. This peculiarity of a free volume structure accounts for both outstanding permeability of thermally rearranged polymers with fast diffusion of gases and their still high permselectivity in the separation of small molecules. The increase of fractional free volume in the polymer matrix up to 30% is comparable to PTMSP (Poly[1-(trimethylsilyl)-1-propyne]), a well-known polymer with the highest gas permeability. The gas permeabilities of almost all TR (Thermally Rearranged) polymer membranes are enhanced by at least two orders of magnitude over those of original polymers and typical glassy polymers, confirming the presence of larger interconnected free volume elements. The gas permeabilities of TR polymer membranes are lower than those of PTMSP membrane, but gas selectivities of CO_2

separations (e.g., CO_2/CH_4 and CO_2/N_2) are two to three times higher than for PTMSP.

In order to enhance the properties of the polymeric membranes, new *mixed matrix membranes* were recently introduced. Their microstructure consists of an inorganic material in the form of micro- or nanoparticles (dispersed phase) incorporated into a polymeric matrix (continuous phase). The inorganic fillers usually are zeolite, silica, TiO_2 and also carbon nanotubes. The use of two materials with different flux and selectivity provides the possibility to better design a membrane, for example, CO_2 capture that allows the synergistic combination of polymers' easy processability and superior GS performance of inorganic materials. Furthermore, the addition of inorganic materials in a polymer matrix offers enhanced physical, thermal and mechanical properties for aggressive environments and represents a way to stabilize the polymer membrane against changes in permselectivity with temperature. These membranes offer very interesting properties; however, their cost, difficulty in commercial-scale manufacturing and brittleness remain as an important challenge to be faced. The main difficulty in mixed-matrix membrane fabrication is related with the control of adhesion between the polymer phase and the fillers' external surface, especially when using zeolites. For this reason, new materials based on metal organic frameworks were recently studied [10]. *Facilitated transport membranes* in which a *carrier* (typically, metal ions) with a special affinity toward a target gas molecule is mixed in the polymeric matrix also showed interesting results in GS [11]. There are several types of facilitated transport membranes; their drawbacks are related to the low fluxes [12].

As already mentioned, almost all industrial GS processes use polymeric membranes, a significant interest is focused on the development of inorganic membranes (e.g., metal, zeolite, ceramic and carbon), particularly for their use in high temperature separations and membrane reactors. *Metal membranes* are specifically used for the purification of specific gaseous streams, for example, Pd-based membranes for hydrogen purification or silver membranes for pure oxygen production. Several metal membranes, including tantalum, niobium, vanadium, copper, golden, iron, cobalt and platinum are used for hydrogen separation. These membranes are extraordinarily selective because they are extremely permeable to hydrogen and essentially impermeable to all other gases. They are generally used for the production of pure stream membranes such as in the case of Pd–Ag membrane reactors that are widely studied for the production of pure hydrogen reactions of high industrial interest (water gas shift, steam reforming of light hydrocarbons, dehydrogenation, etc.) [13]. However, they must be operated at high temperatures (>300 °C) to obtain useful permeation rates and to prevent embrittlement and cracking of the metal by adsorbed gas. Additional drawbacks are high cost, low permeability in the case of highly selective dense membranes (e.g., metal oxides at temperatures below 400 °C) and difficultly in sealing at high temperatures (greater than 600 °C).

Zeolite membranes
These show high thermal stability and chemical resistance compared with those of polymeric membranes. They are capable of separating mixtures continuously on the basis of differences in the molecular size and shape [14], and/or on the basis of different adsorption properties [15–19], since their separation ability depends on the interplay of the adsorption equilibrium mixture and the mixture. Different types of zeolites were studied (e.g., MFI, LTA, MOR and FAU) for the membrane separation. In the field of GS, they are still used in laboratory-scale, as catalytic membranes in CO clean-up, water gas shift, methane reforming and so on [20, 21]. The reason for this limited application in industry might be because of economic feasibility (development of higher flux membranes should reduce both cost of membranes and modules). However, currently they found large application in pervaporation.

Among the inorganic membranes, *carbon molecular sieve membranes* show good transport properties that allow their applications in GS. They are obtained by pyrolysis (at a high temperature in an inert atmosphere) of polymeric precursors that are already processed in the form of membranes [22]. These membranes combine good gas transport properties for light gases (gases of molecular sizes smaller than 0.4–0.45 nm) with thermal and chemical stability. However, the major disadvantages that hinder their commercialization are their brittleness, which means that they require careful handling and when compared to polymeric membranes, the cost of carbon-based membranes 1 to 3 orders higher per unit area.

Ion transport membranes
These are new dense inorganic membranes that are capable of permeating only by oxygen (or hydrogen) ions. They show good performance in terms of permeability and selectivity at a very high temperature (>600 °C); however, the main problem is related to their durability. Once the time of operation is passed, the formation of micro-pinholes depletes the membrane properties, thus significantly reducing the selectivity.

A more detailed analysis on the state of the art of the field of membrane GS in which the mass transport properties of many different membrane materials are reported that can be found in [3].

4.3.1 Membrane modules

In order to apply membranes on a technical scale, large membrane areas are normally required. The smallest unit into which the membrane area is packed is called "module" and is the central part of membrane installation [23].

In designing membrane module, some general requirements need to be fulfilled:

1. Good mechanical, thermal and chemical stability
2. Good flow distribution (no dead zones, no bypass)
3. High packing density
4. Low pressure drop
5. Possibility of cleaning
6. Ease of maintenance and operation
7. Cheap manufacturing
8. Compactness of the system scale
9. Possibility of membrane replacement

The importance of each characteristic with respect to the others is strictly correlated to the final application. Usually, the main interest of module design is a high packing density, which implies lower manufacturing costs. A number of module designs are possible and all are based on two types of membrane geometry: (a) flat sheet and (b) cylindrical [23]. Typical dimensions are shown in Figure 4.4.

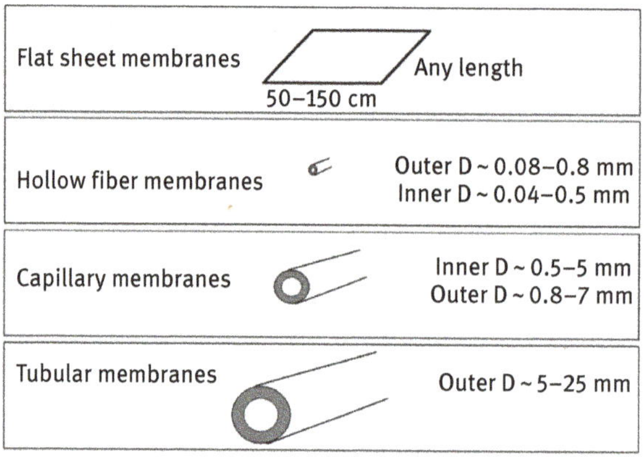

Figure 4.4: Available types and geometry of membranes. Published from Scholz et al. [23] with permission of The Royal Society of Chemistry.

There are three major types of module configurations for GS membrane modules: (1) plate-and-frame, (2) spiral wound and (3) hollow fiber.

In a *plate-and-frame module* configuration, two membranes are placed in a sandwich-like shape with their feed sides facing each other. A spacer is placed between each feed and permeate compartment. Several sets of two membranes constitute a stack. The packing density (membrane surface per module volume) of such modules is low and about 100–400 m^2/m^3 [23]. *Spiral-wound modules* are made from flat membrane envelopes, wrapped around a central tube. The feed distributed by the central tube passes along the module spiral and the permeate passes into a membrane envelope and then goes out via the peripheral collector. Modern modules tend to contain multiple membranes that are all attached to the same central tube. Usually in spiral-wound membrane modules, the size of the feed channel is greater than the active membrane area. In particular, the feed channel spacer is typically about 20% wider than the membrane envelope. This additional width may promote flow perpendicular to the bulk flow (along the spiral). Currently, the spiral wound contains around 1–2 m of rolled sheets, for 20–40 m² of membrane area [23].

Hollow-fiber modules contain a large number of membrane fibers housed in a shell. The free ends of the fibers are potted with agents such as epoxy resins, polyurethanes, silicone rubber, thermoplastics or inorganic cements. Feed can be introduced on either the fiber side or the shell side. Permeate is usually withdrawn in a cocurrent or countercurrent manner, with the latter being generally more effective (Figure 4.5).

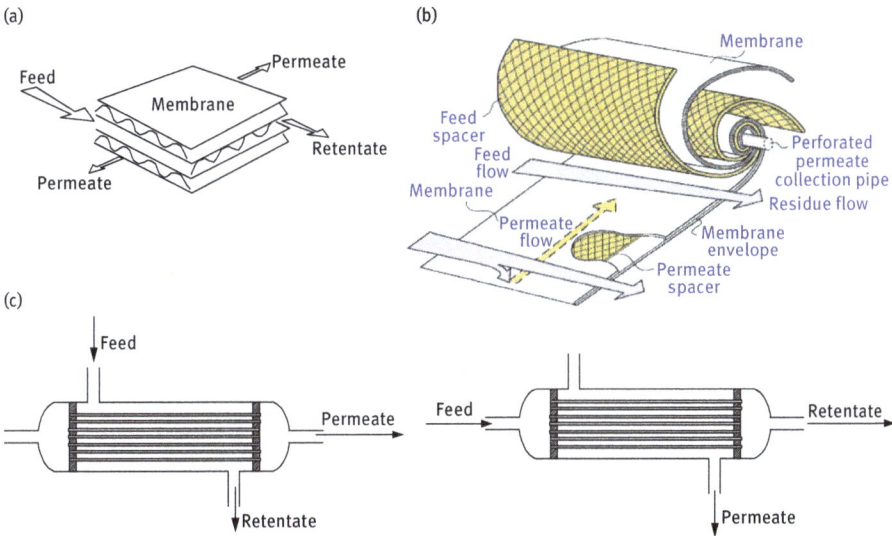

Figure 4.5: Schematic diagram of membrane module configurations: (a) plate and frame (Reproduced from Scholz et al. [23] with permission from The Royal Society of Chemistry) and (b) spiral wound (published from Brunetti et al. [24] with permission from Wiley, (c) hollow fiber.

Each membrane module configuration has advantages and drawbacks. Hollow fibers are the cheapest on a per-square-meter basis (with the highest membrane area to module volume ratio); however, to make very thin selective layers on the hollow fiber is harder than making in flat sheet configuration. This implies that the permeance of hollow fibers is generally lower than that of a flat sheet membrane prepared with the same material and, thus, having the same permeability. Therefore, larger membrane area is required for achieving the same separation by hollow fiber modules. Hollow fiber modules also require more pretreatments of the feed than is usually required by spiral wound modules for removing particles, oil residue and other fouling components. These factors strongly affect the cost of the hollow fiber module design;

therefore, currently, spiral wound modules are employed in several separations (e.g., in natural gas processing), particularly for those separations that cannot support the costs associated with the hollow fiber modules. Spiral-wound modules are also used where pressure drop has to be considered and when countercurrent flow is not needed to maximize the separation efficiency. The choice of the membrane module is also determined by economic considerations, even if it is in industrial plants; especially in refinery and petrochemical operations, the module costs are only 10% to 25% of the total costs, so that significant reductions in membrane costs might not markedly change the cost of the whole plant [25]. In a hollow fiber spinning plant operating continuously, the membrane costs are in a range of US$2–5 m^{-2} of membrane area. An equivalent of spiral-wound modules would cost 10 to US$100 m^{-2}.

A membrane GS production process can be realized by assembling the membrane modules in several configurations, depending on the particular type of separation. The single stage, double stage and multistage with recycling constitute the main design solutions. Each module can be operated with different flow patterns (Figure 4.6) in which the feed and the permeate stream flow through the module [23]:

1. Cocurrent
2. Countercurrent

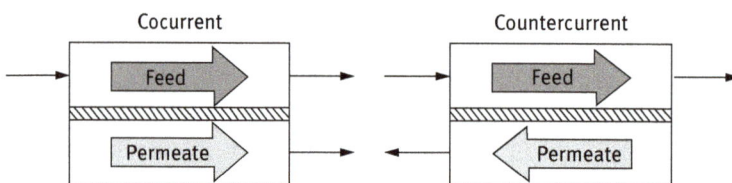

Figure 4.6: Membrane module flow patterns. Published from Scholz et al. [23] with permission from The Royal Society of Chemistry.

In cocurrent flow, the feed and the permeate flow are parallel in the same direction. In countercurrent flow, the feed and the permeate flow parallel but in the opposite direction. The overall driving force in countercurrent flow is higher compared to cocurrent and thus, it is preferred in most of the applications, in particular those with retentate products.

4.4 Conventional technologies for separation of gases

Cryogenic distillation, absorption and adsorption are the traditional technologies for GS at commercial scale. Cryogenic separation is universally used for the large-scale separation of atmospheric gases. It requires the liquefaction of the gases that are distilled at cryogenic temperatures for separating gaseous mixture into its components. The presence of water gives serious problems of cavitation. Absorption technologies are well established for the scrubbing of carbon dioxide and removal of water from natural gases. In gas absorption process, one or more components of the gas mixture are chemically and physically absorbed in a solvent in a gas–liquid contacting device. The separation of the nonabsorbing/nonreacting gases from the gas mixture is thus achieved. Both these methods are complex and capital intensive but cost competitive. The adsorption process consists of selective adsorption of gaseous species on the high surface area of solid particles to contact with the gas mixture, where one component of the gas mixture adsorbs leaving others behind. The main disadvantage of this technology is of being a cyclic process; therefore, for continuous separation of gases, one needs to use more than one adsorbent bed depending on the required regeneration stage after saturation (or breakthrough). Adsorbent processes are best suited to produce high purity gases, especially for the removal of trace gas impurities from a gas mixture.

Currently, membrane technology for GS is a well-consolidated technique in various cases that are competitive with traditional operations. The greatest asset to membrane separation is simplicity. While PSA (Pressure Swing Adsoprtion) requires equipment for swinging pressure, cryogenic distillation must endure extreme temperatures and absorption requires huge amount of sorbent; the only equipment necessary for GS are the membrane and fans. There are almost no moving parts, and the construction is fairly simple. The gaseous stream to be separated generally requires a compression, but this is much smaller than that necessary for PSA.

4.5 Current applications of membranes in gas separation

Since 1950, Weller and Steiner [26] considered membrane processes as feasible for the separation of hydrogen from hydrogenation tail gas, enrichment of refinery gas and air separation. However, commercial-scale GSs using membrane systems were applied for the first time in late 1970s to early 1980s. Their applications were limited to separation of hydrogen from petroleum refineries and in the H_2/CO ratio adjustments for the synthesis gas. In 1980, Permea, with its hydrogen separating prism membrane, launched the first large industrial application of GS membranes [27]. Since then, membrane-based operations, substituting or to be integrated with the traditional ones, had a rapid growth, with many companies, such as Cynara-Natco, Separex-UOP, GMS, Generon, Praxair, AirProducts and UBE that are involved in this field [28, 29]. Numerous membrane GS systems are in operation today for a wide variety of separation applications. Although a lot of research is underway in developing novel membrane materials, until now, only polymeric membranes are commercially used in large scale.

4.5.1 Hydrogen recovery

Currently, the growing request of hydrogen in many industrial processes pushes toward new H_2 sources and production. However, the separation of hydrogen is the most important issue in its production cycle. The first widespread commercial application of membranes in GS was the hydrogen separation in the ammonia purge stream, by using Prism^{TM} systems equipped with Polysulfone (PSF) membranes with silicon (PDMS) "caulking" layers and belonging to Permea. Since that time, other membrane-based processes using the same system for hydrogen recovery were developed by UOP with the Polysep systems and Monsanto [30]. The PRISM system uses hollow fiber membranes and is today the most important membrane technology for hydrogen separation. More than 500 PRISM membrane systems for GS application are worldwide used. Still today, the most important application is hydrogen recovery in ammonia purge stream with 230 plants installed worldwide [31]. The PRISM membrane systems treat the purge stream of the ammonia reactor. Usually, the system has a water scrubber unit for ammonia recovery. The gas stream is fed into a membrane GS unit with hollow fibers operating at 110–130 bar. The gas composition includes high concentration of hydrogen (about 66.5%) and nitrogen (about 22.2%). Membrane system is able to recover a stream at a lower pressure (25–70 bar) with a hydrogen concentration up to 94% and a hydrogen recovery up to 90%. The latter is fed into a compressor unit and is recycled into the reactor. Usually, the membrane module height is about 3 m and is 10 cm or 20 cm in diameter.

The PRISM technology is also applied in other separations, for example, methanol production or syngas ratio adjustment. The methanol/hydrogen stream coming out from the reactor can be further treated in a GS unit for downstream processing to enrich methanol stream in the retentate with hydrogen separated in the permeate that can be recycled to the reactor. The very first PRISM membrane system for the SynGas (H_2/CO) ratio adjustment was built in 1977. Generally, membranes are used for stripping hydrogen out of the syngas in order to reduce the H_2/CO ratio with a feed stream at 48% of hydrogen and 51% of carbon monoxide, a permeate stream with 88% of hydrogen and about 11% of carbon monoxide is obtained, with a retentate stream at a very high CO concentration (about 95%) [32].

The demand of hydrogen recovery in refineries is also rapidly increasing for environmental regulations. The hydrogen content in the various refinery purges and off-gases ranges between 30%–80%, with hydrogen mixing with light hydrocarbons (C_1–C_5); 90%–95% hydrogen purity is required for recycling it to a process unit. A typical refinery operation is the separation of the hydrogen contained in the stream coming out from the hydrocracker. The membranes can be used alone or together with an absorber system, at a reduced capital cost and better process efficiency. At the moment, the Prism^{TM} system (using polysulfone hollow fibers with a thin silicone film on it) is dominant on the market for this type of separation, showing interesting selectivities.

Large room of operation was recently found by the VaporSep technology of Membrane Technology & Research Inc. (MTR) [33]. With this solution, the ammonia production increases by 4%–5% with respect to the traditional system, without increasing gas feed to the reformer. The decreasing gas consumption, easy to operate, room-temperature operation and simple installation are the most important benefits of this system, together with the compact gross dimension (6 m × 3 m × 2.5 m [length × wide × height]). VaporSep technology can be used in the same application of the PRISM technologies: (1) SynGas ratio adjustment [34] and (2) enrichment of methanol stream [35].

4.5.2 Air separation

Separation of air into nitrogen- and oxygen-enriched streams by using membrane has grown fast in the past decades. The most used membranes are oxygen selective; therefore, the nitrogen-rich stream is recovered in the high pressure side (retentate), whereas O_2-enriched stream is obtained as permeate at a low pressure. As reported by Baker [25], the first membranes used for this separation showed an O_2/N_2 selectivity of ca. 4. Approximately the same selectivity was obtained for asymmetric PVTMS (poly(trimethylvinylsilane)) membrane, the first GS membrane manufactured in industrial scale. Currently, nitrogen separation by membrane systems is the largest GS process in use. Membrane selectivity does not need to be high in order to produce a relatively pure nitrogen stream, thus they became the dominant technology instead of PSA or cryogenic distillation. Thousands of compact onsite membrane systems generating nitrogen gas are currently installed in the offshore and petrochemical industry.

Air Products Norway has delivered more than 670 PRISM® systems producing N_2 for different ship applications, and more than 160 PRISM® systems for offshore installations [36]. In December 2006, Air Products started with another PRISM® production plant in Missouri (U.S.) [37]. Another new air separation unit with a capacity of 550 ton/day of oxygen was installed by Air Liquide in Dalian (China) [38]. In Japan, Ube Industries [39] is increasing the production of polyimide hollow fibers for nitrogen separation to introduce a number of ethanol refining plants, mainly in the United States and Europe, driven by the rapid increase in the demand for bioethanol as an additive for oil products.

Among the various, PermSelect® [40] technology is one of the most important membrane systems for air separation. It uses silicone membrane, mostly PDMS. Oxygen is roughly twice more permeable than nitrogen in this membrane, so the permeate is oxygen rich (nitrogen poor) and the retentate stream is nitrogen enriched. The system allows to recover a high nitrogen concentration in the retentate stream (>99.9%). In addition, the recovery of an oxygen-enriched stream with less nitrogen than in the air content brings advantages in any process where inert nitrogen has a ballast effect.

Other silicone membranes were recently introduced in the market by the GRASYS company that developed many system processes for chemical industries, such as Exxon Mobile, ENI, Shell, Gazpronand so on. The modular membrane systems cover a considerable range in terms of nitrogen purity from 90% to 99.9% and nitrogen production capacity –10 to 3,150 Nm^3/h (Figure 4.7).

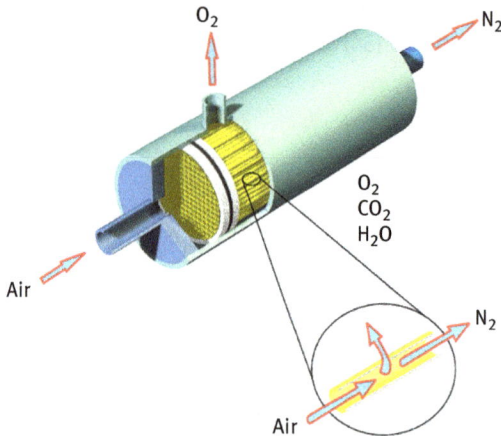

Figure 4.7: PermSelect module for the oxygen capture. (Source: https://en.wikipedia.org/wiki/Membrane_technology. CC0-1.0 public domain). Last access: 14/02/2018.

The hollow fiber membrane consists of a porous polymer layer with a rubbery polymer layer, which is the real selective separation layer. Mostly, the composition of the fiber is the same as that of the PermSelect® technologies.

4.5.3 Air drying

The removal of water vapors from atmospheric air is commercially practiced by the use of refrigeration or adsorption methods. These processes are not energy efficient as a significant energy amount is consumed in condensing water vapor. In some cases, adsorbents are used to capture water from air and are very effective, especially for the production of very dry air. However, they need regeneration step (heat, vacuum or sweep) to remove adsorbed water. Membrane systems are very attractive for air drying applications since almost all the polymers have higher water permeability than air permeability, even though to make a membrane air drying system with low air loss, the selectivity of the membrane in the module should be >1,000. Various commercial products use either a glassy polymer such as polysulfone, polyimide or ionomers that are fluorine-containing polymer membranes. In these membrane systems, wet air is fed on one side of the membrane and the water vapor permeates through it retaining dry air in the retentate. Air drying membrane modules are produced by the major membrane manufacturing companies such as Air Products, Air Liquide, Ube and Asahi Glass. However, most of the air dryer business is handled by other equipment manufacturers (OEMs). Commercial membrane modules can handle flows from 0.1 to 2.3 m³/min. The membrane system is capable of supplying oil and particulate free, dry compressed air to dew point as low as −40 °C at air pressure of 2 to 20 bar [41].

4.5.4 Hydrocarbons separation

Silicon containing membranes are currently dominant in the membrane GS technologies market for hydrocarbon separation and the most important companies on this process are MTR (United States) and licensees of GKSS technology (Europe). The main industrial applications are as follows [42]:
1. Ethylene recovery
2. Polyolefin plant resin degassing
3. Gasoline vapor recovery systems at large terminals
4. Polyvinyl chloride manufacturing vent gas
5. Natural gas processing/fuel gas conditioning

In all the cases, most of commercial plants use silicone rubbery membranes (e.g., PDMS), owing to the high permeability, which allows smaller membrane area with respect to conventional glassy polymers to be used, combined with the adequate vapor/inert gas selectivity for most of the applications.

The *recovery of hydrocarbon monomers from ethylene in polyethylene and polypropylene* plants is actually the largest application of vapor membrane-based separation. After the production of the polyolefin resin, unreacted monomer and hydrocarbon solvents that are dissolved in the resin powder must be separated in order to reuse the polymer. The traditional application involves stripping with hot nitrogen in a column known as a "degassing bin." The value of nitrogen and monomer are both high; therefore, the recovery and reuse of these components is of great interest. For this scope, a membrane operation is profitably used. It consists of two membrane units in series where the off-gas from the "bin" is compressed at 200 bar. The first membrane unit produces a permeate stream enriched with propylene and a purified residue stream containing 97%–98% nitrogen. The vapor-enriched permeate stream is recycled to the compressor. The nitrogen-rich residue can often be directly recycled to the degassing bin without further treatment. The residue gas is passed to a second membrane unit to upgrade the nitrogen more than 99% purity. The hydrocarbon-rich stream of the second unit is sent to flare. The spiral-wound membrane modules are allocated in the horizontal tubes around the compressor. This unit recovers 500 kg/h of hydrocarbons. During the past 20 years, more than 50 of these systems were installed all over the world (Figure 4.8) [43].

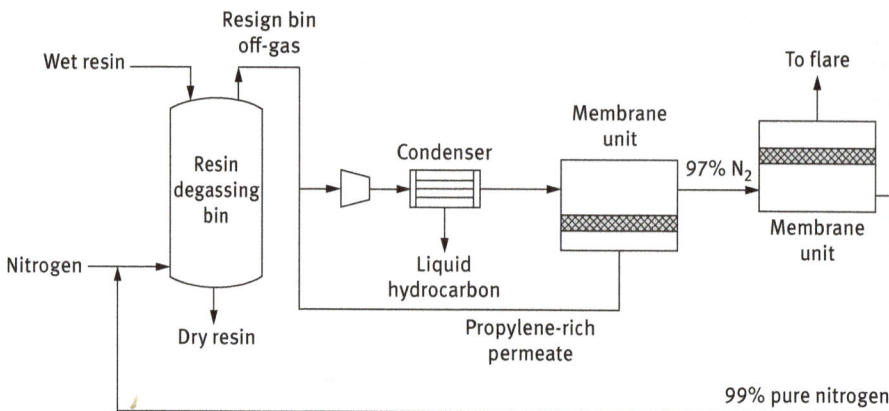

Figure 4.8: Scheme of hydrocarbon/nitrogen membrane plant for hydrocarbon recovery http://www.mtrinc.com/publications/MT01%20Fane%20Memb%20for%20VaporGas_Sep%202006%20Book%20Ch.pdf. (accessed 14/02/2018) [43].

Ethylene oxide is produced through the catalytic oxidation of ethylene with 99.6% pure oxygen; carbon dioxide and water are byproducts. The mixture of products is sent to a water-based scrubber to recover the ethylene oxide. Carbon dioxide is then absorbed with hot potassium carbonate; fresh ethylene and oxygen are added to the unreacted gases and the mixture is recycled back to the reactor. Owing to the presence of argon in the incoming oxygen and ethane in the incoming ethylene, part of the gases in the reactor loop must be purged for keeping the concentration of these inerts under control. The purge gas for a typical ethylene oxide plant contains approximately 20%–30% ethylene, 10%–12% argon, 1%–10% carbon dioxide, 1%–3% ethane, 50% methane and 4%–5% oxygen. This purge gas can be treated in a membrane-based recovery unit: ethylene preferentially permeates through the membrane, producing an ethylene-enriched permeate stream and an argon-enriched residue stream.

The *gasoline vapor recovery* became an important field for membrane application in the past few years. Several hundred retail gasoline stations, in fact, have installed small membrane systems for the recovery of the hydrocarbon vapors during the transfer of hydrocarbons from tankers to holding tanks and then to trucks. Generally, the hydrocarbon concentration in the emitted gas is in the range of 10%–30%. In the range of 3%–15%, the hydrocarbon/air mixture is dangerous because of hydrocarbon explosion. In the membrane system, the vapor hydrocarbon stream is fed in a membrane unit for separation. GKSS licences have installed about 30 gasoline vapor recovery systems at fuel transfer terminals, mostly in Europe (www.gkss.de). MTR and OPW fueling components have developed a membrane vapor recovery system for fuel storage tanks of retail gasoline stations. The OPW Vaporsaver™ system, fitted with MTR's PDMS-based membranes, reduces hydrocarbon emissions by 95%–99% and pays for itself with the value of the recovered gasoline (Figure 4.9).

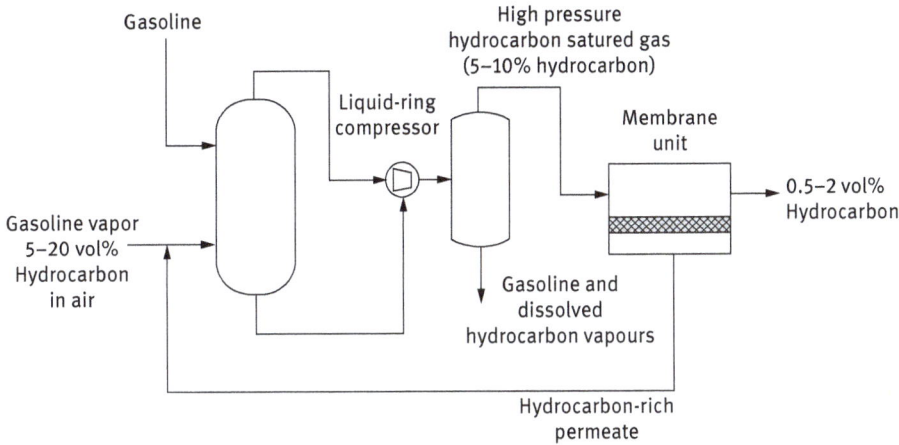

Figure 4.9: Scheme of membrane recovery of hydrocarbon in a gasoline plant. From http://www. mtrinc.com/publications/MT01%20Fane%20Memb%20for%20VaporGas_Sep%202006%20Book% 20Ch.pdf. (accessed 14/02/2018) [43].

In the *polymerization of polyvinyl chloride*, side reactions generate unwanted gas and some small amounts of air leak into the reactors. These inert gases must be vented from the process. However, the vented gas stream, although small, may contain monomers of high value, such as vinyl chloride. Feed gas containing vinyl chloride monomer and air is sent to the membrane system. The vinyl chloride monomer-enriched permeate from the membrane system is compressed in a liquid-ring compressor and cooled to liquefy the vinyl chloride monomer. The nonconden-sable gases are mixed with the feed gas and returned to the membrane section. Vinyl chloride monomer recovery is more than 99%. The first unit of this type was installed by MTR in 1992. Since then, about 40 similar systems were installed.

4.5.5 Volatile organic compound separation

The recovery of volatile organic compounds (VOCs) is an important application in the petrochemical industries. Currently, various industrial-scale plants are designed and built by MTR [44], OPW and Vaporsaver by using silicon rubber (PDMS) and PTMSP polymer membranes that exhibit preferential selectivity versus VOCs than air.

The system is designed to remove VOCs from the air and vapor stream, producing a concentrated VOC liquid phase and a clean air and vapor stream with less than 10 ppm by weight [44]. In particular, the MTR membranes have a composite structure constituted by a microporous layer with dense permselective coating in spiral-wound modules that can be connected in series or in a parallel flow arrangement to meet the flowrate and separation requirements of a particular application [44]. The feed air typically includes water vapor (1%–2%) and VOCs (0.2%). The stream is compressed (15 bar) and it is fed in a cooler. Here, water vapor and some of the VOCs condenses. The air leaving the air cooler enters the membrane GS units where VOCs are separated from the rest of the stream and recovered in the permeate side of the membrane modules. The retentate stream is fed in the second membrane module for obtaining a clean air stream. The concentrated VOCs stream, recovered in the permeate of the first membrane module, is fed into the heat exchanger for condensation, and the gas stream is fed in a third membrane GS system. Here, VOCs-enriched stream returns into the heat exchanger in a concentration loop for improving the VOCs condensation. The air stream returned into the feed stream of the plant is recovered as liquid VOCs (Figure 4.10).

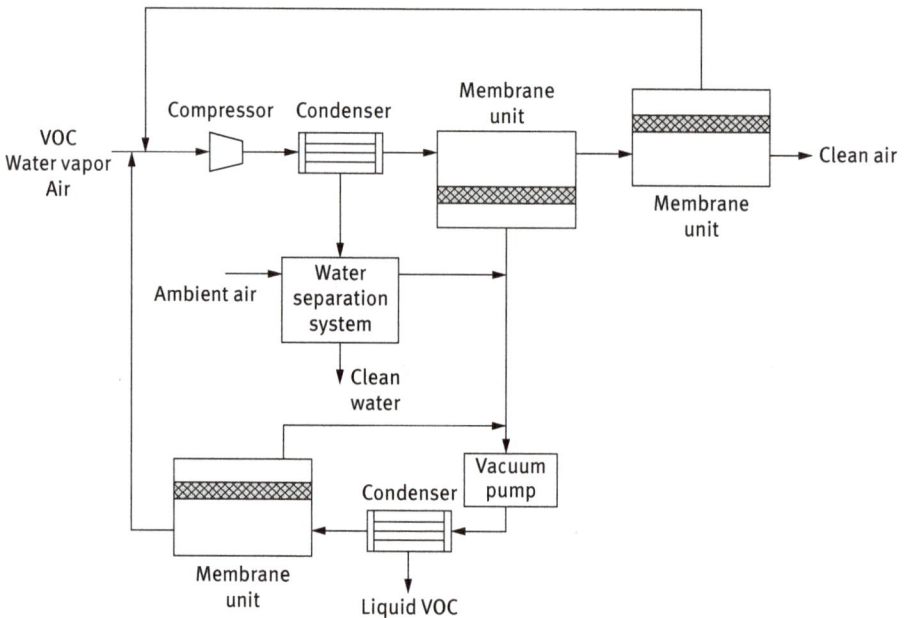

Figure 4.10: Scheme of membrane recovery of VOCs for clean air production by air/VOCs mixture. From https://www.dndkm.org/DOEKMDocuments/ITSR/TRUMixedWaste/Membrane_System_for_ the_Recovery_of_Volatile_Organic_Compounds_from_Remediation_of_Off-Gases.pdf. (accessed: 14/02/2018) [44].

4.5.6 CO_2 separation

Carbon dioxide is produced in huge quantities in various sectors; power and hydrogen production, heating systems (e.g., in steel and cement industries), natural gas and biogas purification and so on are some examples. CO_2 separation from hydrogen and methane streams was used because of high value of these streams [45–49]. Recent constrains and regulations on CO_2 emissions have focused on its separation from flue gas streams where

N_2 is the more relevant species (ca. 80%), whereas CO_2 concentration range is 5%–20% [50–58]. The U.S. Department of Energy (DOE) estimates that post-combustion capture using conventional solvents will increase the cost of electricity by about 80% and incur a US\$68/ton avoided cost for CO_2 [59]. Considering that CO_2-containing streams coming from power plants or heating systems are waste with no "profit" margin involved in their treatment, a significant separation cost (no less than US\$20–25 per ton) would significantly affect the final cost (e.g., electricity) (Table 4.1).

Table 4.1: Typical sources of CO_2 emissions.

	Source	Separation	Feed composition	Temperature and pressure	Ref.
Flue gas streams	Power plants	CO_2/N_2	5%–25% CO_2 65%–80% N_2 3%–5% O_2	35–100 °C and 1 bar	[14, 60]
	Coal gasification plants Steel factories Cement factories Transportation		Rest N_2, SO_x, H_2S, H_2O		
Natural gas	Natural gas pipes	CO_2/CH_4	1%–8% CO_2 70%–90% CH_4 0%–20% C_2H_6,	25 °C–30 °C and 1.2 bar	[61, 62, 63]
	Sweetening of natural gas and so on		C_3H_8, C_4H_{10} Rest O_2, N_2, H_2S, Ar, Xe, He		
Biogas	Various		34%–40% CO_2 50°C–70% CH_4 Rest N_2, O_2, H_2S, H_2O	25 °C–35 °C and 1 bar	[64]

Membrane operations are now being explored for CO_2 capture from power plant emissions and other fossil fuel-based flue gas streams, owing to their interesting engineering and economic advantages over competing separation technologies.

Various materials can be considered suitable for the separation of CO_2 from flue gas or methane streams [65–77], and many advances were made in the maximization of their mass transport properties.

Natural gas membrane processing
Natural gas is mainly (from 75% to 90%) composed of methane; it also contains undesired components such as acid gaseous impurities like carbon dioxide or hydrogen

sulfide that should be removed to prevent pipeline corrosion, condensable compounds (higher C2$^+$ hydrocarbons and water) that must be removed in order to prevent condensation troubles, hydrate formation or corrosion when flowing natural gas in pipelines, inert gases, such as nitrogen that lower the calorific value of natural gas.

Removal of carbon dioxide (*natural gas sweeting*) increases the calorific value and transportability of the natural gas stream. Carbon dioxide content in the natural gas obtained from the gas or oil well can vary from 4% to 50%. It has to be reduced down to ca. 2%–5%. This goal is typically achieved by means of absorption with an aqueous alkanolamine solution that has as the main drawback of tendency to corrode equipment and to lose amine properties by degradation, as well as the amine emissions [78]. Membrane GS systems is an alternative technology for separation of carbon dioxide from the natural gas, particularly for offshore applications [79]. The current level of CO_2/CH_4 selectivity of commercial membranes ranges between 12 and 25 in field conditions [62]. As a consequence, the relatively moderate values of CO_2/CH_4 selectivity offered result in a partial loss of the treated methane in the low-pressure permeate. For this reason, process optimization calculation led to the introduction of two-stage configurations, where in the first stage the permeate was pressurized and fed to a second permeation stage, recovering a part of the methane. This generally leads to a significant improvement of the overall methane recovery (higher than 95%), though introducing extra cost related to the interstage compression unit [62].

Membrane systems can also be integrated with traditional units. The design of a hybrid membrane separation system depends on several aspects, such as membrane permeance and selectivity, CO_2 concentration of the inlet gas and the target required, the gas value (per ca. 30 Nm^3, the price of gas in 2007 was US$6–7 in the United States, whereas in Nigeria, which is far from being called as a well-developed gas market may be as low as US$0.50 if the gas can be used at all) and the location of the plant (on an offshore platform, the weight, footprint and simplicity of operation are critical; onshore, total cost is more significant) [80].

Several natural gas reserves are considered as subquality because of the high nitrogen content. The gas pipeline specifications for inert gases, in fact, fix the upper nitrogen content to a 4% limit [81]. Currently, the cryogenic distillation is used for this separation; however, membrane technology could be used here. The only challenge is the reduction of the methane loss in the permeate. However, methane-permeable membranes can be used conveniently in combination with a cryogenic plant (Figure 4.11). The feed gas, containing 15% nitrogen, is separated by a membrane into two streams: (1) a retentate stream containing 30% nitrogen to be sent to the cryogenic tower and (2) a permeate stream containing 6% nitrogen to be sent to the product pipeline gas. The membrane unit reduces the volume of the gas to be treated by the cryogenic unit by more than half. Simultaneously, the concentrations of water, C$_3$$^+$ hydrocarbons and carbon dioxide are brought to very low levels, because these components also preferentially permeate through the membrane. The removal of these components prior to cryogenic condensation is required to avoid freezing and cavitation in the plant. The savings produced by using a smaller, simpler cryogenic plant more than the offset cost of the membrane unit [82].

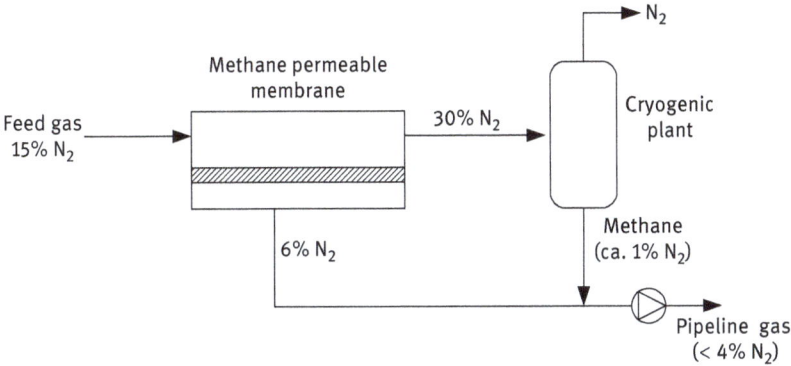

Figure 4.11: Scheme of a hybrid membrane/cryogenic distillation plant for removal of nitrogen from natural gas. Reprinted from Baker [25] with permission from American Chemical Society.

Biogas purification

Biogas is obtained from the anaerobic digestion of domestic, agricultural or sewage wastes. The composition of biogas is variable ranging from 25% to 75% in methane, 15% to 40% in carbon dioxide, from few hundreds ppm up to 50% nitrogen, oxygen, hydrogen sulfide, ammonia, water vapor and VOCs (such as halogenated or aromatic compounds). The biogas can be used in a wide range of applications [83], for example, to cogenerate thermal energy or it can be burned to generate heat energy in boilers. It is also important as a direct fuel for automotive applications or in reforming processes to generate hydrogen to be further supplied to fuel cells [84–86]. However, the presence of gases such as CO_2 and H_2S strongly lowers the fuel calorific value and reduces the possibility to compress and transport over long distances because of the corrosive nature of these gases. In addition, the presence of fouling traces including, for example, siloxanes can induce the formation of fouling in engines and turbines. The biogas upgrading is currently one of the most studied options in biogas treatment leading to the production of biomethane that can be directly supplied to natural gas grids.

Cellulose acetate (CA) is used since 1980 for CO_2–CH_4 separation and covers ca. 80% of the market of membranes for natural gas processing. As an alternative to CA, polyimides show interesting separation properties toward CO_2/CH_4 mixtures along with a good thermal and chemical stability. As CA, polyimides are subjected to plasticization. However, new studies have demonstrated that such an effect can be reduced by cross-linking and more important, polyimides do not exhibit problems in the presence of humidified gaseous streams. The last class of polymers suitable for biogas treatment is represented by the perfluoropolymers, which exhibit a great chemical, thermal and plasticization resistance. In addition, they can be used in separations where a significant amount of water vapor is present as they are hydro-phobic. The main hurdles are concerned with the high fabrication costs because of the expensive nature of the precursors.

The main aspect that currently limits the development of biogas upgrading plants is related to the transportation costs of the material needed for digestion in large plants. As suggested by Scholz [23], the market should move on exploring solution for biogas upgrading for small upgrading plants ($< 100 \, \mathrm{Nm^3/h}$), where (a) membrane-integrated systems are particularly efficient and (b) the main assets of membrane technology (modularity, low plant size, etc.) become still more and more important. De Hullu et al. [87] carried out comparative cost analysis of different biogas upgrading techniques and estimated that the upgrading costs were within the range of € 0.13–0.44 per $\mathrm{Nm^3}$ biogas, and lower operating cost of €0.12 per $\mathrm{Nm^3}$ biogas could be achieved by membrane technology, despite the initial capital cost and membrane fouling. All details on the state of the art on upgrading techniques for biogas can be found in Salihu and Alam 2015 [88].

First attempts to value biogas with membranes were carried out at the beginning of the 1980s by Envirogenics, Permea and Separex using a single-stage configuration and by Permea or Membratek-Envig with a two-stage solution. Today, according to the recently published EBA Biogas Report, there are already more than 15,000 biogas plants in Europe [89], and this number is continuously growing.

After many applications of small/medium size in biogas from agricultural waste to produce biomethane to be injected in grid or as a vehicle fuel, a large-size plant was installed from organic civil waste in Italy. The problem of disposal of organic waste produced by each family or restaurants is now solved by anaerobic digestion with biogas production intended to electric power generators. These power generators release flue gases containing CO_2 and many impurities. The first large commercial-scale upgrading plant installed in Italy was designed for the treatment of more than $6,000 \, \mathrm{Nm^3 \, h^{-1}}$ biogas [90].

CO_2 capture

Today, all the existing coal-fired power plants present over the world emit more than 2–3 billion tons of CO_2 per year. The regulation of the carbon dioxide emissions implies the development of specific CO_2 capture technologies that can be retrofitted to existing power plants as well-designed new plants with the goal to achieve 90% of CO_2 capture limiting the increase in cost of electricity to not more than 35% [91].

Therefore, CO_2 recovery from large emission sources is a formidable technological and scientific challenge that has received considerable attention since several years. Currently, the main strategies for the carbon dioxide capture in a fossil fuel combustion process are as follows:

1. **Oxy-fuel combustion:** This option consists of performing the oxygen/nitrogen separation on the oxidant stream, so that a CO_2/H_2O mixture is produced through the combustion process.

2. **Pre-combustion capture:** This solution is developed in two phases: (i) the fuel conversion in a mixture of H_2 and CO (syngas mixture) through, for example, partial oxidation, steam reforming or autothermal reforming of hydrocarbons, followed by water–gas shift; ii) the separation of CO_2 (at 30%–35%) from the H_2 that is then fed as a clean fuel to turbines. In these cases, CO_2 is separated at very high pressures (up to 80 bar of pressure difference) and high temperatures (300 °C –700 °C) [92, 93].

3. **Post-combustion capture:** In this case, the CO_2 is separated from the flue gas emitted after the combustion of fossil fuels (from a standard gas turbine combined cycle or a coal-fired steam power plant). CO_2 separation is realized at relatively low temperature, from a gaseous stream at atmospheric pressure and with low CO_2 concentration (ca. 5%–25% if air is used during combustion). SO_2, NO_2 and O_2 may also be present in small amounts.

The post-combustion capture is by far the most challenging process since a diluted, low pressure, hot and wet CO_2/N_2 mixture has to be treated. Nevertheless, it also corresponds to the most widely applicable option in terms of industrial sectors (power, kiln and steel production, for instance). Moreover, it shows the essential advantage of being compatible to a retrofit strategy (i.e., an already existing installation can be, in principle, subjected to this type of adaptation).

Membranes are most often listed as potential candidates for their application in post-combustion capture. However, the main problem related to their limited application is the low CO_2 concentration and pressure of the flue gas, which requires the use of membranes with high selectivities (ca. 100) for fitting the specification delivered by the *International Energy Agency*, that is, a CO_2 recovery of 80% with a concentration of at least 80%. The commercial membranes (CO_2/CH_4 selectivity ca.50) that are currently used to separate CO_2 from natural gas at high pressures are not suited for one-stage operation, implying a large membrane area and high compression costs.

In 2008, Baker [94] proposed the possibility of using membranes with CO_2/N_2 selectivity of ca. 50 (already commercial) as integrated multistage solutions. In this case, in fact, the appropriate choice of which membrane type can be used in each separation stage can make this application already feasible [58] (Figure 4.12).

In 2010, Brunetti et al. [50] introduced some general guidelines to rightly drive the application of membrane GS technology as suitable operation for CO_2 capture from flue gas emissions. Considering as case study, a flue gas stream containing 13%

of CO_2, some general maps of CO_2 recovery versus CO_2 purity were introduced, taking into account the membrane characteristics, the flue gas conditions and the desired output to be obtained as useful tools for an immediate and preliminary analysis of the membrane technology suitability for CO_2 separation from flue gas.

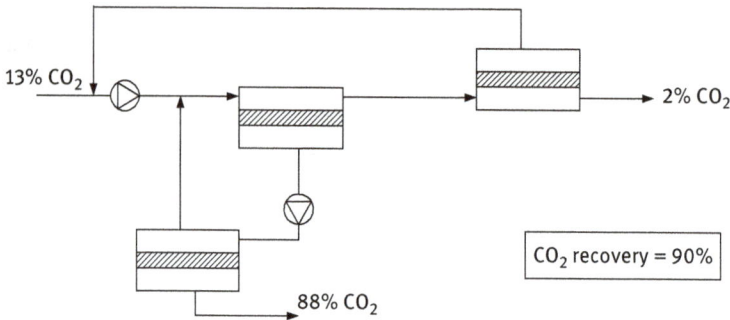

Figure 4.12: New applications: CO_2 from coal power plant flue gas (scheme elaborated from the oral presentation of Ref. Ciferno et al. [91]).

The results showed that with currently available membranes (selectivity up to 50), it is not possible to get, simultaneously, the desired CO_2 recovery and purity (80% CO_2 in permeate stream). To fit into this target, a fundamental role was demonstrated to be played by the operating pressure ratio more than selectivity. In fact, with a selectivity of 100 (value already reached in the lab by some membrane materials), shifting the pressure ratio from 10 to 20 or 50, the CO_2 recovery passes from 22% to more than 60% or 80%, respectively. A high pressure ratio is also necessary when high selective (100–150) membranes are operated (Figure 4.13).

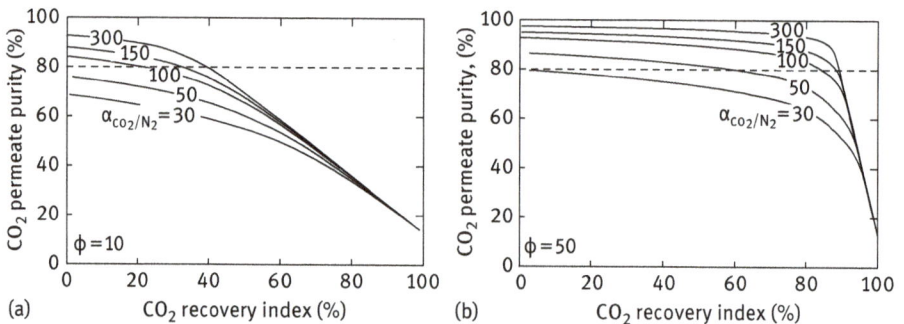

Figure 4.13: CO_2 permeate purity versus recovery index for pressure ratio of (a) 10 and (b) 50 at different CO_2/N_2 ideal selectivity from 30 to 300. Reprinted from Brunetti et al. [50] with permission from Elsevier.

For the pre-combustion and oxy-fuel capture processes, membranes based on alumina, zeolites, silica and carbon that show stability up to 300 °C are generally proposed. Conductive materials that transport CO_3^{2-} ions (e.g., molten Li_2CO_3 formed from the reaction of Li_2ZrO_3 with CO_2) were also studied for CO_2/CH_4 separation up to 600 °C. These membranes may be economically efficient, stable and robust in applications where excess heat/energy is readily available to melt the carbonate. In addition, their use in high-temperature membrane reactors for integration in power generation cycles with CO_2 capture was proposed. However, significant design optimization would be required to identify efficient, feasible and environmentally sound technical solutions. In addition, further development and validation of performance of these membranes in real applications are needed.

Enhanced oil recovery

This technique aims at increasing the yield of oil fields recovery because of high pressure injection of gaseous carbon dioxide. This leads to the maintenance of high pressures in the reservoir and to an improvement of oil displacement.

Since the volumes of concerned carbon dioxide are very large (from 140 to 280 Nm^3 per extracted barrel), it is necessary to separate carbon dioxide from the hydrocarbon gaseous phase to have it recycled to the reservoir (after pressurization). A typical Enhanced oil recovery (EOR) process starts with >50% CO_2 and high pressure (up to 140 bar); CO_2 contents increase significantly over time. Natco (former Cynara) designed the first plant for reducing CO_2 concentration from 45% down to 28% CO_2 for processing 60,000 Nm^3/h of gas. This plant was later expanded and is processing 120,000 Nm^3/h of gas, decreasing CO_2 concentration from 80% to less than 10%.

4.5.7 Commercially available membranes for CO_2 separation

Commercially available membranes for CO_2 separation applications are usually based on polymeric materials forming a dense ultrathin layer as either asymmetric or composite structures [95].

Table 4.2 summarizes some of the most important commercially available membranes, companies and principal membrane materials [96, 97]. These membranes are based on a few polymeric materials that have dominated the industry for the past few decades, mainly owing to the ability of polymers to form low-cost membranes with stable thin active layers that can be processed into modules.

In the recent years, the price of GS membranes was settled to US$~50/$m^2$ that was mainly associated with the polymeric materials and fabrication method. Baker and Lokhandwala [80] estimated that the additional costs led to a membrane skid costing US$500/$m^2$, which is an order greater than the estimated membrane price,

Table 4.2: Important commercially available membranes for CO_2 separation. From Scholes [98].

Membrane name	Supplier	Material
Cynara	Cameron	Cellulose acetate
Prism	Air Products	Polysulfone
Medal	Air Liquide	Polyimide/polyaramid
Separex	UOP	Cellulose acetate
IMS	Praxair	Polyimide
Grace	Kvaerner	Cellulose acetate
UBE	Ube Industries	Polyimide

and demonstrated that for high pressure applications, membrane price is a small variable in the overall cost. For low-pressure applications, such as post-combustion carbon capture, these additional expenses are not necessary because there is no need for pressure vessels and extensive instrumentation for pressure control. This will yield membrane module costs for low-pressure carbon capture comparable to those of reverse osmosis at US\$30–50 for per m^2 of membrane area.

4.6 Selection guidelines for gas separation

The choice of the technology suitable for specific separation is related to different parameters such as economics, stream conditions, product target and also to design considerations. In this logic, new design parameters were introduced by Miller and Stoker [98] used for H_2 separation technologies; however, they can be considered valid, in general, for GS.

The recovery of hydrogen is one of the most common operations in refineries and pressure swing adsorption or cryogenic separations are the means that are generally used for carrying out such an operation. Owing to low capital costs, low energy required and modularity involved in the use of membrane systems are becoming more and more usual in the application of hydrogen separation. Table 4.3 summarizes the comparison among the project parameters that are described above for the three operations considered in the case of H_2 separation.

Membrane systems can maintain the product purity even at reduced capacity (down to 10% of the original design). They are quite capable of operating under variable feed quality conditions, either on a short- or long-term basis; however, the increase of impurity concentrations in the feed can cause a lowering in the level of product purity. Membrane systems can be considered very reliable with respect to the onstream factor, as the membrane separation process is continuous and has few

Table 4.3: Comparison among some important project parameters. From Brunetti et al. [24].

	Operating flexibility	Response to variations	Start-up after the variations	Turndown	Reliability	Control requirement	Byproduct value
Membrane	Moderate	Instantaneous	Extremely short	Down to 10%	100%	Low	Moderate
PSA	High	Rapid (<15 minutes)	1 h	Down to 30%	95%	High	Not economical
Cryogenic	Low	Slow	8–24 h	Down to 30%–50%	Limited	High	High

control components, which can cause a shutdown. Typically, the response to unscheduled shutdowns is rapid.

As it could be seen, membrane systems, owing to their high flexibility, reliability, modularity and ease of control, can be a competitive alternative to the other two classical technologies, in particular, for some specific applications typical of refinery hydrogen upgrading.

References

[1]   Y. Yampolskii. Fundamental science of gas and vapour separation in polymeric membranes, Advanced Membrane Science and Technology for Sustainable Energy and Environmental Applications, Woodhead Publishing Series in Energy, 2011, Pages 22–55.
[2]   W. J. Koros and R. Mahajan., Pushing the limits on possibilities for large scale gas separation: which strategies?, J. Mem. Sci. 2000; 175: 181–196.
[3]   P. Bernardo, E., Drioli, and Golemme G., Membrane Gas Separation: A Review/State of the Art, Ind. Eng. Chem. Res. 2009; 48(10): 4638–4663.
[4]   L.M. Robeson. Correlation of separation factor versus permeability for polymeric membranes J. Mem. Sci. 1991; 62: 165–171.
[5]   L.M. Robeson. The upper bound revised J. Mem. Sci. 2008; 320: 390–400.
[6]   S. H. Han, Y. M. Lee, Recent High Performance Polymer Membranes for CO_2Separation in E. Drioli and G. Barbieri, eds. Membrane Engineering for the treatment of gases, The Royal Society of Chemistry, Cambridge, The United Kingdom, 2011, pages 84–124, ISBN 978-1-84973-239-0
[7]   H. Lin and B. D. Freeman., Gas Permeation and Diffusion in Cross-Linked Poly(ethylene glycol diacrylate), Macromol. 2006; 39: 3568–3571.
[8]   S. R. Reijerkerk, M. H. Knoef, K. Nijmeijer and M. Wessling., oly(ethylene glycol) and poly(dimethyl siloxane): Combining their advantages into efficient CO_2 gas separation membranes, J. Membr. Sci. 2010; 352: 126–133.
[9]   H. B. Park, C. H. Jung, Y. M. Lee, A. J. Hill, S. J. Pas, S. T. Mudie, E. Van Wagner, B. D. Freeman and D. J. Cookson., Polymers with cavities tuned for fast selective transport of small molecules and ions, Sci. 2007; 318: 254–257.
[10]   C.S. Scholes, G. W. Stevens, S. E. Kentish., Membrane gas separation applications in natural gas processing, Fuel. 2012; 96: 15–28.
[11]   S. Hess, C.S. Bickel and R.N. Lichtenthaler., Propene/propane separation with copolyimide membranes containing silver ions, J. Mem. Sci. 2006; 275: 52–59.
[12]   T.-J. Kim, B. Li and M.B. Hägg., Novel fixed-site–carrier polyvinylamine membrane for carbon dioxide capture, J. Polym. Sci., Part B: Polym. Phys. 2004; 42(23): 4326–4332.
[13]   R. Dittmeyer, V. Höllein, and K. Daub., Membrane Reactors for Hydrogenation and Dehydrogenation based on Supported Palladium, J. Mol. Cat. A: Chemical. 2001; 173: 135–184.
[14]   J. Caro, M. Noack, P. Kolsch and R. Schafer., Zeolite membranes - state of their development and perspective, Micropor. Mesopor. Mater. 2000; 38: 3–16.
[15]   A. Caravella, P.F. Zito, A. Brunetti, E. Drioli, G. Barbieri., Evaluation of Pure-Component Adsorption Properties of DD3R Based on the Langmuir and Sips Models, J. Chem. Eng. Data. 2015; 60(8):2343–2355.
[16]   A. Caravella, P.F. Zito, A. Brunetti, E. Drioli, G. Barbieri., Evaluation of pure-component adsorption properties of silicalite based on the Langmuir and Sips models, AIChE J. 2015; 61 (11): 3911–3922.

[17] A. Caravella, P.F. Zito, A. Brunetti, E. Drioli, G. Barbieri., A novel modelling approach to surface and Knudsen multicomponent diffusion through NaY zeolite membranes, Microporous and Mesoporous Mater. 2016; 235: 87–99.

[18] P.F. Zito, A. Caravella, A. Brunetti, E. Drioli, G. Barbieri., Research articleAbstract only Knudsen and surface diffusion competing for gas permeation inside silicalite membranes, J. Membr. Sci. 2017; 523(1): 456–469.

[19] P.F. Zito, A. Caravella, A. Brunetti, E. Drioli, G. Barbieri., Light Gases Saturation Loading Dependence on Temperature in LTA 4A Zeolite, Microporous and mesoporous mater. 2017; 249: 67–77.

[20] P. Bernardo, C. Algieri, G. Barbieri, E. Drioli., Hydrogen purification from carbon monoxide by means of selective oxidation using zeolite catalytic membranes Sep. Purif. Technol. 2008; 62: 629–635.

[21] P. Bernardo, C. Algieri, G. Barbieri. E. Drioli., Catalytic (Pt-Y) membranes for the purification of H2-rich streams, Cat. Tod. 2006; 118: 90–97.

[22] P.S. Tin, Y.C. Xiao and T.S. Chung., Polymide-carbonizedmembranes for gas separation: structural composition and morphological control of precursors, Sep. Purif. Rev. 2006; 35, 285–292.

[23] M. Scholz, M. Wessling, J. Balster., Design of membrane modules for gas separation Chapter 5 in E. Drioli and G. Barbieri, eds. Membrane Engineering for the treatment of gases, The Royal Society of Chemistry, Cambridge,
The United Kingdom, 2011, pages 125–149, ISBN 978-1-84973-239-0

[24] Brunetti, P. Bernardo, E. Drioli, G. Barbieri, Membrane engineering Progresses and Potentialities, in Yampolskii Y., Freeman B. eds.
"Membrane Gas Separation", Wiley & Sons, 2010, pp.281–312. (ISBN 978 0 470 74621-9)

[25] R.W. Baker., Future Directions of Membrane Gas Separation Technology, Ind. Eng. Chem. Res. 2002; 41: 1393–1411.

[26] S.M. Weller and W.A. Steiner., Separation of gases by fractional permeation through membranes J. Appl. Phys., 1950; 21: 279.

[27] R.W. Baker. Ind. Eng. Chem. Res., Future Directions of Membrane Gas Separation Technology, 2002; 41: 1393–1404.

[28] E. Sanders, D.O. Clark, J.A. Jensvold, H. N. Beck, G.G. Lipscomb and F.L. Coan., Process for preparing POWADIR membranes from Tetrahalobisphenol A Polycarbonates U.S. Patent 4,772,392, (1988).

[29] O.M. Ekiner, R.A. Hayes and P. Manos., Novel multicomponent fluid separation membranes, U.S. Patent 5,085,676. (Feb. 1992).

[30] "Advanced Prism® Membrane Systems" For Cost Effective Gas Separations", http://www.airproducts.com/NR/rdonlyres/81FB384C-3DB5-4390-AED0-C2DB4EEDDB18/0/PrismPGS.pdf. (accessed 21/01/2016)

[31] http://www.airproducts.com/~/media/Files/PDF/products/supply-options/prism-membrane/en-process-gas-membrane-systems-ammonia-plants.pdf. (accessed 04/02/2018)

[32] Ockwig N.W., Nenoff T.M., Membranes for hydrogen separation, Chem Rev. 2007; 107: 4078–4093.

[33] http://www.mtrinc.com/pdf_print/refinery_and_syngas/MTR_Brochure_Hydrogen_Recovery_from_Ammonia_Plant_Purge_Gas.pdf. (accessed 14/02/2018)

[34] http://www.mtrinc.com/hydrogen_separation_in_syngas_processes.html. (accessed 04/02/2018)

[35] http://www.mtrinc.com/hydrogen_recovery_from_methanol_plant_purge_gas.html. (accessed 04/02/2018)

[36] www.airproducts.no. (accessed 14/02/2018)

[37] www.airproducts.com. (accessed 14/02/2018)

[38] http://www.cn.airliquide.com/en/who-we-are/air-liquide-in/main-business-lines-in-china.html. (accessed 14/02/2018)

[39] SERCK, B. Ube expands gas separation membrane production. Filtration Industry Analyst. 2006; 9(3).

[40] https://permselect.com/markets/oxygen-enrichment. (accessed 14/02/2018)

[41] P. Pushpinder P., in E. Drioli and G. Barbieri, Commercial Applications of Membranes in Gas Separations, eds. Membrane Engineering from the treatment of gases, The Royal Society of Chemistry, Cambridge, The United Kingdom, 2011, pages 215–244, ISBN 978-1-84973-239-0

[42] R.W. Baker. Membranes for vapor/gas separation, 2006. http://www.mtrinc.com/publications/MT01%20Fane%20Memb%20for%20VaporGas_Sep%202006%20Book%20Ch.pdf. (accessed 04/ 05/2016)

[43] http://www.mtrinc.com/publications/MT01%20Fane%20Memb%20for%20VaporGas_Sep%202006%20Book%20Ch.pdf. (accessed 14/02/2018)

[44] https://www.dndkm.org/DOEKMDocuments/ITSR/TRUMixedWaste/Membrane_System_for_the_Recovery_of_Volatile_Organic_Compounds_from_Remediation_of_Off-Gases.pdf. (accessed: 14/02/2018)

[45] L. Shao, B. T. Low, T.-S. Chung, A.R. Grenberg., Polymeric membranes for the hydrogen economy: contemporary approaches and prospects for the future, J. Mem. Sci. 2009; 327 (1–2): 18–31.

[46] E. Favre, R. Bounaceur, D. Roizard., Biogas, Membranes and Carbon Dioxide Capture, J. Mem. Sci. 2009; 328: 11–14.

[47] S. Basu, A. L. Khan, A. Cano-Odena, C. Liu, I. F. J. Vankelecom., Membrane-based technologies for biogas separations, Chem. Soc. Rev. 2010; 39: 750–768.

[48] H. Lin, E. Van Wagner, R. Raharjo, B. D. Freeman, I. Roman., High Performance Polymer Membranes for Natural Gas Sweetening, Adv. Mater. 2006; 18: 39–44.

[49] C. A. Scholes, J. Bacus, G. Q. Chen, W. X. Tao, G. Li, A. Qader, G. W. Stevens, S. E. Kentish, Pilot plant performance of rubbery polymeric membranes for carbon dioxide separation from syngas, J. Mem. Sci. 2012; 389: 470–477.

[50] A. Brunetti, F. Scura, G. Barbieri, E. Drioli. J. Mem. Sci., Membrane Technologies for CO_2 separation 2010; 359: 115–125.

[51] J.P. Ciferno, T. E. Fout, A. P. Jones, J.T. Murphy., Capturing carbon form existing coal-fired power plants, Chem. Eng. Prog., 2009; 105, 33–41.

[52] H. Herzog. Env. Sci. Tech., What Future for Carbon Capture and Sequestration?, 2001; 35(7): 148–153.

[53] C. M. White, Separation and capture of CO_2 from large stationary sources and sequestration in geological formations J. Air Waste Management Association. 2003; 53: 645–715.

[54] E. Favre., Carbon dioxide recovery from post-combustion processes: can gas permeation membranes compete with absorption?, J. Mem. Sci. 2007; 294; 50–59.

[55] T. C. Merkel, H. Lin, X. Wei, R. Baker. J. Mem. Sci., Power plant post-combustion carbon dioxide capture: An opportunity for membranes, 2010; 359: 1–2, 1 126–139.

[56] B. Li, Y. Duan, D. Luebke, B. Morreale., Advances in CO_2 capture technology: A patent review, App. En. 2013; 102: 1439–1447.

[57] L. Peters, A. Hussain, M. Follmann, T. Melin, M. B. Hägg., CO_2 removal from natural gas by employing amine absorption and membrane technology - A technical and economic analysis, Chem. Eng. J. 2011; 172: 952–960.

[58] L. Daal, L. Claassen, R. Bruns, B. Schallert, G. Barbieri, A. Brunetti, K. Nijmeijer., Field tests of carbon dioxide removal from flue gasses using polymer membranes, VGB Powertech. 2013; 6: 78–84.

[59] A.Y. Ku, P. Kulkarni, R. Shisler, W. Wei. J. Mem. Sci., Membranes performance requirements for carbon dioxide capture using hydrogen-selective membranes in integrated gasification combined cycle (IGCC) power plants 2011; 367: 233–239.

[60] M. Pidwirny. (2006). Fundamentals of Physical Geography, 2nd Edition. http://www.physical-geography.net/fundamentals/7a.html

[61] http://naturalgas.org/overview/background.asp

[62] A. Baudot, Gas/Vapor Permeation Applications in the Hydrocarbon-processing Industry, in E. Drioli and G. Barbieri, eds. Membrane Engineering for the treatment of gases, The Royal Society of Chemistry, Cambridge, The United Kingdom, 2011, pages 150–195, ISBN 978-1-84973-239-0.

[63] Y. Xiao, B.-T. Low, S.S. Hosseini, T.S. Chung, D.R. Paul, The strategies of molecular architecture and modification of polyimide-based membranes for CO_2 removal from natural gas-A review, Prog. Polym. Sci. 2009; 34, 561–580.

[64] L. Deng, M-B. Hagg., Techno-economic evaluation of biogas upgrading process using CO_2 facilitated transport membrane, Int. J. Greenhouse Gas Control. 2010; 4: 638–646.

[65] C. E. Powell, G. G. Qiao, Polymeric CO_2/N-2 gas separation membranes for the capture of carbon dioxide from power plant flue gases J. Mem. Sci. 2006; 279: 1–49.

[66] P. Luis, T. Van Gerven, B. Van der Bruggen., Recent developments in membrane-based technologies for CO_2 capture, Progr,. En. Com. Sci. 2012; 38(3): 419–448.

[67] K. Ramasubramanian, WS W. Ho., Recent developments on membranes for post-combustion carbon capture Curr. Op. Chem. Eng. 2011; 1(1): 47–54.

[68] H. B. Park, C. H, Jung, Y. M. Lee, A. J. Hill, S. J. Pas, S. T. Mudie, E.Van Wagner, B. D. Freeman, D. J. Cookson., Polymers with cavities tuned for fast selective transport of small molecules and ions, Sci. 2007; 318: 254–258.

[69] C. H. Jung, J. E. Lee, S. H. Han, H. B. Park, Y. M. Lee., Highly permeable and selective poly (benzoxazole-co-imide) membranes for gas separation, J. Mem. Sci. 2010; 350: 301–309.

[70] M. Calle, Y. M. Lee., Thermally Rearranged (TR) Poly(ether–benzoxazole) Membranes for Gas Separation Macromol. 2011; 44: 1156–1165.

[71] R.T. Adams, J. S. Lee, T.-H. Bae, J. K. Ward, J. R. Johnson, C. W. Jones, S. Nair, W. J. Koros ., CO_2-CH_4 permeation in high zeolite 4A loading mixed matrix membranes, J. Mem. Sci. 2011; 367: 197–203.

[72] R, Adams, C, Carson, J, Ward, R, Tannenbaum, W, Koros, Metal organic framework mixed matrix membranes for gas separations, Micr. Mes. Mat. 2010; 131(1): 13–20.

[73] F. Falbo, F. Tasselli, A. Brunetti, E. Drioli, G. Barbieri, Polyimide hollow fiber membranes for CO_2 separation from wet gas mixtures, Brazilian J. Chem. Eng. 2014; 31: 1023–1034.

[74] M. Cersosimo, A. Brunetti, F. Fiorino, E. Drioli, G. Dong, K. T. Woo, J. Lee, Y. M. Lee, G. Barbieri, Separation of CO_2 from humidified ternary gas mixtures using thermally rearranged polymeric membranes J. Membr. Sci. 2015; 492: 257–262.

[75] A. Brunetti, M. Cersosimo, G. Dong, K.T. Woo, J. Lee, J.S. Kim, Y.M. Lee, E. Drioli, G. Barbieri, In situ restoring of aged thermally rearranged gas separation membranes J. Membr. Sci. 2016; 520: 671–678.

[76] F. Falbo, A. Brunetti, G. Barbieri, E. Drioli, F. Tasselli ., CO_2/CH4 separation by means of Matrimid hollow fibre membranes Applied Petrochemical Research Journal. 2016; 6(4): 439–450.

[77] A. Brunetti, M. Cersosimo, J. S. Kim, G. Dong, E. Fontananova, Y. M. Lee, E. Drioli, G. Barbieri, Thermally rearranged mixed matrix membranes for CO_2 separation: An aging study Int. J. Greenhouse Gas Control. 2017; 61: 16–26.

[78] R. Bounaceur, N. Lape, D. Roizard, C. Vallieres, E. Favre, Membrane Processes for Post-Combustion Carbon Dioxide Capture: A Parametric Study, Energy. 2006; 31: 2556–2570.

[79] K. Datta and P. K. Sen, Optimization of membrane unit for removing carbon dioxide from natural gas., J. Mem. Sci. 2006; 283: 291–300.

[80] R. W. Baker, T. Hofmann, J. Kaschemekat, K. A. Lokhandwala, Field Demonstration of a Membrane Process to Recover Heavy Hydrocarbons and to Remove Water from Natural Gas, Annual Report, Contract Number DE-FC26-99FT40723, Report Period Ending September 29, 2001.

[81] C. C. Tannerhill, Nitrogen Removal Requirement from Natural Gas, Gas Research Institute Topical Report GRI-99/0080; Gas Research Institute: Washington, D.C., May 1999.

[82] K. A. Lokhandwala, Membrane-augmented cryogenic methane/nitrogen separation U.S. Patent 5,647,227, (1997).

[83] M. Poschl, S. Ward, P. Owende, Evaluation of energy efficiency of various biogas production and utilization pathways, Appl. Energy. 2010; 87: 3305–3321.

[84] J.V. Herle, Y. Membrez, O. Bucheli, Biogas as a fuel source for SOFC co-generators, J. Power Sou. 2004; 127: 300–312.

[85] D. Papadias, S. Ahmed, R. Kumar, Fuel quality issues with biogas energy – An economic analysis for a stationary fuel cell system Energy. 2012; 44: 257–277.

[86] E. Ryckebosch, M. Drouillon, H. Vervaeren, Techniques for transformation of biogas to bio-methane, Biomass and bioenergy. 2011; 35: 1633–1645.

[87] J. De Hullu, J.I.W. Maassen, P.A. van Meel, S. Shazad, J.M.P. Vaessen. Comparing different biogas upgrading techniques, Eindhoven University of Technology, The Netherlands, (2008).

[88] A. Salihu and Md. Z. Alam, Upgrading strategies for effective utilization of biogas, Environmental Progress & Sustainable Energy. Vol.34(5), 1512-1520.

[89] European Biogas Association. Biogas. Brussels. EBA; c2013[cited]. Available from: http://european-biogas.eu/biogas/. (accessed: 14/02/2018).

[90] U. Moretti, Polymeric Membrane-based Plants for Biogas Upgrading, Chapter 11 in "Membrane engineering for the treatment of gases", Volume 1 "Gas-separation issues with membranes", Editors E. Drioli, G. Barbieri, A. Brunetti. The Royal Society of Chemistry, Cambridge, The United Kingdom, 2017.

[91] J.P. Ciferno, T. E. Fout, A. P. Jones, J.T. Murphy, Capturing Carbon from Existing Coal-Fired Power Plants, Chem. Eng. Prog. April 2009, 33.

[92] Barbieri G.; Brunetti A.; Caravella A.; Drioli E, HYPERLINK "C:\Users\A. Brunetti\Desktop \BRUNETTI\ITM-CNR\altro\Barbieri\AppData\Roaming\Barbieri\AppData\Roaming\Microsoft \PortFolio\04_Papers\01_Papers_ISI\0050__2011_RSC_Advance_WGS.pdf.lnk"Pd-based Membrane Reactors for one-stage Process of Water Gas Shift RSC Adv. 2011; 1 (4): 651–661.

[93] Brunetti A.; Caravella A.; Fernandez E.; Pacheco Tanaka D.A.; Gallucci F.; Drioli E.; Curcio E.; Viviente J.L.; Barbieri G, Syngas upgrading in a membrane reactor with thin Pd-alloy supported membrane Int. J. Hydrogen Energy. 2015; Vol. 40(34): 10883–10893.

[94] R. W. Baker, Recent developments and future directions in membrane modules, Proceedings of 12th Aachener membran kolloquim, October 29–30, 2008, Aachen (Germany).

[95] R. Spillman. In Membrane Separations Technology, eds. R. D. Noble and S. A. Stern, Elsevier Science, Eastbourne, 1995, pp. 589–667.

[96] C. A. Scholes. Chapter 8 in "Membrane engineering for the treatment of gases", Volume 1 "Gas-separation issues with membranes", Editors E. Drioli, G. Barbieri, A. Brunetti. The Royal Society of Chemistry, Cambridge, The United Kingdom, 2017.

[97] A. G. Fane, R. Wang and Y. Jia. In Membrane and Desalination Technologies, eds. L. K. Wang, J. P. Chen, Y.-T. Hung and N. K. Shammas. Springer, New York, 2011, pp. 1–46.

[98] Miller G. Q., Stöcker J. NPRA Annual Meeting, March 19–21, 1989, San Francisco, California (USA). 1989.

Francesca Macedonio, Enrico Drioli

5 Membrane contactors

5.1 Definition

Membrane contactors (MCs) represent relatively new membrane-based devices that are gaining wide consideration. MCs are systems in which microporous membranes are used not as selective barriers but as a tool for interphase mass transfer operations. Such extremely compact devices are able to use the microporous membrane as a fixed interface between two different phases without dispersing one phase into another, and to create a large contact area for promoting an efficient mass or energy transfer. Gas/liquid MCs have been tested in a large variety of systems, including (i) absorption into aqueous or organic solutions of CO_2, NH_3 and so on, (ii) oxygen removal in a semiconductor industry for the production of ultrapure water, (iii) ozonation for water treatment, (iv) in dehumidification processes as absorption air-handling systems working with liquid desiccants and (v) in concentration and crystallization processes to be carried out at low temperature (i.e., membrane distillation [MD], osmotic distillation [OD] and membrane crystallization [MCr]). In short, traditional stripping, scrubbing, absorption and liquid–liquid extraction operations, as well as condensation, dehydration, crystallization and phase transfer catalysis, can be carried out according to MC configuration.

This chapter illustrates the working principles, the fundamental concepts and the transport phenomena through microporous membranes in MD, MCr, membrane condenser, membrane dryer and membrane emulsification operations.

5.2 Membrane contactor technology

MD, membrane extraction, pertraction, perstraction, gas adsorption, membrane-based solvent extraction, liquid–liquid extraction, membrane-based gas absorption and stripping, membrane-assisted crystallization, membrane-assisted condensation, membrane dryer and membrane emulsification are generally referred to as MCs. The separation performance in these processes is determined by the distribution coefficient of a component in two phases, and the membrane acts only as an interface. In general, it is not the enhanced mass transfer but rather the large area per volume (that can be found in hollow fiber and capillary modules) that makes this process more attractive than conventional dispersed-phase contactors.

Francesca Macedonio, Enrico Drioli, National Research Council of Italy, Institute for Membrane Technology (ITM-CNR), University of Calabria, Rende, Italy

https://doi.org/10.1515/9783110281392-005

MCs can be gas–liquid (G–L) and liquid–liquid (L–L). In the G–L contactors, one phase is a gas or a vapor and the other phase is a liquid, whereas in the L–L contactors both phases are liquids. In general, if a component is transferred from the feed phase to the permeate phase, three steps must be considered: (1) transport from the feed phase to the membrane, (2) diffusion through the membrane and (3) transfer from the membrane to the permeate phase.

The flux J of a component i is conveniently expressed in terms of an overall mass transfer coefficient k_{ov} as follows:

$$J_i = k_{ov,i} \Delta c_i \text{ where}$$

$$\frac{1}{k_{ov,i}} = \frac{1}{k_i(\text{feed})} + \frac{1}{k_i(\text{membrane})} + \frac{1}{k_i(\text{receiving phase})}$$

In general, the mass transfer resistance in the boundary layers cannot be neglected and these must be calculated or estimated from mass transfer correlations.

5.2.1 Gas-liquid membrane contactor

The membranes used in MCs are in general porous and act as a barrier between the phases. Two concepts are possible: the membrane pores are either filled with the gas phase or with the liquid phase. If a hydrophobic membrane is used, then the pores of the membrane are filled with the gas phase whereas the liquid phase is an aqueous solution that does not wet the membrane. The liquid must be prevented from wetting, which means that the wetting pressure (or liquid entry pressure – LEP_w) should not be exceeded.

The Laplace (Cantor) equation allows estimating the LEP_w. It provides the relationship between the membrane's largest allowable pore size (d_{max}) and the related operating conditions:

$$LEP_w = \frac{-B\gamma_L \cos\theta}{d_{max}} \tag{5.1}$$

where B is a geometric factor determined by pore structure, γ_L the liquid surface tension and θ is the liquid/solid contact angle.

When the hydrostatic pressure on the feed side of the membrane exceeds LEP_w, liquid penetrates the pores and it is able to pass through the membrane. Once a pore has been penetrated, it is said to be "wetted" and the membrane must be completely dried and cleaned before the wetted pores can once again support a vapor–liquid interface.

On the other hand, if a hydrophilic membrane is used, the aqueous phase will wet the membrane.

Both if hydrophobic and hydrophilic membranes are used, the mass transfer resistance is normally located in the liquid phase [1].

G–L and L–G MCs may be applied, for example, (i) in oxygen transfer systems in fermentation processes and aerobic waste water treatment without bubble formation; (ii) in carbon dioxide transfer to beverages; (iii) for the separation of saturated/unsaturated hydrocarbons (e.g., paraffin olefin separation); (iv) for the removal of acid gases (such as CO_2, H_2S, CO, SO_2 and SO_x) from flue gas, biogas and natural gas; (v) for the separation of volatile bioproducts (alcohols and aroma compounds) and (vi) for the removal of O_2 from water.

5.3 Membrane distillation

MD is a thermal membrane separation process that belongs to the class of MCs. MD involves the transport of vapor through microporous hydrophobic membranes and operates on the principle of vapor–liquid equilibrium as a basis for molecular separation.

This technique allows the separation of volatile components from solutions. If the solutions contain nonvolatile components, it is possible to remove solvent by concentrating the solutions. Since its invention, MD has had a gamut of reported applications including desalination; brine concentration; concentration of fruit juices, radioactive solutions, acids and VOCs; removal of heavy metals and dyes; wastewater treatment and so on.

The first patent on MD was filed by Bodell on 3 June 1963 [3] and the first MD paper was published in 1963 by Findley [2]. Intense interests in MD process began in early 1980s with the advent of new membrane manufacturing techniques, and membranes became available with porosities as high as 80% and thickness as low as 50 μm [4]. Improvements in module design and a better understanding of temperature and concentration polarization phenomena also contributed to the renewed interest in MD, in particular, within the academic community; however, MD still needs to be developed for its widespread industrial implementation. Academic interest in MD is fueled by the process versatility.

In MD, one side (feed side) of a hydrophobic membrane is brought into contact with a heated, aqueous feed solution. The hydrophobic nature of the membrane prevents penetration of the aqueous solution into the pores, resulting in a vapor–liquid interface at each pore entrance. Here, volatile compounds evaporate, diffuse and/or convect across the pores, and are condensed on the opposite side (permeate) of the system (Figure 5.1). The driving force of the process is supplied by a partial pressure difference between both sides of the membrane, caused by temperature gradient imposed between the liquid–vapor interfaces.

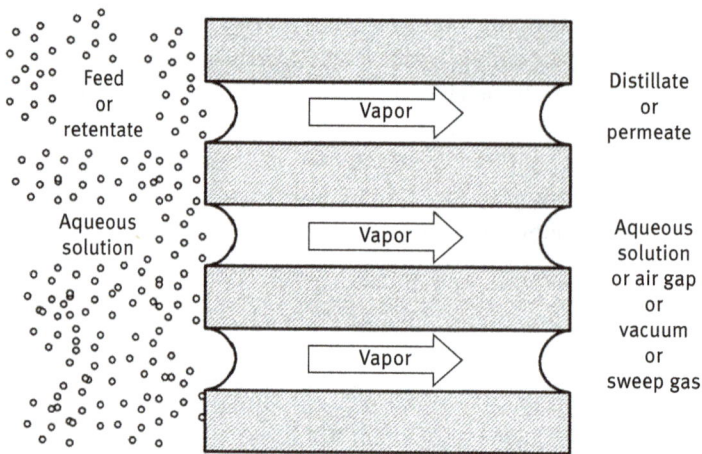

Figure 5.1: A general scheme of the MD process: aqueous solution on feed side, whereas four different solutions can be realized on permeate side (aqueous solution or air gap or vacuum or sweeping gas). From Curcio and Drioli [5]. Reprinted with Permission.

5.3.1 Types and versions

A variety of methods may be employed to impose the vapor pressure difference across the membrane to drive flux and, according to the nature of the permeate side of the membrane, MD systems can be classified into four basic configurations (Figure 5.2):
1. direct contact membrane distillation (DCMD-Figure 5.2a), in which the membrane is in direct contact only with liquid phases (e.g., saline water on one side and fresh water on the other);

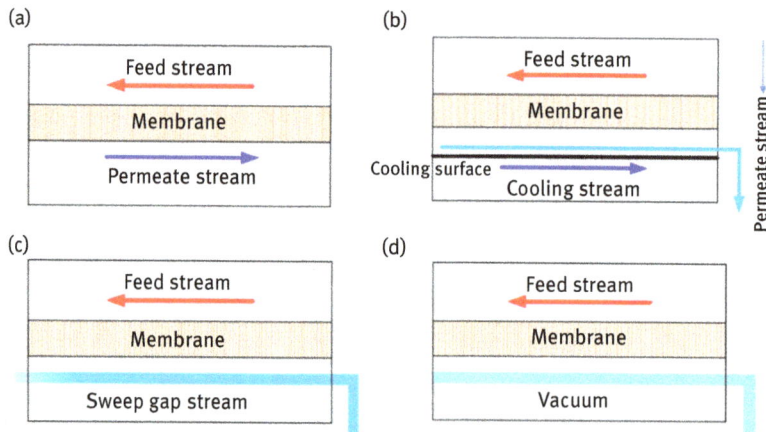

Figure 5.2: Common configurations of MD process that may be utilized to establish the required driving force. (Adapted from Wang and Chung [8]. Reprinted with Permission.)

2. vacuum membrane distillation (VMD-Figure 5.2d), in which the vapor phase is vacuumed from the liquid through the membrane, and condensed, if needed, in a separate device;
3. air gap membrane distillation (AGMD-Figure 5.2b), in which an air gap is interposed between the membrane and the condensation surface and
4. sweeping gas membrane distillation (SGMD-Figure 5.2c), in which a stripping gas is used as a carrier for the produced vapor instead of vacuum as in VMD.

The selection of a specific configuration depends upon feed and permeate compositions as well as upon requested productivity. In general, DCMD (the cheapest and the simplest to operate) is the best choice for applications in aqueous environments in which water is the major permeate component; SGMD and VMD are typically used to remove volatile organic or dissolved gas from aqueous solutions and AGMD is the most versatile MD configuration, which can be applied to almost any application whenever high fluxes are not required.

Some new configurations with improved energy efficiency, better permeation flux or smaller foot print have been proposed such as material gap membrane distillation (MGMD), multieffect membrane distillation (MEMD), multieffect VMD, permeate gap membrane distillation (PGMD) and hollow fiber MEMD [6].

PGMD is an enhancement of DCMD in which a third channel is introduced by an additional nonpermeable foil (Figure 5.3).

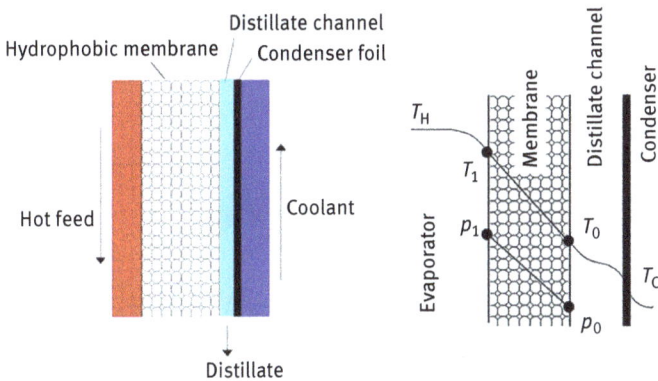

Figure 5.3: A basic channel arrangement and temperature profile for PGMD. (From Winter et al. [7]. Reprinted with Permission.)

One significant advantage of PGMD is the separation of the distillate from the coolant. Therefore, the coolant can be any other liquid, such as cold feed water. In module development, this opens the opportunity to integrate an efficient heat recovery system. The presence of the distillate channel reduces sensible heat losses due to an additional heat transfer resistance. An additional effect is the reduction of the effective temperature difference across the membrane, which slightly lowers the permeation rate.

Multistage MD and MEMD arise from the concept of AGMD module with internal heat recovery as illustrated in Figure 5.4.

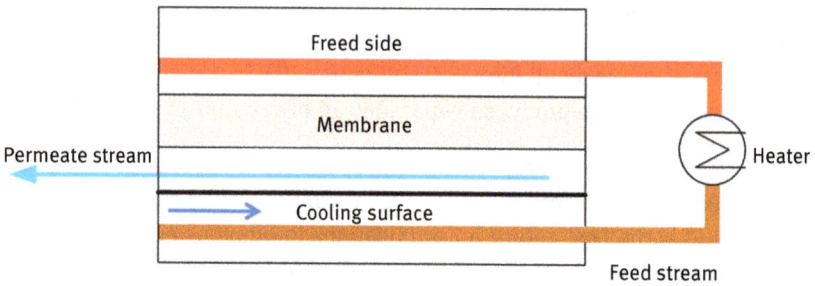

Figure 5.4: Illustration of an AGMD configuration with internal heat recovery. (From Wang and Chung [8]. Reprinted with Permission.)

The cold feed solution is placed beneath the condensation surface as a coolant to condense the permeated vapors as well as to gain heat. The preheated feed solution is further heated before it enters the feed channel.

Vacuum-multieffect membrane distillation (V-MEMD) is a modified form of VMD that integrates the concept of multieffect distillation (MED) into the VMD (Figure 5.5). The typical V-MEMD consists of a heater, multiple evaporation–condensation stages and an external condenser [9]. A vacuum condition is employed at the air gap region to remove the excess air/vapor. Distillate is hence produced in both condensation stages and inside the condenser. The feed in each stage recovers the condensation heat and a multiple-effect characteristic.

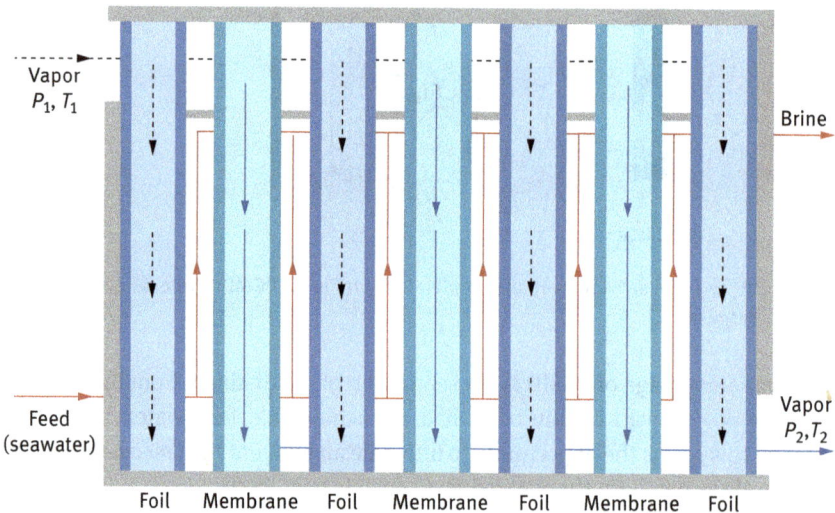

Figure 5.5: Illustration of a V-MEMD configuration. (From Zhao et al. [9]. Reprinted with Permission.)

MGMD represents a development of AGMD. The latter normally shows a lower permeation flux as compared with other MD configurations due to the presence of a layer of stagnant air between the membrane and the condensation surface. To abate this disadvantage, Francis et al. [10] developed a new MD module design called MGMD. The air gap in the module was filled with different materials like sponge (polyurethane) and polypropylene (PP) mesh. As a result, an increase of 200–800% in water vapor flux was observed during the MGMD [8].

Hollow fiber MEMD consists of a multieffect AGMD hollow fiber module with internal heat recovery [8, 11]. The process was patented by Qin et al. [11] from Chembrane Research & Engineering, Inc. Unlike the PGMD design, the feed solution is preheated to 90 °C before entering the MD module. At the exit of the MD module, the concentrated feed solution at a reduced temperature is further cooled down by an external cooler. This cooled feed solution is fed back into the MD module and serves as the coolant to condensate vapor in the permeate side. As a multistage design, the effluent stream can serve as the feed solution for the next membrane module to enhance heat recovery and efficiency.

OD represents another extension of the MD concept: a microporous hydrophobic membrane separates two aqueous solutions that are kept in contact at different solute concentrations; this difference in activity causes a vapor-pressure difference that activates mass transport through the membrane.

OD is not a purely mass transfer operation: transport involves an evaporation at the feed side and a condensation at the stripping side. A temperature difference at the membrane interfaces is thus created, even if the bulk temperatures of the two liquids are equal. In general, the temperature difference in aqueous systems is lower than 1 °C, leading to a negligible decrease of the vapor flux. The salts chosen as osmotic pressure agents are in general NaCl, $MgCl_2$, $CaCl_2$ and $MgSO_4$, due to their relatively low cost; in some cases, organic liquids (glycerol and polyglycols) are preferred [12]. In osmotic evaporation, transmembrane fluxes rise at higher stripping solution concentrations and feed temperatures.

Because OD operates essentially at room temperature, it is appropriate for applications in the agro-food industry (such as in integrated membrane system for the clarification and the concentration of citrus and carrot juices that have been proposed as an alternative and efficient approach to the traditional techniques currently in operation), in pharmaceutical biotechnology and medicine [13, 14].

Further MD applications include the following:
- Membrane distillation bioreactor (MDBR): This is a system able to produce high-quality product water with simultaneous biodegradation of organics [15, 16]. The MDBR combines a thermophilic bioprocess with the MD process, which works by transferring water vapor across a thermal gradient through a hydrophobic, microporous membrane to produce water.
- Photocatalytic membrane reactor systems (coupling photocatalysis with MD) for textile dye effluent treatment [17].

5.3.2 Benefits and drawbacks of MD technology

The advantages of MD compared to conventional distillation processes (such as multistage flash [MSF] and MED) from one side and compared to reverse osmosis (RO; i.e., the membrane technique usually utilized in conventional desalination process) from another side are as follows:

1. *lower operating temperatures and vapor space required* than MSF and MED. Because the process can be conducted at temperatures typically below 70 °C, and driven by low-temperature difference (20 °C) of the hot and the cold solutions, low-grade waste and/or alternative energy sources such as solar and geothermal energy can be coupled with MD systems for a cost-and energy-efficient liquid separation system.
2. *Lower operating pressure than RO.* The MD process can be performed at operating pressures generally near the atmospheric pressure. This allows using equipment made of plastic material reducing or avoiding corrosion problems.
3. *Complete rejection of nonvolatile solutes.* Since MD operates on the principle of vapor–liquid equilibrium, 100% (theoretical) of ions, macromolecules, colloids, cells and other nonvolatile constituents are rejected.
4. *Performance not limited by high osmotic pressure or concentration polarization.* This means that MD can be preferentially employed whenever elevated permeate recovery factors or high retentate concentrations are requested.
5. *Less demanding membrane mechanical properties.* Since MD membranes act merely as a support for a vapor–liquid interface, they can be fabricated from almost any chemically resistant polymers with hydrophobic intrinsic properties, such as polytetrafluoroethylene (PTFE), PP and polyvinylidenedifluoride (PVDF). This characteristic increases the membrane's life.
6. *Less membrane fouling.* Membrane fouling in MD is less problematic than in other membrane separations because (a) the pores are relatively large compared to RO/UF pores, (b) the process liquid cannot wet the membrane; therefore, fouling layers can be deposited only on the membrane surface but not in the membrane pores and (c) due to the low operating pressure of the process, the deposition of aggregates on the membrane surface would be less compact and only slightly affect the transport resistance.

The disadvantages of MD include the following:

1. lack of membranes and modules designed specifically for MD. Compared to other membrane separation processes including RO, only few research groups have considered the possibility of designing and manufacturing novel membranes for MD applications. The few commercial available membrane modules are still expensive.
2. Risk of membrane pore wetting.

3. Temperature polarization, similar to concentration polarization, arises from heat transfer through the membrane and it is often the rate-limiting step for mass transfer.

5.3.3 MD mass and heat transfer phenomena

In MD, the driving force of the process is the vapor pressure difference across the membrane. The vapor–liquid equilibrium for nonideal mixtures is described as follows:

$$p_i = P y_i = p_i^0 a_i = p_i^0 \xi_i x_i \tag{5.2}$$

where p is the total pressure, x_i and y_i are the liquid and vapor mole fraction, respectively, and ξ_i is the activity coefficient.

The vapor pressure p^0 of a pure substance varies with temperature according to the Clausius–Clapeyron equation:

$$\frac{dp^0}{dT} = \frac{p^0 \lambda}{RT^2}$$

where λ is the latent heat of vaporization ($=9.7$ cal/mole for water at $100\,°C$), R is the gas constant and T is the absolute temperature. At the pore entrance, the curvature of the vapor–liquid interface is generally assumed to have a negligible effect on the equilibrium; however, possible influences on the vapor pressure value can be estimated by Kelvin equation [4]:

$$P^0_{\text{convex surface}} = P^0_\infty \exp \left(\frac{2 \cdot \gamma_L}{r \cdot c \cdot R \cdot T} \right) \tag{5.3}$$

where $P^0_{\text{convex surface}}$ is the pure liquid saturation pressure above a convex liquid surface with radius of curvature r, P^0_∞ is the pure liquid saturation pressure above a flat surface, γ_L is the liquid surface tension, c is the liquid molar density, R is the gas constant and T is the temperature.

Activity coefficients ξ_i can be deduced by a large number of equations aiming at evaluating the excess Gibbs function of mixtures; the most popular of them are listed in Table 5.1. The Margules equation is the simplest one, and has been found to give similar results to Van Laar equation for several organic solutions. For asymmetric systems, showing large positive deviation from ideality, the Van Laar equation is preferentially used. The Wilson equation is a powerful tool for systems that do not exhibit L–L phase splitting. A most flexible and complex approach to the determination of activity coefficients is given by UNIQUAC model; it is based on a combinatorial term that contains pure-components parameters, and a residual contribute depending on adjustable parameters that are characteristic for each binary system.

Table 5.1: Empirical expressions for the activity coefficients (γ_i^∞ is the activity coefficient at infinite dilution).

Margules equation	$\ln \gamma_1 = x_2^2 [A_{12} + 2(A_{21} - A_{12})x_1]$ $\ln \gamma_2 = x_1^2 [A_{21} + 2(A_{12} - A_{21})x_2]$	$\ln \gamma_1^\infty = A_{12}$ $\ln \gamma_2^\infty = A_{21}$
Van Laar equation	$\ln \gamma_1 = \dfrac{A'_{12}A'_{21}x_1x_2}{(A'_{12}x_1 + A'_{21}x_2)^2}$ $\ln \gamma_2 = \dfrac{A'_{21}A'_{12}x_1^2}{(A'_{12}x_1 + A'_{21}x_2)^2}$	$\ln \gamma_1^\infty = A'_{12}$ $\ln \gamma_2^\infty = A'_{21}$
Wilson equation	$\ln \gamma_1 = -\ln(x_1 + \Lambda_{12}x_2) + x_2 \left(\dfrac{\Lambda_{12}}{x_1 + \Lambda_{12}x_2} - \dfrac{\Lambda_{21}}{x_2 + \Lambda_{21}x_1} \right)$ $\ln \gamma_2 = -\ln(x_2 + \Lambda_{21}x_1) + x_1 \left(\dfrac{\Lambda_{21}}{x_2 + \Lambda_{21}x_1} - \dfrac{\Lambda_{12}}{x_1 + \Lambda_{12}x_2} \right)$	$\ln \gamma_1^\infty = -\ln \Lambda_{12} + 1 - \Lambda_{21}$ $\ln \gamma_2^\infty = -\ln \Lambda_{21} + 1 - \Lambda_{12}$

The expression for activity coefficient in diluted aqueous ionic solutions can be derived from the Debye–Hückel theory:

$$\log \xi_\pm = -|z_+ z_-| \Psi \sqrt{I} \tag{5.4}$$

Here $\xi_\pm$ is the activity coefficient of the electrolyte, Ψ is a constant that depends on the temperature and solution permittivity, z is the ion valence and I the ionic strength of the solution given as follows:

$$I = \frac{1}{2} \sum_i z_i^2 c_i \tag{5.5}$$

In an aqueous solution at 25 °C the constant Ψ is 0.509 $(mol/kg)^{1/2}$.

Mass transfer for MD can be described in terms of resistances in series upon transfer between the bulks of two phases contacting the membrane (Figure 5.6).

Mass transfer boundary layers adjoining the membrane generally result in a negligible contribution to the overall mass transfer resistance, whereas molecular diffusion across the polymeric membrane often represents the controlling step. Resistance to mass transfer on the distillate side is omitted whenever MD operates with pure water as condensing fluid in direct contact with the membrane, or if the configuration used to establish the required driving force is based on vacuum. The resistances within the membrane are associated with Knudsen, molecular and surface diffusion mechanisms and convective transport.

A mass balance across the feed side boundary layer allows to derive a relationship between molar flux J, mass transfer coefficient k_x and solute concentrations c_m and c_b at the membrane interface and in the bulk, respectively:

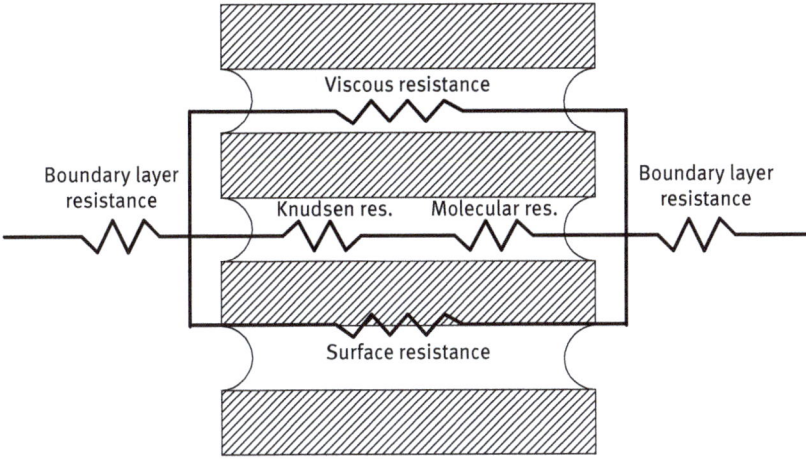

Figure 5.6: Serial and parallel arrangement of resistances to mass transport in MD.

$$\frac{J}{\rho} = k_x \ln \frac{c_m}{c_b} \tag{5.6}$$

ρ is the density of the solution. Literature provides several correlations [18], often derived by analogy with those evaluated for heat transport, that are practical for determining the mass transfer coefficient. These empirical relationship are usually expressed in the following form:

$$Sh = \alpha \, Re^\beta \, Sc^\gamma \tag{5.7}$$

where Sh is the Sherwood number $Sh = \frac{k_x d_h}{D}$ (d_h: hydraulic diameter, D: diffusion coefficient); Re is the Reynolds number $Re = \frac{\rho v d_h}{\mu}$ (ρ: fluid density, v: fluid velocity, μ: fluid viscosity); Sc is the Schmidt number $Sc = \frac{\mu}{\rho D}$.

A brief list of specific predictive equations for mass transfer coefficients is given in Table 5.2.

As consequence of solvent permeation through the membrane, the solute concentration c_m at the feed solution/membrane interface becomes higher than that in the bulk solution, c_b. This phenomenon, known as concentration polarization, is quantified by the CPC coefficient, defined as follows:

$$CPC = \frac{c_m}{c_b} \tag{5.8}$$

In a porous medium, if surface diffusion is assumed negligible, mass transfer can be affected by viscous resistance (resulting from the momentum transferred to the supported membrane), Knudsen diffusion resistance (due to collisions between

Table 5.2: Examples of specific predictive equations for mass transfer coefficients in MD.

Correlation	α	β	γ	k_x (10^{-5} m/s)	Comment	Reference
$k_x = \beta Q^\gamma$	–	4.02×10^{-5}	0.38	3.5–7.6	Q: volumetric feed flowrate (L/min) VMD Stirred cell Feed side	[60]
$Sh = \alpha Re^\beta Sc^\gamma$	2.0	0.48	0.33	–	Stirred cell Stirring rate: 200–800 rpm With aqueous LiBr solution (0–55% w/w)	[61]
$Sh = \alpha Re^\beta Sc^\gamma$	1.86	0.38	0.38	–	Tangential flux	[62]
$Sh = \alpha Re^\beta Sc^\gamma$	0.96–0.45φ	0.55	0.33	17.5	Helicoidal hollow fibers φ: angle of inclination With oxygen $50<R_e<400$ Feed side	[63]
$Sh = \alpha Re^\beta Sc^\gamma$	0.023	0.33	0.33	6.6–7.4	Tubular fibres With water and NaCl aqueous solutions (2 and 4% w/w)VMD Feed side	[64]

molecules and membrane walls) or ordinary diffusion (due to collisions between diffusing molecules) [19]. Predominance, coexistence or transition between all of these different mechanisms are estimated by comparing the mean free path ι of diffusing molecules to the mean pore size of the membrane (Knudsen number). Kinetic theory of ideal gases calculates ι as follows:

$$\iota = \frac{k_B T}{P\sqrt{2\pi\sigma^2}} \tag{5.9}$$

where k_B is the Boltzmann constant (1.380×10^{-23} J/K) and σ is the collision diameter of the molecule (2.7 Å for water).

In the continuum region, the free mean path of a gas is small when compared with the average membrane pore diameter, and molecule–molecule collisions predominate over molecule–wall collisions. The Knudsen number, defined as the ratio of the free path of the gas to the pore diameter (Kn = ι/d_{pore}) is < 1 and the flux can be described by Darcy's law. In the Knudsen region this situation is reversed: the mean free path of a gas is large with respect to the average membrane pore diameter (Kn > 1), molecule–wall collisions predominate over molecule–molecule collisions and the mass transport can be described by Knudsen's law. In many practical cases, ι is comparable to the typical pore size of MD membranes and no simplifications can be done when modeling MD mass transfer operations. Dusty gas model (DGM) is frequently used for describing gaseous molar fluxes through porous media; the most general form (neglecting surface diffusion) is expressed as follows:

$$\frac{J_i^D}{D_{ie}^k} + \sum_{j=i\neq i}^{n} \frac{y_j J_i^D - y_i J_j^D}{D_{ije}^0} = -\frac{1}{RT}\nabla p_i \qquad (5.10)$$

$$J_i^v = -\frac{\varepsilon r^2 p_i}{8RT\tau\mu}\nabla P \qquad (5.11)$$

$$D_{ie}^k = \frac{2\varepsilon r}{3\tau}\sqrt{\frac{8RT}{\pi M_i}} \qquad (5.12)$$

$$D_{ije}^0 = \frac{\varepsilon}{\tau}D_{ij}^0 \qquad (5.13)$$

where J^D is the diffusive flux, J^v the viscous flux, D^k the Knudsen diffusion coefficient, D^0 the ordinary diffusion coefficient, y the molar fraction in gaseous phase, p the partial pressure, M_i the molecular weight, μ the gas viscosity, r the pore radius, ε the membrane porosity and τ the membrane tortuosity. The subscript e indicates the "effective" diffusion coefficient, calculated by taking into account the structural parameters of the membrane as shown in eqs. (5.12) and (5.13).

Although DGM was derived for an isothermal system, it is successfully applied in MD working under relatively small thermal gradients by assuming an average value of temperature across the membrane. The adoption of empirical correlations is in some cases preferred. The transmembrane flux is often expressed as a linear function of the vapor pressure difference across the membrane [4]:

$$J = C\Delta p \qquad (5.14)$$

where C is the MD coefficient, and Δp the partial pressure gradient evaluated at the membrane surfaces. In eq. (5.14), the MD coefficient C is a function of the structural membrane properties (pore size, thickness, porosity and tortuosity), physical and chemical properties of the vapor transported across the membrane (molecular weight and diffusivity) and operative conditions.

Regarding heat transfer, Figure 5.7 illustrates the possible heat transfer resistances in MD using an electrical analogy.

Heat is first transferred from the heated feed solution of uniform temperature T_f across the thermal boundary layer to the membrane surface at a rate $Q = h_f \cdot \Delta T_f$. At

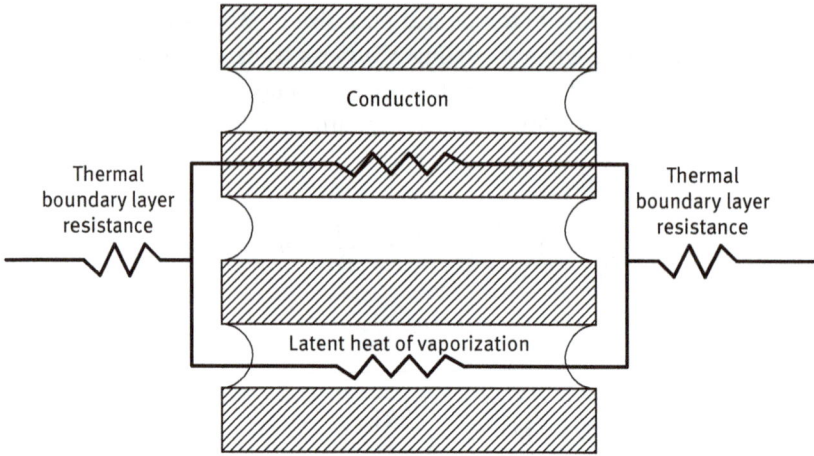

Figure 5.7: Serial and parallel arrangement of resistances to heat transport in MD.

the surface of the membrane, liquid is vaporized and heat is transferred across the membrane at a rate $Q_V = h_V \cdot \Delta T_m = N \cdot \Delta H_V$ (where N is the rate of mass transfer and ΔH_V is the heat of vaporization). Additionally, heat is conducted through the membrane material and the vapor that fills the pores at a rate $Q_m = h_m \cdot \Delta T_m$ where $h_m = \varepsilon \cdot h_{mg} + (1 - \varepsilon)h_{ms}$ (ε is the membrane porosity, and h_{mg} and h_{ms} represent the heat transfer coefficients of the vapor within the membrane pores and the solid membrane material, respectively). Conduction is considered as a heat loss mechanism, because no corresponding mass transfer takes place. Total heat transfer across the membrane is $Q = Q_V + Q_m$. Finally, as vapor condenses at the liquid–vapor interface, heat is removed from the cold-side membrane surface through the thermal boundary layer at a rate $Q = h_p \cdot \Delta T_p$.

The overall heat transfer coefficient of the MD process is given as follows:

$$\frac{1}{U} = \frac{1}{h_f} + \frac{1}{h_m + h_v} + \frac{1}{h_p} = \frac{1}{h_f} + \frac{1}{\left(\frac{K_g \cdot \varepsilon + K_m(1-\varepsilon)}{\delta}\right) + \left(\frac{N \cdot \Delta H_V}{T_{fm} - T_{pm}}\right)} + \frac{1}{h_p} \tag{5.15}$$

where each h and each T represent the corresponding heat transfer coefficients and temperatures shown in Figure 5.7.

In eq. (5.15), since K_g is generally an order of magnitude smaller than K_m (Table 5.3), heat lost by conduction through the membrane can be reduced by increasing the membrane porosity ε.

The total heat transferred across the membrane is given as follows:

$$Q = U \cdot \Delta T \tag{5.16}$$

Equation (5.15) illustrates the importance of minimizing the boundary layer resistances (maximizing the boundary layer heat transfer coefficients). A commonly

Table 5.3: Thermal conductivity of various polymers, air and water.

	Thermal conductivity, W/mK
PP	0.11–0.16
PVDF	0.17–0.19
PTFE	0.25–0.27
Air	$2.72 \times 10^{-3} + 7.77 \times 10^{-5} T$
Water	$2.72 \times 10^{-3} + 5.71 \times 10^{-5} T$
Water at 60 °C	0.022

used measure of the magnitudes of the boundary layer resistances relative to the total heat transfer resistance of the system is given by the temperature polarization coefficient (TPC):

$$\text{TPC} = \frac{T_{\text{fm}} - T_{\text{pm}}}{T_{\text{f}} - T_{\text{p}}} \qquad (5.17)$$

- if TPC → 1, the MD system is well designed and it is limited by mass transfer;
- if TPC → 0, the MD system is poorly designed and it is limited by heat transfer through the boundary layers.

The boundary layer heat transfer coefficients are estimated from empirical correlations usually expressed in the form:

$$\text{Nu} = \alpha \cdot \text{Re}^{\beta} \text{Pr}^{\gamma}$$

where Nu is the Nusselt number, Re is the Reynolds number and Pr is the Prandtl number (more details are available in [20, 21]).

The heat transfer across the membrane has already been described. For what concerns the heat transferred by convection within the membrane pores, this can be also considered but is negligible because convection accounts for, at most, 6% of the total heat lost through the membrane and only 0.6% of the total heat transferred across the membrane [4].

In the MD process, the low-to-moderate flow rates and high heat transfer coefficients reduce the impact of concentration polarization, which is lower than that of the temperature polarization effect [22]. In fact, boundary layers next to the membrane can contribute substantially to the overall transfer resistance: heat transfer across the boundary layers is often the rate-limiting step for mass transfer in MD because a large quantity of heat must be supplied to the membrane surface to vaporize the liquid, and because the membrane fabrication technology has improved so much in the last decades that the MD process has shifted away from being limited by mass transfer across the membrane to being limited by heat transfer through the boundary layers on either side of the membrane.

5.3.4 Membranes for MD technology

The membranes for MD need to be:
- porous and hydrophobic for serving their intended function,
- with low thermal conductivity to minimize conduction losses and achieve better thermal efficiency,
- with good thermal and chemical stability, as well as sufficient mechanical strength for maintaining a good long-term operation.

In most of the MD experiments, membranes typically fabricated from PTFE, PP or PVDF have been used, possessing a high porosity (70–80%), a membrane thickness of 10–300 µm and providing microfiltration properties with pore sizes of 0.2–1 µm (Table 5.4).

Table 5.4: Common commercial membranes for use in MD (Modified from Khayet [24]).

Membrane type	Trade name	Manufacturer	Material	Pore size (µm)	Porosity
Flat sheet	TF200	Gelman	PTFE supported	0.20	80
	TF450		by PP	0.45	80
	TF1000			1.0	80
	3MA	3M Corporation	PP	0.29	66
	3MB			0.40	76
	3MC			0.51	79
	3MD			0.58	80
	3ME			0.73	85
	FGLP	Millipore	PTFE supported	0.20	70
	FHLP		by PE	0.50	80
	GVHP		PVDF	0.22	75
	HVHP			0.45	75
Capillary	Accurel S6/2	AkzoNobel Microdyn	PP	0.20	70
	Accurel BFMF	Enka AG		0.20	

Information about hydrophobicity is obtained by contact angle (Θ) measurements: a droplet of water deposited on a hydrophobic surface gives a contact angle greater than 90°. According to the Young equation,

$$\gamma_{LV} \cos \theta = \gamma_{SV} - \gamma_{SL}$$

where γ_{LV}, γ_{SV} and γ_{SL} are the surface tension for liquid–vapor, the surface energy of the polymer and the solid–liquid surface tension, respectively.

The effect of surface heterogeneity on contact angle is generally established by relation that allows predicting the contact angle Θ^* of a rough surface from the contact angle Θ of the equivalent smooth surface [23]:

$$\cos \Theta^* = f_1 \cos \Theta - f_2$$

where f_1 and f_2 are the fractions of liquid–solid and liquid–air surfaces, respectively.

Microporous polymeric membranes are prepared by various techniques: sintering, stretching and phase inversion.

Sintering is a simple technique for the production of microporous structure having a porosity in the range of 10–40% and rather irregular pore sizes, ranging from 0.2 to 20 μm.

Stretching allows producing microporous membranes with a relatively uniform porous structure, pore size distribution in the range of 0.1–3 micron and porosity of about 90%.

MD membranes can be prepared (and often happened) by phase inversion technique from polymers that are soluble at a certain temperature in an appropriate solvent or solvent mixture, and that can be precipitated as a continuous phase by changing temperature and/or composition of the system. These changes aim to create a miscibility gap in the system at a given temperature and composition; from a thermodynamic point of view, the free energy of mixing of the system becomes positive.

The formation of two different phases, that is, a solid phase forming the polymeric structure (symmetric, with porosity almost uniform across the membrane cross section, or asymmetric, with a selective thin skin on a sublayer) and a liquid phase generating the pores of the membrane, is determined by few and conceptually simple actions:

1. by changing the temperature of the system (cooling of a homogeneous polymer solution which separates in two phases) – temperature-induced phase separation technique and
2. by adding nonsolvent or nonsolvent mixture to a homogeneous solution – diffusion-induced phase separation.

The first step to fabricate a high-performance membrane is to choose a correct membrane material. Among the hydrophobic materials applied in fabrication of membranes for MCs, fluoropolymers constitute a unique class of materials with a combination of interesting properties that attracted significant attention from material researchers over the past few decades [25]. Generally, these polymers have high thermal stability, improved chemical resistance and lower surface tension because of the low polarizability and the strong electronegativity of the fluorine atom, its small van der Waals radius (1.32 °A) and the strong C–F bond (485 kJ/mol). The important fluoropolymers for membrane operations are PVDF, PTFE, poly(ethylene chlorotrifluoroethylene), poly(chlorotrifluoroethylene), poly(vinyl fluoride), poly(fluorenyl ether), Hyflon® AD, Teflon® AF and Cytop®. Among the technological developments, fullerene, graphene membranes, carbon nanotubes and biomimic membranes as well as thermally rearranged polymers [26] and two-dimensional materials [27] are emerging as developed membranes with superior permeability, durability and selectivity.

5.4 Membrane crystallization technology

MCr is an extension of MD; it is a hybrid membrane separation–crystallization process where a solution first becomes saturated, then supersaturated and finally the crystals are obtained.

In its current conception, a membrane crystallizer is a system in which a solution containing a nonvolatile solute that is likely to be crystallized (defined as the crystallizing solution or feed or retentate) is in contact with, by means of a microporous membrane, a solution on the distillate side. The membrane might be made of polymeric or inorganic materials or by a combination of both in a hybrid or composite configuration. Hollow fibers as well as flat-sheet membranes can be employed in a similar manner.

When the membrane is prevented from becoming wet due to the adjacent solutions, no mass transfer through its porous structure is observed directly in liquid phase, but the two subsystems, which are in contact, are subjected to mass interexchange in the vapor phase. As in MD, wetting of the membrane, with the consequent deleterious direct passage of liquids, can be avoided when the pressure of the solutions facing it is lower than the entry limit (P_{entry}).

The gradient of vapor pressure between the two subsystems induces the evaporation of the volatile component from feed solution, migration through the porous membrane and, finally, the recondensation at the distillate side (Figure 5.8(a)). The continuous removal of solvent from the feed solutions in a membrane crystallizer increases solute concentration, thus generating supersaturation. Accordingly, the membrane in MCr does not act as a sieving barrier for the selective transport of specific components, but as a physical support able to generate and to sustain a controlled supersaturated environment in which crystals can nucleate and grow.

Depending on the chemical–physical properties of the membrane and on the process parameters (temperature, concentration, flowrate, etc.), the solvent evaporation rate, and hence supersaturation degree and supersaturation rate, might be regulated very precisely. The effect would be the control of the nucleation and growth rate by choosing a broad set of available kinetic trajectories in the thermodynamic phase diagram, which are not readily achievable in conventional crystallization methods, and which would lead to the production of specific crystalline morphologies and structures [29–31] (Figure 5.9). Furthermore, the generation of an extremely homogeneous supersaturation over the whole solution, due to the numerous points for solvent removal (pores), allows the production of homogeneous distribution of initial aggregates, which, in turn, will produce macroscopic crystals with uniform size distribution and controlled morphology. This aspect would be of undoubted benefit for industrial production of organic crystals.

Moreover, in a membrane crystallizer, the crystallizing solution is in direct contact with the membrane surface; therefore, a solute–membrane interaction is likely to occur, depending on the fluidynamic regime. This effect can be due to both the

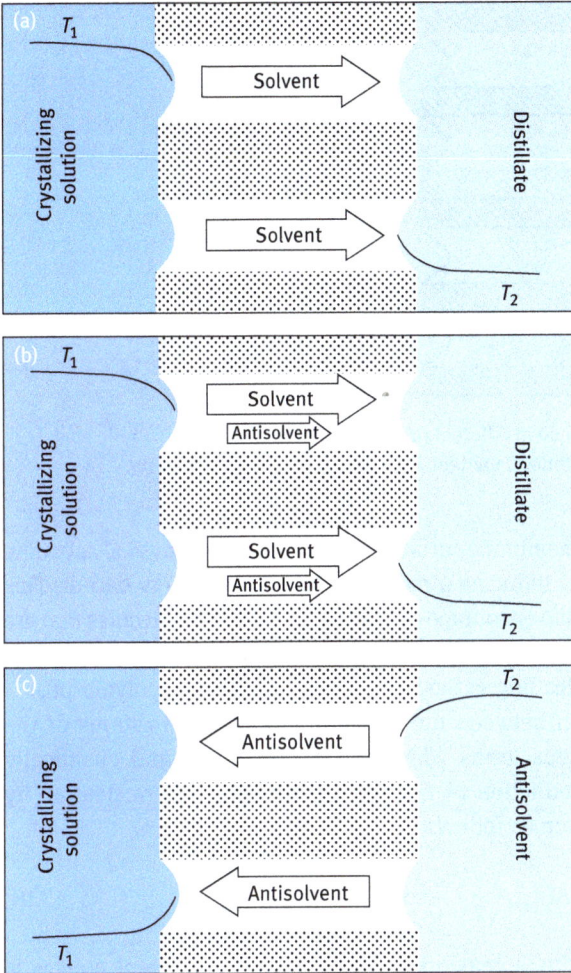

Figure 5.8: Basic principle of a membrane crystallizer: (a) solvent removal MCr, where solvent is removed from the crystallizing solution under a temperature gradient ($T_1 > T_2$); (b) solvent/antisolvent demixing MCr, in which the preferential evaporation of the solvent induces the increase of the antisolvent volume fraction, thus reducing solubility ($T_1 > T_2$) and (c) antisolvent addition MCr, where an antisolvent is evaporated into the crystallizing solution in vapor phase from the other side of the membrane ($T_1 < T_2$) Drioli et al. [28]. Reprinted with Permission.

structural and chemical properties of the membrane surface: first, the porous nature of the surface might supply topographical heterogeneities where solute molecules are physically entrapped leading, locally, to enhanced levels of supersaturation; secondly, the nonspecific and reversible chemical interaction between the membrane and the solute can allow to concentrate and reorient molecules on the surface, without loss of mobility, thus facilitating effective interaction among them, which is apt for

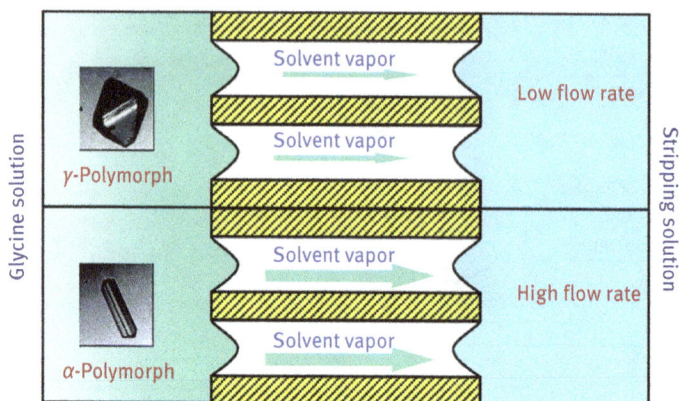

Figure 5.9: Glycine polymorphs obtained in different conditions of flow rate. From Di Profio et al. [29]. Reprinted with permission from Di Profio. Copyright 2007 American Chemical Society.

crystallization. In this way, the membrane surface might operate as a physical substrate for heterogeneous nucleation by inducing a reduction in the free energy barrier. This effect would be extremely useful to encourage crystallization of such molecules that are reluctant to crystallize, as is generally the case for biomacromolecules, or to facilitate specific molecular interactions leading to the formation of preferential polymorphs.

Considering the interaction between the solute and substrate in terms of the contact angle α that the nucleus forms with the ideally smooth and chemically homogeneous substrate, the reduction of the activation energy for nucleation by heterogeneous activation is given as follows:

$$\Delta G^*_{het} = \Delta G^*_{hom} \left(\frac{1}{2} - \frac{3}{4}\cos\alpha + \frac{1}{4}\cos^3\alpha \right) \tag{5.18}$$

Figure 5.10 graphically shows the aforementioned expression for different polymeric materials used as heterogeneous nucleants. If the nucleus wets the substrate completely ($\alpha = 180°$), $\Delta G^*_{het} = \Delta G^*_{hom}$; when the contact angle is 90° (limit between hydrophobic and hydrophilic behavior), $\Delta G^*_{het} = \frac{1}{2}\Delta G^*_{hom}$, and the smaller the contact angle α, the smaller the value of the activation energy for nucleation, which is zero for $\alpha = 0$.

When using MCr, eq. (5.18) is no longer applicable because nucleation takes place on a porous substrate. In this case, a modified version of the equation that takes into account the porous structure of the surfaces has to be considered [33]:

$$\frac{\Delta G^*_{het}}{\Delta G^*_{hom}} = \frac{1}{4}(2 + \cos\alpha)(1 - \cos\alpha)^2 \left[1 - \varepsilon\frac{(1 + \cos\alpha)^2}{(1 - \cos\alpha)^2} \right]^3 \tag{5.19}$$

where ε is the overall surface porosity, defined as the ratio of the total pore areas on the whole geometrical surface. If $\varepsilon = 0$, eq. (5.19) reduces to the form reported in the literature (eq. 5.18) for heterogeneous nucleation on nonporous surfaces.

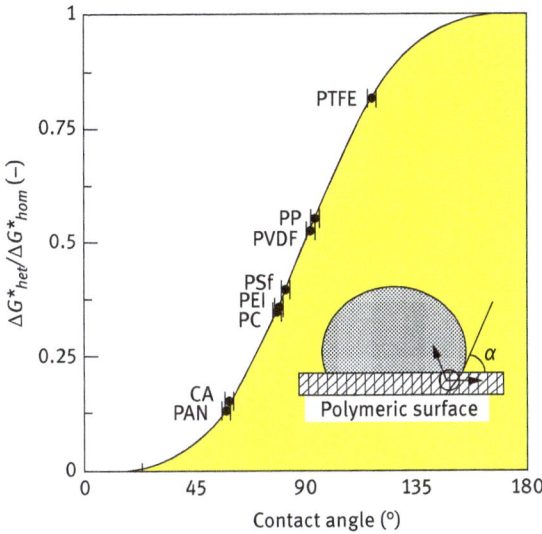

Figure 5.10: Reduction in the free energy of the nucleation barrier due to heterogeneous nucleation as a function of the water contact angle with the polymeric surface (CA, cellulose acetate; PAN, polyacrylonitrile; PC, polycarbonate; PET, polyetherimide; PES, polyethersulfone; PP, polypropylene; PSf, polysulfone; PTFE, polytetrafluoroethylene; PVDF, polyvinylidene fluoride). (From Di Profio et al. [32].) Reprinted with permission from Di Profio. Copyright 2010 American Chemical Society.

5.4.1 Membrane crystallizer types and versions

From a process design point of view, the membrane can be applied for a membrane-assisted operation (i.e., on a mixture recirculating loop), or directly for in situ crystallization purposes. The first case can be seen as a typical hybrid process approach and it is shown in Figure 5.11(a). The membrane module is here used to generate the supersaturation, or simply to concentrate the solid phase, but the nucleation and the crystal growth take place in the crystallizer [34]. In the second case (Figure 5.11(b)), the crystallization takes place directly in the membrane module where the supersaturation is generated [35, 36].

In a recent development of the process, Di Profio et al. [37] proposed a new design of the MCr process in which crystallization is induced by using antisolvent. This new approach operates in two configurations: first, solvent/antisolvent demixing (Figure 5.8(b)), and second, antisolvent addition (Figure 5.8(c)). In both the cases, solvent/antisolvent migration occurs in the vapor phase, according to the general concept of MCr and, unlike the aforementioned configuration, not by forcing it in liquid phase through the membrane.

The selective and precise dosing of the antisolvent, controlled by the porous membrane, allows a finer control of the solution composition during the process

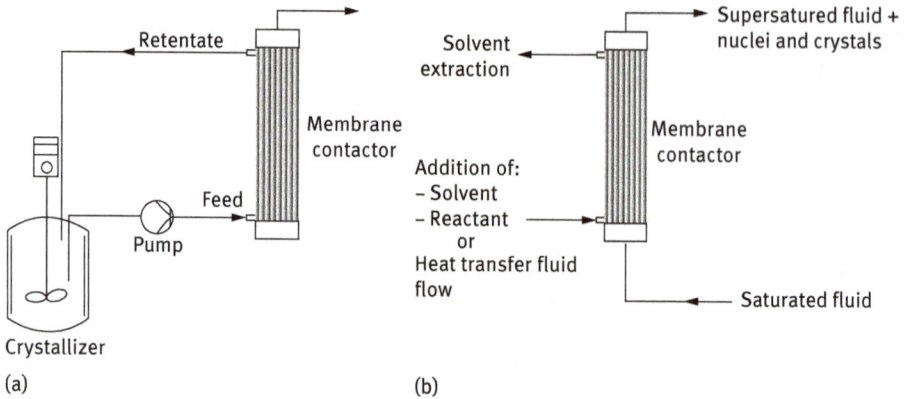

Figure 5.11: Schematic representation of the two process designs: (a) hybrid MCr process and (b) the crystallization takes place directly in the membrane module.

and at the nucleation point, with consequent improvement of the final crystal characteristics.

According to the considerations above, crystallization by using membranes can be classified depending on the different working principles [38]:

1. MD/OD-based processes where diffusion of solvent molecules in vapor phase through a porous membrane, under the action of a gradient of chemical potential as driving force, generates supersaturation in the crystallizing solution;

2. membrane-assisted crystallization in which pressure-driven membrane operations (MF, NF and RO) are used to concentrate a solution by solvent removal in liquid phase, while crystals are recovered in a separate tank, often operated at lower temperature and with seeding;

3. solid (nonporous) hollow fibers used as heat exchanger to generate supersaturation by cooling;

4. antisolvent (or crystallizing solution) forced directly in the liquid state into the crystallizing solution (or into the antisolvent) through the pores of a membrane under a pressure gradient and

5. antisolvent MCr, where dosing of the antisolvent in the crystallizing solution is carried out by means of a membrane, according to the working principle of point 1, in the two solvent/antisolvent demixing and antisolvent addition configurations.

5.4.2 Membrane crystallization: transport phenomena

Detailed relations and models for heat and mass transport through the membrane in MCr can be described by using the same concepts developed for MD. As a general description, heat and mass transport through membranes occurs only if the overall system is not in thermodynamic equilibrium. For the mass transport, it can be separated into three steps: mass transfer in feed boundary layer, mass transfer *through the membrane pores* and mass transfer in permeate boundary layer. The mass transfer in permeate boundary layer is not taken into account since the mole fraction of the transporting species in the permeate stream is approximately equal to one. The mass transfer in boundary layers is analyzed by film theory, whereas DGM is usually employed to describe the mass transfer across the membrane. DGM elucidates mass transfer in porous media by four possible mechanisms: viscous flow, Knudsen diffusion, molecular diffusion and surface diffusion. It is general for MD in direct contact configuration to neglect surface diffusion and viscous flow, and to employ a Knudsen-molecular diffusion transition model [4, 39]:

$$N' = \frac{\varepsilon P D_{ij}}{\tau \delta RT} \ln \left(\frac{2 \frac{2r}{3}\left(\frac{8RT}{\pi M_i}\right)^{1/2}}{p_a} + P D_{ij}}{\frac{1 \frac{2r}{3}\left(\frac{8RT}{\pi M_i}\right)^{1/2}}{p_a} + P D_{ij}} \right) \tag{5.20}$$

where N' is the molar transmembrane flux, ε is the porosity, P is the total pressure, D_{ij} is the diffusivity, τ is the membrane tortuosity, δ is the membrane thickness, R is the ideal gas constant, T is the temperature, r is the pore size, M_i is the molecular weight and p_a^1 and p_a^2 are the partial pressure of air at feed and membrane surface, respectively. The model can be further simplified in the specific cases. Knudsen diffusion model is suitable for the system where the collision between molecule and pore wall dominates the mass transport. On the other hand, molecular diffusion model is preferred when the collision between the molecules plays main role in the mass transfer across the membrane. Nevertheless, if both molecule–pore wall and molecular–molecular collisions occur frequently, the Knudsen-molecular diffusion transition model must be employed. Both molecular diffusion limit and Knudsen-molecular diffusion transition model were successfully applied to describe the flux in DCMD system [39–42]. In both cases, the transmembrane flux is proportional to membrane porosity ε, whereas it is inversely proportional to membrane thickness δ. Therefore, membrane structural properties will strongly affect MCr performance in terms of both solvent evaporation rate and crystals nucleation and growth. In fact, a crystallizing solution can be imagined as a certain number of solute molecules moving among the molecules of solvent and colliding with each other, so that a number of them converge forming clusters. According to Volmer [43], the critical size n^*, which an assembly of molecules must have in order to be stabilized by further growth, is given as follows:

$$n^* = \frac{32\pi v_0 \, \gamma^3}{3(k_B T)^2 \ln^3 S} \tag{5.21}$$

where v_0 is the molecular volume, γ is the interfacial energy, k_B is Boltzmann's constant and S is the supersaturation. Equation (5.21) shows how critical size n^*

depends on supersaturation S: the higher the operating level of supersaturation, the smaller is the size (typically a few tens of molecules). Therefore, the proper choice of membrane chemical–physical properties and process parameters (temperature, concentration, etc.) allow regulating flowrate, supersaturation degree, supersaturation rate, nucleation and growth.

5.5 Membrane distillation and membrane crystallization in zero liquid discharge system

Among the different and various MD and MCr applications, an important field where these technologies are expected to give a fundamental contribution is seawater desalination. The latter is the most economically competitive way to resolve the potable water demand in regions with high deficiencies. Growing global demand for water made membrane filtration the prominent technology in desalination: the global cumulative contracted capacity, dominated by reverse osmosis, reached 99.8 million m^3/day in 2017 [44], and membrane desalination technologies account for more than 90% of all desalination plants [45]. However, the management of brine is becoming one of the main problem of seawater and brackish water desalination. Current practice of handling these concentrates is to discharge them into the coastal waters, which could have detrimental effects on the aquatic life and coastal environment. To mitigate major environmental concerns related to brine/concentrate discharges, concentrates should be prediluted with the seawater to minimize the effects related to high salt concentrations. Removal or recovery of substances from the concentrates by implementing alternative treatment methods is an attractive option that would provide both environmental benefits (in reducing the magnitude and environmental impact of disposal) and economic benefits (in production of valuable metals) [46].

As described earlier, MD and MCr are not limited by concentration polarization phenomena. Therefore, they can be utilized in integrated membrane-based desalination

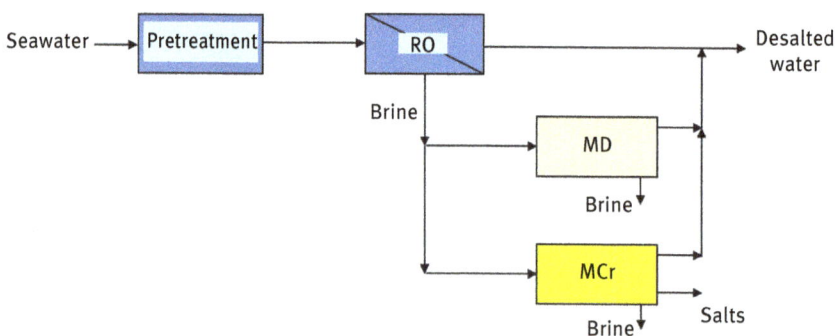

Figure 5.12: Possible integrated membrane-based desalination process.

processes (Figure 5.12), downstream of RO, for the recovery of water and the various chemicals present in the concentrated streams.

The studies carried out by Drioli and coworkers [47–50] showed that the introduction of a MCr unit on NF and RO retentate streams of an integrated membrane-based desalination system constituted of MF/NF/RO increases plant recovery factor so much to reach 92.8%, higher than that of an RO unit (about 45%) and much higher than that of a typical MSF (about 10–20%). Therefore, integrated membrane-based desalination systems with MD and MCr units offer the possibility to reduce brine disposal problem and increase water recovery factor of desalination plants, thus approaching a zero liquid discharge, or near zero liquid discharge, system.

5.6 Membrane condenser

A membrane condenser is an innovative membrane operation where microporous hydrophobic membranes are used to promote water condensation and recovery.

Water condensation is a common phenomenon in natural and industrial settings. When condensation takes place on a surface that is not wet by the condensate, water beads up into droplets and rolls on the surface. This process is referred to as dropwise condensation. Water vapor preferentially condenses on solid surfaces rather than directly from the vapor because of the reduced activation energy of heterogeneous nucleation in comparison to homogeneous nucleation [51, 52].

The working principle of membrane condenser has been recently introduced by Macedonio et al. [55] and consists of condensing and recovering the water contained in a gaseous stream on the retentate side of the membrane module by exploiting the hydrophobic nature of the membrane, whereas the dehydrated gases pass through the membrane in the permeate side. Figure 5.13 schematizes the membrane condenser principle.

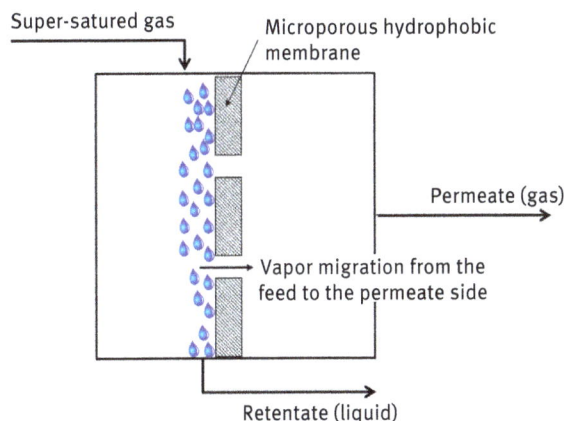

Figure 5.13: A scheme of the membrane condenser process for the recovery of evaporated "waste" water from a gaseous stream. From Macedonio [53]. Reprinted with Permission.

In particular, the feed (e.g., a gaseous stream such as flue gas) at a certain temperature and, in most of the cases, water saturated is fed to the membrane condenser kept at a lower temperature for cooling the gas up to a supersaturation state. The water condenses in the membrane module, and once this stream is brought into contact with the retentate side of the microporous membranes, their hydrophobic nature prevents the penetration of the liquid into the pores letting pass the dehydrated gases through the membrane. Therefore, the liquid water is recovered at the retentate side, whereas the other gases at the permeate side of the membrane unit.

The technologies until now proposed for the capture of evaporated water from gaseous streams are cooling with condensation, liquid and solid sorption, dense membranes or porous hydrophilic membranes. Each of them has its own advantages and disadvantages:

– Traditional condensers represent the easiest process even if corrosion phenomena due to the presence of acid pollutant in the waste gases stream are their main limitation.
– Adsorption of water by a liquid or solid desiccant is another valid alternative despite desiccant losses, cost and regeneration of adsorbent, and low quality of water are the main drawbacks.
– High-energy consumption due to the high-pressure requirements is associated with the utilization of dense membrane for the recovery of water vapor from the gaseous streams. On the contrary, its main advantage is the preferential transport of water vapor with respect to gases through the membrane via sorption–diffusion mechanism.

Advantages of a membrane condenser are higher water recovery, cleaner operation, lower energy consumption and no corrosion phenomena [54]. In fact, the membranes to be utilized in a membrane condenser can be fabricated from chemically resistant polymers such as polytetrafluoroethylene, PP and PVDF, highly resistant to the acid compounds that can be present in the flue gas streams. On the contrary, water quality can be limited by the possible condensation of contaminants, if the latter are present in the gaseous stream. However, as it will be described in the following sections, in a membrane condenser, the condensation of contaminants in the recovered liquid water can be controlled by opportunely tuning the operating conditions.

5.6.1 Transport phenomena in a membrane-assisted condensation system

The vapor flux across the membrane of a membrane condenser follows the reduced Knudsen-molecular diffusion transition form of the DGM [55, 56]:

$$N_i^v = -\frac{1}{RT_{avg}} \left(\frac{D_w^k D_{w-a}^0}{D_{w-a}^0 + p_a D_w^k} \right) \frac{\Delta p}{\delta} M$$

with $D_w^k = \frac{2\varepsilon r}{3\tau} \sqrt{\frac{8RT_{avg}}{\pi M}}$

and $D_{w-a}^0 = 4.46 {*}10^{-6} \frac{\varepsilon}{\tau} T_{avg}^{2.334}$

where Δp is the partial pressure gradient of water through both membrane surfaces generated by a temperature gradient and/or a concentration difference (i.e., it is the driving force to mass transfer in the proposed process), N^v is the viscous flux, D^k is Knudsen diffusion coefficient, D^0 is the ordinary diffusion coefficient, M is the molecular weight, r is the membrane pore radius, ε is the membrane porosity, τ is the membrane tortuosity, δ is the membrane thickness, R is the gas constant, the subscript "w" is indicative of water, the subscript "a" is indicative of air and the subscript "avg" is indicative of average value.

When the transmembrane flux N is known, the amount of water that can be recovered from the fed gaseous stream can be calculated with the following equation:

$$\text{Fraction of recovered water} = \frac{\left[n_{H_2O, feed} - \left(\frac{P_{H_2O}(T_{out}) \cdot \left(n_{feed} - n_{H_2O, feed} \right)}{P - P_{H_2O}(T_{out})} - N \cdot a \right) \right]}{n_{H_2O, feed}}$$

where $n_{H_2O, feed}$ is the number of water moles in the fed gaseous stream, n_{feed} is the total number of fed gaseous moles, $P_{H_2O}(T_{out})$ is the partial pressure of water at the temperature at the exit of the condenser T_{out} and a is the membrane area.

The concentration c of each contaminant i exiting from the system (i.e., $c_{i, OUT, liquid}$) with the recovered water can be estimated through a mass balance

$$c_{i, OUT, liquid} = \frac{moli_{i, FEED} - moli_{i, OUT, vapor}}{\text{recovered water}}$$

where the solubility of the different gases in water can be estimated by Henry's law:

$$k_H = k_H^0 {*}\exp\left(\frac{-\Delta_{soln}H}{R} \left(\frac{1}{T} - \frac{1}{T^0} \right) \right)$$

with

– $\frac{\Delta_{soln}H}{R} = \frac{-d \ln k_H}{d(1/T)}$

– k_H^0 and $T^0 \exp$ refer to standard condition (298.15K)

The results of the simulation give, therefore, indications about the amount and quality of recovered water. An example can be found in Figure 5.14, showing the amount of water that can be recovered at increasing temperature difference between the fed gaseous stream and the membrane module, at various feed temperatures.

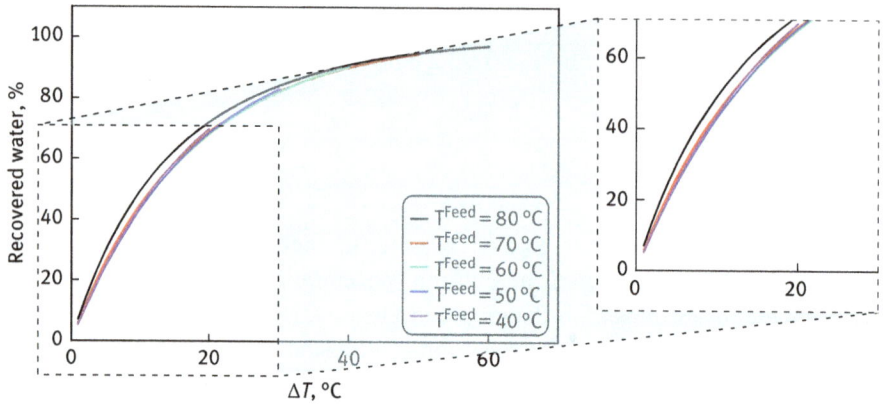

Figure 5.14: Recovered water vs ΔT (i.e., temperature reduction between fed flue gas and membrane module) at various feed temperature (Q^{Feed}=380.8 SCCM, relative humidity [RH] = 100%). Reprinted with permission from Macedonio et al. [56].

It has been estimated [56] that, in general, ΔT lower than 20 °C are sufficient to recover more than 65% of the water present in the gaseous waste stream (Figure 5.14). Moreover, the amount of recovered water increases more than proportionally with the increasing ΔT and T^{Feed} due to the exponential dependence of partial pressure of water on temperature. One more parameter most influencing the process is the feed flow rate Q^{Feed} and interfacial membrane area $A^{Membrane}$ ratio: a low value of this ratio means that the membrane area is more than sufficient to treat the feed; on the contrary, a high value of the ratio implies that the feed flow rate is too high with respect to the membrane area available in the module. As a consequence by keeping the membrane area constant and increasing Q^{Feed}, the amount of water recovered will not increase proportionally at the increasing Q^{Feed} (Figure 5.15).

The contaminants concentration in the recovered liquid water strongly depends on the temperature. In particular, it was proved that the concentration of contaminants in the recovered liquid water increases with the increasing temperature difference ΔT (Figure 5.16) between the fed flue gas and the membrane module (that is when the temperature of the fed flue gas is constant, whereas the temperature of the membrane module decreases). In fact, Henry's law constant increases reducing the temperature and, as a consequence, the solubility in aqueous solution increases.

For the energy consumption, in a membrane condenser (i.e., the *condenser heat duty*) it mainly constitutes of two terms: 1) the power required to drive an eventual compression (e.g., through a fan or blower) and 2) the heat duty required to cool the gaseous stream and condense the vapor or part of it. However, it was proved [54] that the energy consumption of the system is mainly owing to the heat required to condense the water vapor (Figure 5.17).

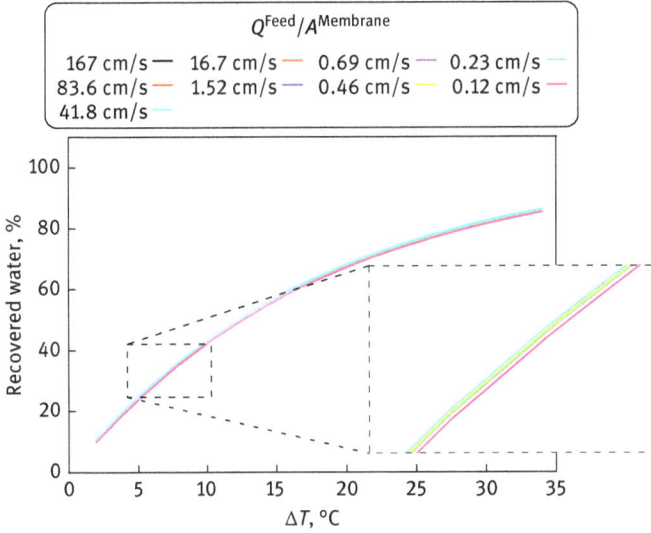

Figure 5.15: Recovered water vs ΔT at various $Q^{Feed}/A^{Membrane}$ ratio (constant feed temperature = 55.5 °C and RH = 100%). Reprinted with permission from Macedonio et al. [56].

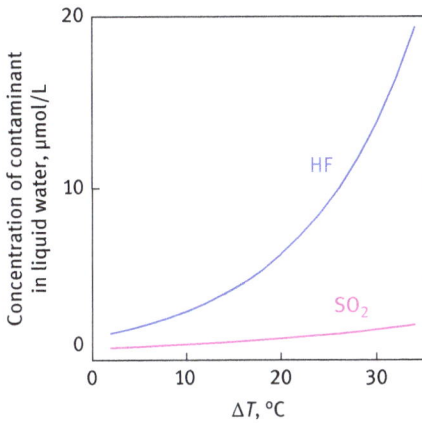

Figure 5.16: Concentration of HF and SO_2 in the recovered water as a function of the temperature difference ΔT between feed and membrane module ($Q^{Feed}/A^{Membrane}$ = 0.69 cm/s, T^{Feed} = 55.5 °C, RH^{Feed} = 100%). From Macedonio et al. [53]. Reprinted with Permission.

With the aim to reduce membrane condenser energy consumption, two more membrane-assisted condenser configurations were analyzed (Figure 5.18): in a second configuration, a cold sweeping gas cools the feed gaseous stream directly inside the membrane module; in the third configuration, the fed waste gas is first partially cooled via an external medium and then a sweeping gas is used for the final cooling of the stream. These two configurations were compared in terms of amount of recovered liquid water and energy consumption (Figure 5.19) with the first above described membrane condenser process (the one where the fed waste gas is cooled via cooling water before entering the membrane module).

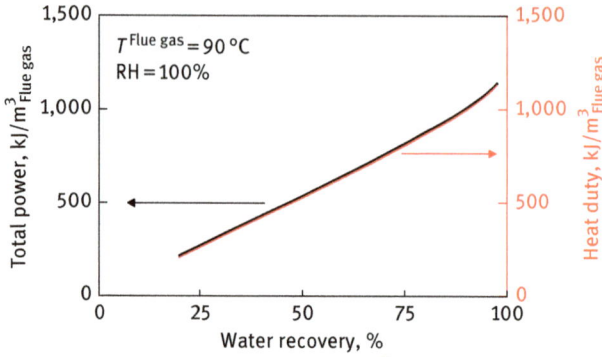

Figure 5.17: Power needed to drive the process per cubic meter of treated flue gas vs recovered water. Feed flue gas with at RH = 100% and T = 90 °C. Reprinted with permission from Macedonio et al. [55].

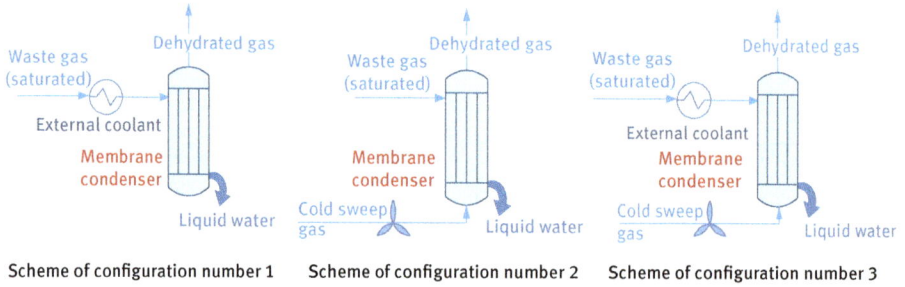

Scheme of configuration number 1   Scheme of configuration number 2   Scheme of configuration number 3

Figure 5.18: Membrane condenser possible configurations.

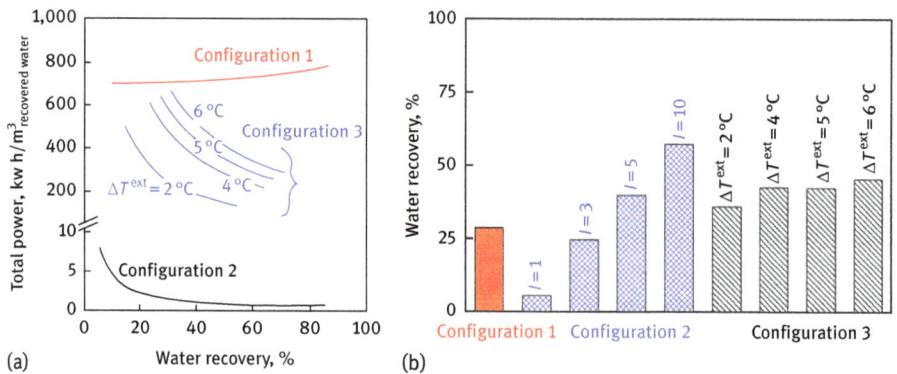

(a)   (b)

Figure 5.19: (a) Energy consumption and (b) maximum amount of recoverable water from the three different proposed configurations (Q^{Feed} =0.03 m^3/h, RH = 100%, 55 °C; sweep gas at 20 °C Q^{Sweep}=0.09 m^3/h). From Macedonio et al. [53]. Reprinted with Permission.

When a cold sweeping gas is utilized, the ratio between its flow rate and the feed flow rate is defined as the sweeping factor, I:

$$I = \frac{\text{cold sweeping gas flow rate}}{\text{feed waste gas flow rate}}$$

Considering waste gas at 55 °C and RH = 100%, and air at 20 °C as cold sweeping gas (for configuration 2 and 3), the highest energy consumption is achieved utilizing configuration 1 (Figure 5.19(a)). On the contrary, configuration 2 is, among those proposed, the one with the lowest energy consumption (Figure 5.19(a)). For water recovery, excluding configuration 2 with $I = 1$ and $I = 10$ (the first one because recovers too low amount (5.42%) of water and the second one because requires too high flow rate of cold sweeping gas), it increases going from configuration 1 to 2 with the increasing sweeping factor I (from about 28% to 39%; Figure 5.19(b)). The highest maximum amount of liquid water can be obtained utilizing configuration 3 whose energy consumption is in between configuration 1 and 2 (Figure 5.19(a)).

5.7 Membrane dryer

A vacuum membrane dryer (VMDr) is an extension of VMD where the transport of water vapor and volatile compounds (from the feed to the permeate side) through the micropores of hydrophobic membranes is applied to streams containing solid particles (Figure 5.20). The result is drying of the particles at the feed side, provided that their size is higher than the pore size of the membrane, while producing a purified permeate. With respect to traditional vacuum dryers, in a VMDr the presence of the membrane prevents the loss of microparticles.

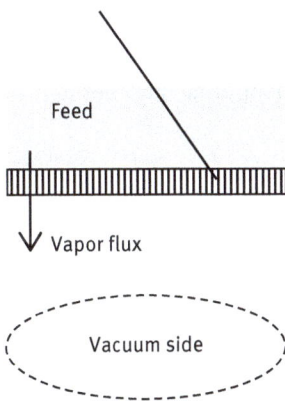

Feed

Vapor flux

Vacuum side

Figure 5.20: Dehydration in a VMDr. From Drioli et al. [57]. Reprinted with Permission.

As a case study, the potentiality of the VMDr was tested for the dehydration of polystyrene microparticles (size ranging from 0.3 to 7 μm) up to a solid residue of 98±0.5 wt.% [57, 58]. The operating feed temperature and vacuum pressure were 30 °C and 4 mbar, respectively. A low temperature was chosen to investigate the efficiency of the system when thermolabile compounds have to be treated (how it might happen in food and pharmaceutical industries). Among different studied configurations, a flat membrane module working with both recirculation and stirring of the feed was identified as the most suitable one for the process. However, the maximum solid residue achieved by recirculating the feed was of about 50±0.5 wt.%. The feed recirculation inside the set-up is, in fact, possible only until the feed is still fluid, and this implies a low solid residue. On the contrary, by loading the feed at one side of the membrane and ensuring its mixing by stirring, the target of a solid residue of 98±0.5 wt.% was obtained. The drying process is affected by different factors, such as the operating temperature and vacuum pressure, the operating time, the amount and the initial solid residue of feed to be dehydrated, the particle size and the membrane properties.

VMDr presents some interesting advantages in comparison with traditional vacuum dryers:
- no need of dust filters to collect particles entrained by the vapor flux;
- ability to efficiently treat a wide range of particle size;
- possibility to homogenize the feed by air bubbling that acts more gently on particles than a delumping bar, a mixer or a crasher.

Moreover, when compared to the other devices also used to dehydrate solids, like press filters, spray dryers and fluidized beds, some further benefits can be added:
- high dehydration efficiency achieved at low operating temperatures and pressures, with saving in energy consumption and better preservation of particle properties, that are not subject to mechanical deformation nor thermal degradation;
- recovery of purified water;
- higher degree of automation.

In order to fulfil all the expectations, membrane dryer needs more systematic analysis, accurate modeling for an easy scale-up and development of membranes appropriate for this operation.

5.8 Membrane emulsification

Membrane emulsification technology is a drop-by-drop emulsification method through a porous membrane, introduced in Japan in 1988. Ever since, there is significant attention given to this method from both the scientific and technological point of views. After around 25 years of research and inventions, membrane emulsification received increasing interest. The method allows the generation of emulsions by a drop-by-drop mechanism through a microporous membrane (Figure 5.21) [59]. The dispersion phase in the form of droplets can be as a pure liquid or an emulsion. In the first case, simple emulsions are produced such as oil-in-water or water-in-oil droplets in which an immiscible liquid is used as continuous phase. In the second case, an emulsion of an emulsion is generated, for example, water-in-oil-in-water and oil-in-water-in-oil emulsions, also termed as multiple or double emulsions. A coarse emulsion may be, alternatively, refined upon passage through a microporous membrane. The process is referred to as premix membrane emulsification to distinguish from the direct process (Figure 5.21).

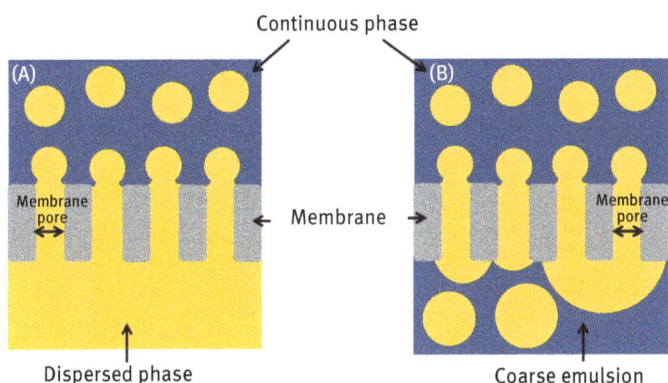

Figure 5.21: Production of particles by direct and premix membrane emulsification. (A) Direct membrane emulsification and (B) premix membrane emulsification. From Piacentini et al. [59]. Reprinted with Permission.

5.9 Concluding remarks

MCs technology can potentially lead to significant innovation in processes and products, thus offering new opportunities in the design, rationalization and optimization of innovative productions.

MD is investigated worldwide as a low cost and energy-saving alternative with respect to conventional separation processes (such as distillation and RO). It is one of the few membrane operations based on a thermal process. However, due to the separation principle different from the traditional pressure-driven membrane processes and due to lesser fouling tendency, a lot of other interesting applications of MD have been explored.

The operation at low temperature makes MD attractive for processing of temperature-sensitive products such as pharmaceutical compounds, juices, dairy products, natural aromatic compounds and so on. Theoretically 100% rejection of nonvolatiles renders MD process ideal for the applications requiring a very high rejection of certain components, such as treatment of nuclear waste or radioactive water and production of water for semiconductor industry. More recently, MD has been also used in MDBR configuration for the treatment of industrial and municipal used waters, in order to effectively retain small size and persistent contaminants. As MD is able to produce highly concentrated brine, the possibility to combine MD with the production of high-quality crystals (e.g., extracted from the brine of RO) is particularly interesting and promising. A MD/MCr process is used for water recovery and to concentrate the feed solution until a desired concentration so that salt crystals can be easily precipitated. Another important advantage of MCr is that the membrane matrix acts as a selective gate for solvent evaporation, thus modulating the final degree and the rate for the generation of the supersaturation. Hence, the final properties of the produced crystals, both in terms of structure (polymorphism) and morphology (habit, shape, size and size distribution), can be modulated by acting on the transmembrane flux (e.g., by changing the driving force of the process). Furthermore, an undoubted benefit of this novel technology, when compared to other traditional techniques, is the ability to speed up crystallization kinetics even for high-molecular weight macromolecules, like proteins, which are characterized by low diffusivity in solution.

More young membrane operations are membrane condenser and membrane dryer. The former is an innovative membrane unit operation for the selective recovery of evaporated waste water from industrial gases and for the control of the composition of the recovered liquid water. The latter allows to recover and dry solid microparticles from liquid suspensions.

References

[1]   Mulder, M. Basic. Principles of Membrane Technology. Kluwer Academic Publishers, London: 1996.
[2]   Findley, M. E. "Vaporization through porous membranes." Ind. Eng. Chem. Process. Des. Dev. (1967); 6(2): 226–230.
[3]   Bodell, B. R. Silicone rubber vapor diffusion in saline water distillation. US Patent. (1963); 285: 032.
[4]   Lawson, K. W., Lloyd, D. R. "Membrane distillation." J. Membr. Sci. (1997); 124(1) 1–25.
[5]   Curcio, E., Drioli, E. Membrane distillation and related operations—a review. Sep. Purif. Rev. (2005); 34(1): 35–86.
[6]   Macedonio, F. Membrane Distillation (MD). 2015. Date: 28 December; pp. 1–9. doi: 10.1007/ 978-3-642-40872-4_361-2.
[7]   Winter, D., Koschikowski, J.,Wieghaus, M. Desalination using membrane distillation: Experimental studies on full scale spiral wound modules. J. Membr. Sci. (2011); 370: 104–112.
[8]   Wang, P., Tai-Shung, C."Recent advances in membrane distillation processes: Membrane development, configuration design and application exploring." J. Membr. Sci. (2015); 474: 39–56.

[9] Zhao, K., Heinzl, W., Wenzel, M., Buttner, S., Bollen, F., Lange, G., Heinzl, S., Sarda, N. Experimental study of the memsys vacuum-multi-effect-membrane-distillation (V-MEMD) module. Desalination. (2013); 323: 150–160.

[10] Francis, L., Ghaffour, N., Alsaadi, A. A., Amy, G. L. Material gap membrane distillation: A new design for water vapor flux enhancement. J. Membr. Sci. (2013); 448: 240–247.

[11] Qin, Y., Liu, L., He, F, Liu, D., Wu, Y. Multi-effect membrane distillation device with efficient internal heat reclamation function and method. China Patent 201010570625. 2013.

[12] Gryta, M. Osmotic MD and other membrane distillation variants. J. Membr. Sci. (2005); 246(2), 145–156.

[13] Drioli, E., Criscuoli, A., Curcio, E. Membrane contactors: Fundamentals, applications and potentialities. Membrane Science and Technology Series. Elsevier, Amsterdam; Boston: 2006; 11.

[14] Drioli, E., Ali, A., Macedonio, F. Membrane distillation: Recent developments and perspectives. Desalination. (2015); 356: 56–84.

[15] J Phattaranawik, A. G., Fane, A. C. S., Pasquier, Bing, W. A novel membrane bioreactor based on membrane distillation. Desalination. (2008); 223: pp. 386–395.

[16] Goh, S., Zhang, J., Liu, Y., Fane, A. G. Membrane distillation bioreactor (MDBR)–A lower greenhouse-gas (GHG) option for industrial wastewater reclamation. Chemosphere. (2015); 140: 129–142.

[17] Yatmaz, H. C., Dizge, N., Kurt, M. S. Combination of photocatalytic and membrane distillation hybrid processes for reactive dyes treatment. Environmental Technology. (2017);1–9.

[18] Gekas, V. Hallstrom, B. J. Membrane Sci. (1987); 30: 153.

[19] Strathmann, H., Giorno, L., Drioli, E. "An Introduction to Membrane Science and Technology," Wiley, Rome: 2011.

[20] Macedonio, F. Membrane Distillation (MD). Date: 28 December 2015; pp 1–9. doi: 10.1007/978-3-642-40872-4_361-2.

[21] Ali, A., Macedonio, F., Drioli, E., Aljlil, S., Alharbi, O. A. Experimental and theoretical evaluation of temperature polarization phenomenon in direct contact membrane distillation. Chem. Eng. Res. Des. (2013); 91(10): 1966–1977.

[22] Macedonio, F. Membrane Distillation (MD). Entry in Drioli, Enrico, and Lidietta Giorno. eds. Encyclopedia of Membranes, Springer-Verlag Berlin Heidelberg, 2016.

[23] Franken, A. C. M., Nolten, J. A. M., Mulder, M. H. V., Bargeman, D., Smolders, C. A. Wetting criteria for the applicability of membrane distillation. J. Membr Sci. (1987); 33: 315–328.

[24] Khayet, M. "Membranes and theoretical modeling of membrane distillation: A review." Adv. Cplloid. Interface. Scie. (2011); 164(1): 56–88.

[25] Cui, Z., Drioli, E., Lee, Y. M. Recent progress in fluoropolymers for membranes. Prog. Polym. Sci. (2014); 39(1): 164–198.

[26] Kim, J. H., Park, S. H., Lee, M. J., Lee, S. M., Lee, W. H., Lee, K. H., Lee, Y. M. Thermally rearranged polymer membranes for desalination. Energy. Environ. Sci. (2016); 9(3): 878–884.

[27] Liu, G., Jin, W., Xu, N. "Two-Dimensional-Material membranes :A new family of high-performance separation membranes minireviews", Angew. Chem. Int. Ed. Engl. 2016; vol. 55. pp. 2–16.

[28] Drioli, E., Di Profio, G., Curcio, E. Progress in membrane crystallization. Curr. Opin. Chem. Eng. (2012); 1(2), 178–182.

[29] Di Profio, G. Tucci, S. Curcio, E., Drioli, E. Selective glycine polymorph crystallization by using microporous membranes. Cryst. Growth. Des. 2007; 7(3), 526–530.

[30] Di Profio, G. Tucci, S. Curcio, E., Drioli, E. Controlling polymorphism with membrane-based crystallizers: Application to form I and II of paracetamol. Chem. Mater. 2007; 19(10), 2386–2388.

[31] Quist-Jensen, C. A., Macedonio, F., Horbez, D., Drioli, E. Reclamation of sodium sulfate from industrial wastewater by using membrane distillation and membrane crystallization. Desalination. (2017); 401: 112–119.

[32] Di Profio, G., Curcio, E., Drioli, E. Supersaturation control and heterogeneous nucleation in membrane crystallizers: Facts and perspectives. Ind. Eng. Chem. Res. (2010); 49(23): 11878–11889.

[33] Curcio, E., Fontananova, E., Di Profio, G., Drioli, E. Influence of the structural properties of poly (vinylidene fluoride) membranes on the heterogeneous nucleation rate of protein crystals. J. Phys. Chem. B. 2006; 110: 12438–12445.

[34] Edwie, F, Chung, T.-S. Development of simultaneous membrane distillation– crystallization (SMDC) technology for treatment of saturated brine. Chem. Eng. Sci. (2013); 98: 160–172.

[35] Kieffer, R.,Mangin, D.,Puel, F.,Charcosset, C. Precipitationofbariumsulphate in a hollow fiber membrane contactor, Part I: Investigation of particulate fouling. Chem. Eng. Sci. (2009); 64: 1759–1767.

[36] Curcio, E., Simone, S., Profio, G. D, Drioli, E., Cassetta, A., Lamba, D. Membrane crystallization of lysozyme under forced solution flow. J. Membr. Sci. (2005); 257: 134–143.

[37] Di Profio, G., Stabile, C., Caridi, A., Curcio, E., Drioli, E. Antisolvent membrane crystallization of pharmaceutical compounds. J. Pharm. Sci. (2009); 98: 4902–4913.

[38] Di Profio, G., Curcio, E., Drioli, E. Membrane crystallization technology, chapter in book: Comprehensive Membrane Science and Engineering, December. (2010); pp. 21–46: DOI: 10.1016/B978-0-08-093250-7.00018-9

[39] Srisurichan, S., Jiraratananon, R., Fane, A. G. Mass transfer mechanisms and transport resistances in direct contact membrane distillation process. J. Membr. Sci. (2006); 277: 186–194.

[40] Phattaranawik, J. Heat and Mass Transfer Models for High Fux Direct Contact Membrane Distillation Process. D. Eng. Thesis. Department of Chemical Engineering, King Mongkut's University of Technology Thonburi, Thailand, 2002.

[41] Bandini, S., Gostoli, C., Sarti, G. C. Role of heat and mass transfer in membrane distillation process. Desalination. (1991); 81: 91–106.

[42] Schofield, R. W., Fane, A. G., Fell, C. J. D. Heat and mass transfer in membrane distillation. J. Membr. Sci. (1987); 33: 299–313.

[43] Volmer, M. Kinetik der Phasenbildung; Steinkopf: Leipzig. 1939.

[44] IDA Desalination Yearbook. Published by Media Analytics Ltd. United Kingdom, 2017–2018.

[45] IDA Desalination Yearbook. Published by Media Analytics Ltd., United Kingdom, 2016–2017.

[46] Macedonio, F., Drioli, E. Membrane engineering for green process engineering. Engineering. (2017); 3(3): 290–298.

[47] Macedonio, F., Curcio. E., Drioli, E. Integrated membrane systems for seawater desalination: Energetic and exergetic analysis,economic Evaluation, Experimental Study. Desalination. (2007); 203: 260–276.

[48] Macedonio, F., Drioli, E. Pressure-driven membrane operations and membrane distillation technology integration for water purification. Desalination. (2008); 223: 396–409.

[49] Macedonio, F., Drioli, E., Curcio, E., Di Profio, G. Experimental and economical evaluation of a membrane crystallizer plant. Desalination and Water Treatment. (2009); 9: 49–53.

[50] Macedonio, F. Drioli, E., Hydrophobic membranes for salts recovery from desalination plants Desalination and Water Treatment. (2010); 18: 224–234.

[51] Enright, R., Miljkovic, N., Al-Obeidi, A., Thompson, C. V., Wang, E. N. Condensation on superhydrophobic surfaces: The role of local energy barriers and structure length scale. Langmuir. (2012); 28(40):14424–14432.

[52] Kashchiev, D. Nucleation: Basic Theory with Applications, Butterworth-Heinemann: Oxford, UK. 1 ed. 2000.

[53] Macedonio, F., Brunetti, A., Barbieri, G., Drioli, E. Membrane condenser configurations for water recovery from waste gases. Sep. Purif. Technol. (2017); 181: 60–68.

[54] Brunetti, A., Santoro, S., Macedonio, F., Figoli, A., Drioli, G., Barbieri, G. Waste gaseous streams: From environmental issue to source of water by using membrane condensers. Clean–Soil, Air, Water. (2013); 42(8), 1145–1153.

[55] Macedonio, F., Brunetti, A., Barbieri, G., Drioli, E. Membrane condenser as a new technology for water recovery from humidified "waste" gaseous streams. Ind. Eng. Chem. Res. (2013); 52(3): 1160–1167.

[56] Macedonio, F., Cersosimo, M., Brunetti, A., Barbieri, G., Drioli, E. Water recovery from humidified waste gas streams: Quality control using membrane condenser technology. Chem. Eng. Process: Process Intensification. (2014); 86: 196–203.

[57] Drioli, E., Carnevale, M. C., Figoli, A., Criscuoli, A. Vacuum Membrane Dryer (VMDr) for the recovery of solid microparticles from aqueous solutions. J. Membr. Sci. (2014); 472: 67–76.

[58] Criscuoli, A., Carnevale, M. C., Drioli, E. Study of the performance of a membrane-based vacuum drying process. Sep. Purif. Techn. (2016); 158: 259–265.

[59] Piacentini, E., Drioli, E., Giorno, L. Membrane emulsification technology: Twenty-five years of inventions and research through patent survey. J. Membr. Sci. (2014); 468: 410–422.

[60] Bandini, S., Gostoli, C., & Sarti, G. C. (1992). Separation efficiency in vacuum membrane distillation. Journal of Membrane Science, 73(2–3), 217–229.

[61] Sudoh, M., Takuwa, K., Iizuka, H., & Nagamatsuya, K. (1997). Effects of thermal and concentration boundary layers on vapor permeation in membrane distillation of aqueous lithium bromide solution. Journal of membrane science, 131(1–2), 1–7.

[62] Tomaszewska, M., Gryta, M., & Morawski, A. W. (1995). Study on the concentration of acids by membrane distillation. Journal of Membrane Science, 102, 113–122.

[63] M.J. Costello, P.A. Hogan and A.G. Fane, Proc. Euromembrane '97, 23–27 Jun 1997, The Netherlands.

[64] Lawson, K. W., & Lloyd, D. R. (1996). Membrane distillation. I. Module design and performance evaluation using vacuum membrane distillation. Journal of membrane science, 120(1), 111–121.

Adele Brunetti, Giuseppe Barbieri, Enrico Drioli,
Rosalinda Mazzei, Lidietta Giorno

6 Membrane reactors and membrane bioreactors

6.1 Introduction

This chapter is composed of two sections. Section 6.2 is focused on membrane reactors using inorganic catalysts, mainly in gas phase and temperature higher than 100 °C. Section 6.3 is dedicated to membrane bioreactors (MBR) that use catalysts of biological origin, mainly in the liquid phase and temperature lower than 100 °C.

6.2 Membrane reactors

Currently, the use of membrane reactors for producing hydrogen is becoming more and more a reality with various pilot installations all over the world and increasing number of studies focused not only on the development of mechanically and chemically stable membranes with high permselectivity but also on integrated processes that are able to maximize the productivity, thus reducing equipment size and energy consumption. The possibility of combining reaction and separation in the same unit, reducing the whole volume of the plant and increasing its efficiency are the main assets that promote the development of this technology. The scope of this chapter is to highlight the main findings about membrane reactors that are used in high-temperature gaseous phase reactions. A detailed discussion is presented about the main membrane reactor configurations that are used for various processes (from packed bed to fluidized bed to microreactors) with a short overview on some representative results in the upgradation of syngas thorugh water-gas shift reaction. In addition, considering water-gas shift as a reference reaction, process intensification metrics are detailed. These parameters together with traditional variables usually used in the evaluation of the process performance supply additional and important information for the selection of the type of technology and the identification of the operating condition windows for making the process more profitable.

Adele Brunetti, Giuseppe Barbieri, Enrico Drioli, Rosalinda Mazzei, Lidietta Giorno, National Research Council of Italy, Institute for Membrane Technology (ITM-CNR), University of Calabria, Rende, Italy

https://doi.org/10.1515/9783110281392-006

6.2.1 Description of the technology

According to IUPAC definition, a membrane reactor (often called multiphase reactor) is a device in which a chemical reaction and a separation can be integrated in a single unit. A typical scheme of a membrane reactor is reported in Figure 6.1. For a tube-in-tube configuration, the outer tube represents the shell of the module and the inner tube the membrane. The latter divides the unit in two zones: (1) the reaction/retentate volume where the reaction occurs and (2) the permeate side where one of the products is selectively recovered when the membrane acts as separator. In case of gaseous phases, the driving force is provided by the pressure gradient between the two membrane sides, whereas by a concentration gradient in the case of a liquid phase reaction.

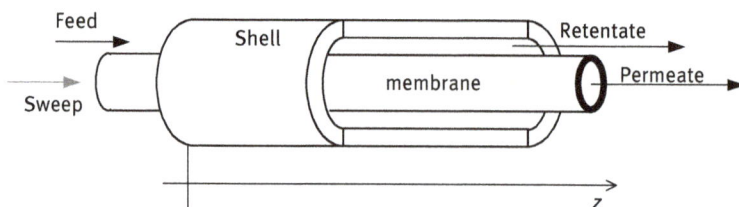

Figure 6.1: Scheme of a tubular membrane reactor.

Membrane reactor functions
The functions of a membrane in a membrane reactor can be basically distinguished as follows:

– "Extractor or separator": This is for selective removal of the products from the reaction mixture
– "Distributor": This is for controlling the addition of reactants to the reactor
– "Contactor": This is for optimizing the contact between reactants and catalyst or the contact between the two phases

Membrane as separator or extractor
In the most common class of membrane reactors, the membrane plays the role of a separator or an "extractor." One or more of the products that are generated by the chemical reaction is continuously removed through the membrane that is recovered in the permeate (Figure 6.2).This selective removal of one of the reaction products from the reaction volume allows the equilibrium conversion to be shifted toward further production of products, thus enhancing the yield of the reaction and in the mean time limiting the undesired side reactions that involve the targeted reaction product.

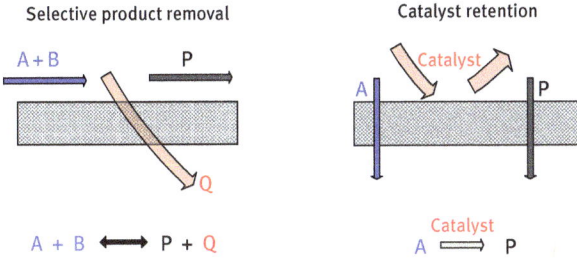

Figure 6.2: Membrane operating as a selective separator or an extractor.

Membrane as distributor

The distribution principle for a membrane is essentially based on controlling the addition of reactants to a reaction mixture through the membrane itself. On the basis of this concept, the membrane can have different functions (Figure 6.3):

– Controlled distribution of limiting reactants in the reaction volume, in order to prevent secondary reactions.
– Upstream separation unit that plays a role in selectively dosing one component from a mixture to another is retained on the other side of the membrane; in this case of coupling of separation and reaction, an increased driving force for permeation is created as the permeating component reacts directly after permeation.

The above-mentioned two functions can be combined.

Most often, the distributor principle is used for coupling of reactions.

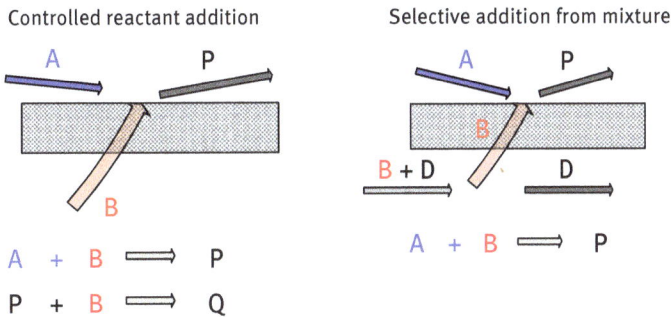

Figure 6.3: Membrane operation as a selective distributor.

Membrane as contactor

The membrane contactor principle is based on the use of microporous membranes that are not selective; however, in some cases they are catalytic. The membrane divides the membrane reactor in two zones: in most cases, the membrane reactor contains reactants in different phases such as liquid–liquid or gas–liquid. The role of the membrane is to act as a support for providing a good contact area between these two phases (Figure 6.4).

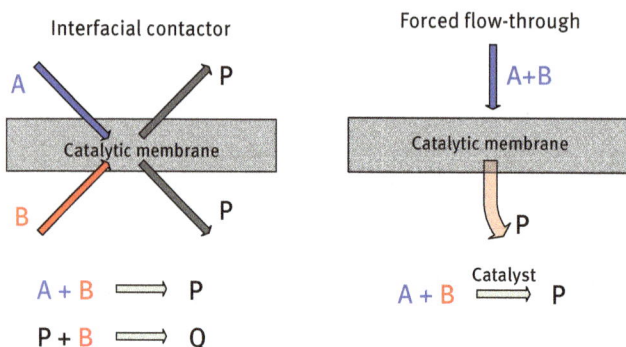

Figure 6.4: Membrane operating as a contactor.

In the case of liquid–liquid reaction, the membrane material can be affined to one phase and not to the other; this is to keep the two phases separate that allows the contact only on the membrane interface. For the gas–liquid reaction, the membrane has to be non-wettable to ensure that the pores are free of liquid even at relatively high liquid pressure and, thus, the operation is stable with a high overall mass transfer coefficient.

The membranes with catalytic properties are most often used in the gas–liquid contactors for reactions that are called "three-phase reactions." The catalytic particles are usually deposited on the pore walls of a thin microporous membrane layer typically supported on an asymmetric thicker layer. One reactant is dissolved in the liquid phase and is sucked into the microporous catalytic layer by capillary forces. The gaseous reactant is fed through the support to the catalytic layer from the other side of the membrane. The pressure on the liquid side can be atmospheric, whereas on the gaseous side, it must be above the bubble point of the support, so that water present there would be forced out and it does not exceed the bubble point of the microporous membrane layer. This membrane reactor concept for gas–liquid reactions is called "catalytic diffuser" (Figure 6.4[a]). As gas–liquid interface is established within the porous membrane structure, the gas phase can be supplied directly to the catalytic region, which consequently increases the concentration of gaseous reactants and enhances the overall reaction rate. Another configuration that has found large application especially in gas–liquid reaction in the "Flow-through catalytic membrane reactor" (Figure 6.4[b]). The difference with the "membrane diffuser" is that in the former the reactants are fed from different sides of the membrane, whereas in forced flow-through mode the premixed reactants are supplied from the same side in a dead-end mode. The function of the membrane is to provide a reaction volume with short and controlled residence time and high catalytic activity. The catalyst placed inside the membrane pores can be better exploited with respect to the conventional fixed bed where channeling phenomena occur. This results in an intensive contact between reactants and the catalyst, thus implying a high catalytic activity.

Membrane reactor configurations

The different types of membrane reactor configurations are basically classified according to the relative placement of the two most important elements of this technology: (1) the membrane and (2) the catalyst. The main configurations are as follows (Figure 6.5):

- Catalyst is physically separated from the membrane and is either packed or finely dispersed on one side of the membrane itself
- Catalyst is dispersed in the membrane
- Membrane is inherently catalytic

The first configuration is often called "inert" membrane reactor by opposition with the two other ones that are "catalytic" membrane reactor.

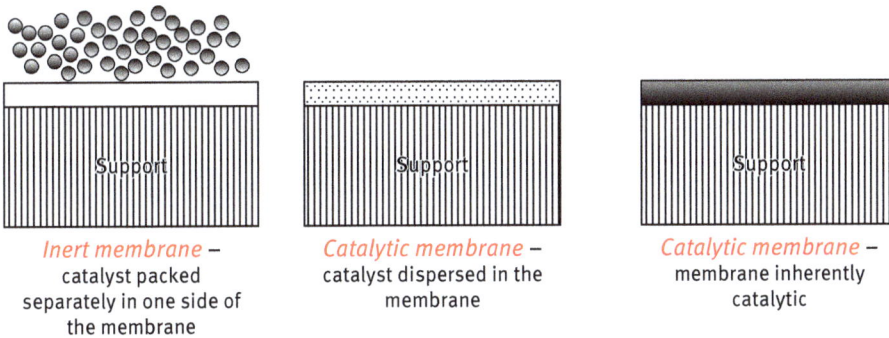

Inert membrane –
catalyst packed
separately in one side of
the membrane

Catalytic membrane –
catalyst dispersed in the
membrane

Catalytic membrane –
membrane inherently
catalytic

Figure 6.5: Main membrane/catalyst combinations.

The membrane can have a cylindrical or flat sheet shape. The cylindrical membranes are subdivided according to their dimensions: "tubular membranes" with a diameter of more than 10 mm, "hollow fibers" (HF) with a diameter of few hundred microns and "capillary membranes" with intermediate sizes (>1 mm). When the membranes are packed closely together in a module, a fiber diameter between 5 and 0.05 mm corresponds to a surface area per volume between 360 and 36,000 m^2/m^3. Tubular membranes are placed inside a pressure-resistant tube. The capillary and hollow fiber membranes are assembled in a module with the free ends of the fibers potted with, for example, epoxy resins or silicone rubber. The flat sheet membranes are usually assembled in a spiral wound or plate and frame configuration, where sets of two membranes are placed in a sandwich-like fashion with their feed sides facing each other and separated by spacers. When such a plate-and-frame module is wrapped around a central collection pipe, a spiral wound module is obtained. In most of the cases, when the membrane acts as a separator with or without catalytic function, the membrane module has a tube-in-tube configuration with the catalytic bed placed in the annulus or in the core in the former case. Same configuration can be assumed if the membrane is catalytic. All these systems can be operated in a continuous mode, with concurrent or countercurrent configuration. Obviously, in addition to the catalyst/ membrane arrangement, the reactor module configuration depends also on the reaction phases and on the function of the membrane as well as on economic considerations, with the correct engineering parameters being employed to achieve this.

6.2.2 Thermodynamic equilibrium in gaseous phase membrane reactors

The nonequilibrium conversion of a Traditional Reactor (TR) depends on several factors, for example, the reactor model considered (continuous stirred tank or plug flow or batch), the thermodynamic variables (temperature and pressure), composition and operating variables (e.g., feed flow rate, catalyst amount and activity and the overall heat exchange coefficient) and so on.

$$\text{Nonequilibrium conversion of a } TR$$

$$= f\left(Kp, \ T^{\text{Reaction}}, \ P^{\text{Reaction}}, \ F^{\text{Feed}}, \ Y_i^{\text{Feed}}, \ W^{\text{Catalyst}}, \ U^{\text{Overall}}\right) \tag{6.1}$$

The upper limit of a chemical reactor is given by the thermodynamics that allows the evaluation of the equilibrium conversion. As it is well known, the calculation of the conversion for a TR starts from reaction equilibrium constants.

$$Kp_j(T) = \prod_i^{N_{\text{Species}}} p_i^{v_{i,j}} \ \forall \text{reaction } j \tag{6.2}$$

The TR equilibrium conversion (TREC) depends only on thermodynamics (temperature, pressure and equilibrium constant) and on the initial concentrations of species.

$$TREC = TR \text{ equilibrium conversion}$$

$$= f\left(Kp, T^{\text{Reaction}}, P^{\text{Reaction}}, Y_i^{\text{Feed}}\right) \tag{6.3}$$

It does not depend on the reaction path.

The nonequilibrium conversion of a membrane reactor (MR) depends on temperature and pressure of the permeate side, sweep gas flow rate and composition and membrane properties, in addition to the other parameters already cited for a TR.

Nonequilibrium conversion of an MR

$$= f\left(Kp, T^{\text{Reaction}}, P^{\text{Reaction}}, F^{\text{Feed}}, Y_i^{\text{Feed}}, W^{\text{catalyst}}, T^{\text{Permeation}}, P^{\text{Permeation}},\right.$$
$$\left. F^{\text{Sweep}}, Y_i^{\text{Sweep}}, U^{\text{Overall}}\right) \tag{6.4}$$

It depends on the reaction and permeation conditions.

The membrane characteristics and its geometry (permeance, thickness and surface) influence the permeation rate and hence the nonequilibrium conversion of an MR.

The equilibrium of an MR is a relatively new concept [1, 2]. The permeation equilibrium has to be reached in an MR in addition to the reaction equilibrium typical of a TR. This extra constraint can no longer be expressed as permeation.

$$J_i^{\text{Permeating}} = 0 \Leftrightarrow f\left(P_i^{\text{Reaction}}, P_i^{\text{Permeation}}\right) = 0 \Leftrightarrow P_i^{\text{Reaction}} = P_i^{\text{Permeation}} \tag{6.5}$$
$$\forall \text{ permeable species } i$$

The condition of the absence of flux is equivalent to zero permeation driving force, that is, when partial pressure of the species on both membrane sides are equal to each other. The said equality means that the equilibrium conversion of an MR is also independent of the permeation law.

Therefore, the MR equilibrium conversion (MREC) is a function of the thermodynamic variables and initial compositions on both sides of the Pd-alloy membranes.

MREC = MR equilibrium conversion

$$= f\left(Kp, T^{\text{Reaction}}, P^{\text{Reaction}}, Y_i^{\text{Feed}}, T^{\text{Permeation}}, P^{\text{Permeation}}, F^{\text{Feed}}/F^{\text{Sweep}}, Y_i^{\text{Sweep}}\right) \tag{6.6}$$

MREC does not depend on the reaction path as the TREC and permeation rate. It is independent of the membrane permeation properties influencing the time-dependent variables (i.e., the residence time for the plug-flow MR) necessary to reach equilibrium, but the final value reached depends on the extractive capacity of the system.

6.2.3 Mathematical modeling of catalytic membrane reactors

A mathematical model allows the description of the evolution of variables defining the system as a function of the external operating conditions (temperature, pressure, composition, etc.). The overall structure of these equations, consisting of momentum, mass and energy balances is the same as that of traditional reactors. However, in the case of MRs, reaction and permeation sides are physically connected through the transmembrane mass and energy flux and, thus, the equation modeling both sides are coupled to each other (Figure 6.6).

In general context, some simplifications can be done to reach an approximated but acceptable representative solution of the target system. For

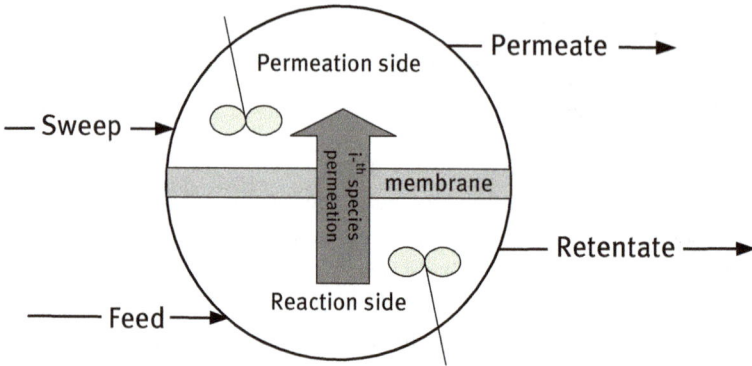

Figure 6.6: Scheme of a catalytic MR. Reprinted from Comprehensive Membrane Science and Engineering Volume 3: Catalytic membranes and catalytic membrane reactors, Edited by E. Drioli and L. Giorno; chapter 3.03; Barbieri, G., Scura F., Brunetti A., Modeling and simulation of membrane reactors and catalytic membrane reactors, Pages 57–79. Copyright 2010, with permission from Elsevier.

example, the local velocity field can be considered known or assigned in most cases.

The mass balance accounts for the net mass flow through an MR:

$$\text{IN} - \text{OUT} + \text{PRODUCTION} = \text{ACCUMULATION}$$

If steady state is considered, the ACCUMULATION term on both reaction and permeation side is zero. Usually, no reaction occurs in the permeation side, even if some examples of coupled reactions in MRs [3, 4] are present in the literature. In such a case, the species produced on one membrane side permeates through the membrane and reacts on the other membrane side [5].

The aforementioned considerations are general and independent of the membrane type. The permeating flux law depends on the mass transfer mechanism determining the permeation (e.g., solution-diffusion in dense polymeric or metallic membranes, and viscous, Knudsen flux and/or surface diffusion in porous membranes).

In the case of solution-diffusion mechanism, the permeating flux of the ith species can be expressed as follows:

$$J_i^{\text{Permeating}} = \frac{S_i D_i}{\delta^{\text{Membrane}}} \left(P_i^{\text{Feed}} - P_i^{\text{Permeate}} \right)\Big|_{\substack{\text{Between} \\ \text{the membrane} \\ \text{interface}}} \tag{6.7}$$

For infinitely selective Pd-based membranes, where the only permeating species is hydrogen, the permeating flux can be expressed by the Sieverts' law in case of internal diffusion-controlling permeation and ideal conditions (infinite dilution in the metal lattice):

$$J_{H_2}^{Sieverts} = \frac{Permeability_{H_2}}{Thickness} \left(\sqrt{P_{H_2}^{Feed}} - \sqrt{P_{H_2}^{permeate}} \right) \tag{6.8}$$

As for the porous membranes, where the pore size is comparable or smaller than the mean free path of molecules (e.g., microporous or zeolite membranes), the permeation is mainly controlled by the Knudsen eq. (6.9a) and/or surface diffusion eq. (6.9b):

$$J_i^{Permeating} = d_{pore} \frac{\varepsilon}{\tau} \sqrt{\frac{1}{3RTM_i}} \frac{\Delta P_i}{Thickness} \tag{6.9a}$$

$$-\rho_{Zeo} \theta_i \frac{\nabla \mu_i}{RT} = \sum_{j=1}^{n_{Species}} \frac{C_{\mu,j}N_i - C_{\mu,i}N_j}{C_{\mu s,i}C_{\mu s,j}D_{ij}} + \frac{N_i}{C_{\mu s,i}D_i} \tag{6.9b}$$

As for the overall reactors, the mathematical expression of MRs may be distinguished as follows:

- Distributed parameter systems, such as, for instance, *tubular and tubes-and-shell MRs*, in which the state variables for both reaction and permeation sides depend on the axial and/or radial position [6]
- Lumped parameter systems such as *completely stirred MRs*, in which both reaction and permeation sides are described by global variables [7].

Tubular membrane reactor

A tubular MR is a tube-in-tube device where the inner tube is usually a permselective membrane promoting the selective mass transfer of reactants/products between the reaction and permeation sides. On both sides, species composition, temperature and pressure can generally change along the reactor length and the radial direction, as well as the permeation rate. Therefore, the behavior of these systems must be described by partial differential equations (PDEs).

However, one-dimensional (1D) mathematical models can provide a satisfactory description in a system for which the radial gradients can be neglected (large radial mixing).

In this section, a simple 1D mathematical model for gas phase reactions in tubular MRs operating in steady state will be presented. The hypotheses of such a model are listed as follows:

- Absence of radial concentration profiles
- Plug flow on both membrane sides
- Isobaric conditions on both membrane sides (i.e., negligible pressure drops in the catalytic bed of lab-scale reactors). However, Ergun's equation can be used for large-scale MRs)
- Ideal gas behavior on both the sides of the membrane

- Pseudo-homogeneous description of the control volumes where the heterogeneous catalytic reactions occur, that is, the void fraction and the specific catalytic surface are included into the reaction rate expressions

Mass balance

The mass balances on both membrane sides for a tubular MR must be written for a differential control volume with a length dz (see Figure 6.6). The MR configuration (reaction in the *Lumen/Annulus*) does not affect the form of the mass balance equations for both the reaction and permeation side, but it utilizes the numerical method to solve them. In fact, we have to deal with initial-value problem (IVP) for the cocurrent configuration – easier to be solved – and boundary-value problem (BVP) for the countercurrent one, for which iterative methods such as shooting and collocation have to be implemented. In this example, a cocurrent configuration is considered, in which equations and corresponding boundary conditions – all defined at the inlets of the system – are reported in eqs. (6.10) and (6.11) (Table 6.1) for the reaction and permeation sides, respectively. All the constitutive equation terms are reported in the Table 6.2 for the reaction side.

The selective removal of one or more products of reaction enhances the conversion as a direct consequence of the *Le Chatelier principle*.

Table 6.1: Mass balance of a tubular MR with cylindrical symmetry. Plug-flow MR (1D – *First-order model*) – Cocurrent flow configuration – *Steady state*.

Reaction side:	$$-\frac{dN_i^{\text{Reaction}}}{dz} + \sum_{j=1}^{N_{\text{Reactions}}} v_{i,j} r_j - \frac{A^{\text{Membrane}}}{V^{\text{Reaction}}} J_i^{\text{Permeating}} = 0$$ $$\text{B.C.} \quad C_i^{\text{Reaction}}\big|_{z=0} = C_i^{\text{Feed}}$$	(6.10)
Permeation side:	$$-\frac{dN_i^{\text{Permeation}}}{dz} + \frac{A^{\text{Membrane}}}{V^{\text{Permeation}}} J_i^{\text{Permeating}} = 0$$ $$\text{B.C.} \quad C_i^{\text{Permeation}}\big|_{z=0} = C_i^{\text{Sweep}}$$	(6.11)

Table 6.2: Constitutive terms.

$N_i^{\text{Reaction}} = C_i v$	Axial convective flux i-th species along the reaction side
$+ \sum_{j=1}^{N_{\text{Reactions}}} v_{i,j} r_j$	Reaction term involving i-th species in all the reactions
$- \frac{A^{\text{Membrane}}}{V^{\text{Reaction}}} J_i^{\text{Permeating}}$	Permeation term of the i-th species through the membrane

The equations governing the permeation side consist of the same terms as those of the reaction side, except for the term relative to the chemical reaction. In the case where the membrane operates as a reactant supplier, the flux goes from the permeation to the feed side and, thus, the flux sign is opposite with respect to the case in which the membrane is used for product separation. Furthermore, when the axial dispersion cannot be neglected (small Péclet's number), a term related to diffusive transport involving the concentration gradients must be included in the equations in addition to the convective flux. This leads to a 1D- second-order model, which requires additional boundary conditions (B.C.) (e.g., Danckwerts' conditions expressing the absence of the concentration gradient just before the reaction inlets and just after the reactor exit).

In a cylindrical symmetry, when the radial profiles are not flat, temperature and pressure depend on the radial coordinate, thus have a differential control volume along both the radial and the axial direction that must be taken into consideration. In this case, a term related to radial diffusion also appears in the balance equations and the mathematical model becomes 2D-second order. As a consequence, the in/out term related to the permeation flux disappearing from the mass balance equation for being taken into account by means of the boundary conditions (B.C.) at the membrane surface can be different from each other depending on whether the membrane thickness is so small that the curvature can be neglected. The required second B.C. over the radial coordinate can be the absence of radial flux on the symmetry axis of the internal tube and at the wall of the external one (shell).

Energy balance

Usually, the heat developed during chemical reactions promotes a temperature profile. Any temperature variation means variations of kinetics (reaction rate) and, specifically for MRs, species permeance and permeating flux. When the heat involved in the reaction produces a sensible temperature change (e.g., the methane steam reforming is a highly endothermic reaction), the energy balances for both the sides of membrane has to be considered as part of the equation set in addition to the mass balances. In cases where there is a low concentration of dissolved species reacting in a liquid phase (e.g., the S-Naproxen methyl ester hydrolysis), then negligible thermal effects can be assumed. In any case, the knowledge of the temperature profiles and the effect that these profiles produce on the membrane properties allows the MRs to be operated as desired.

Equations (6.12) and (6.13) (see Table 6.3) are the energy balances and the corresponding initial and boundary conditions written for the 1D tube-in-tube systems (Table 6.3). These equations have to be coupled to the mass balance eqs. (6.10) and (6.11). The energy balance contains heat exchange between the two sides of the membrane and is transported by permeated species. The annulus exchanges heat with the furnace and lumen side, whereas the stream in the lumen exchanges heat only with the annular volume. The heat generated by the reaction, Ψ (eq. 6.15), see Table 6.4, is present in the equation of the annulus or lumen side, depending on the configuration used. The contribution of transport by permeated species, φ (eq. 6.14), is different from zero on the permeate side (it contributes to increasing temperature), but it is null on the other side because the permeating stream leaves the reaction side at the same temperature. The system behaves as a splitting point of a stream, where only the extensive variables (such as flow rate and stream enthalpy) but not the intensive one (e.g., temperature) undergo variations.

Table 6.3: Energy balance equations of a tubular MR. Plug-flow MR (1D-first order model) – steady-state.

Annulus	$-\sum\limits_{i=1}^{N\,\text{Species}} N_i Cp_i \dfrac{\partial T^{\text{Annulus}}}{\partial z} + \dfrac{U^{\text{Shell}} A^{\text{Shell}}}{V^{\text{Annulus}}}\left(T^{\text{Furnace}} - T^{\text{Annulus}}\right) +$ $-\dfrac{U^{\text{Membrane}} A^{\text{Membrane}}}{V^{\text{Annulus}}}\left(T^{\text{Annulus}} - T^{\text{Lumen}}\right) + \Psi + \varphi \dfrac{A^{\text{Membrane}}}{V^{\text{Annulus}}} = 0$ (6.12)
	I.C $\qquad T^{\text{Annulus}}\big|_{t=0} = T^{\text{Annulus, Initial}}$
	B.C $\qquad T^{\text{Annulus}}\big|_{z=0} = T^{\text{Feed}}$ *or* T^{Sweep}
Lumen	$-\sum\limits_{i=1}^{N\,\text{Species}} N_i Cp_i \dfrac{\partial T^{\text{Lumen}}}{\partial z} + \dfrac{U^{\text{Membrane}} A^{\text{Membrane}}}{V^{\text{Lumen}}}\left(T^{\text{Annulus}} - T^{\text{Lumen}}\right) +$ $+ \Psi + \varphi \dfrac{A^{\text{Membrane}}}{V^{\text{Lumen}}} = 0$ (6.13)
	I.C $\qquad T^{\text{Lumen}}\big|_{t=0} = T^{\text{Lumen, Initial}}$
	B.C $\qquad T\big|_{z=0} = T^{\text{Feed}}$ *or* T^{Sweep}

Table 6.4: Characteristic terms of energy balance in MRs.

$\varphi = \begin{cases} 0 \text{ on reaction side} \\ J_i^{\text{Permeating}}\left(h_i^{T\,\text{Reaction}} - h_i^{T\,\text{Permeation}}\right) \text{ on permeation side} \end{cases}$	Temperature variation owing to enthalpy flux associated with i-th species permeation	(6.14)
$\Psi = \begin{cases} \sum\limits_{j=1}^{N\,\text{Reaction}} r_j\left(-\Delta H_j\right) \text{ on reaction side} \\ 0 \text{ on permeation side} \end{cases}$	Heat produced by chemical reactions	(6.15)

6.2.4 Case study: High temperature water gas shift reaction in a membrane reactor

Successful examples of the use of Pd-based MRs are high temperature reactions for hydrogen production. The possibility of selectively removing hydrogen from the reaction volume leads to significant advantages such as production of pure H_2 stream, the enhancement of the conversion, the deletion of secondary reactions, the increase in residence time of reactants, the reduction of reaction volume and so on [8]. The syngas upgrading by means of water–gas shift (WGS) reaction was widely investigated both experimentally and by simulation [9–22].

WGS reaction is industrially carried out in two fixed bed adiabatic reactors connected in series with a cooler (heat exchanger) between them. The first reactor operates at a high temperature (HT–WGS) ranging from 300 °C to 500 °C employing Fe–Crbased catalysts. The second reactor (LT–WGS, low temperature WGS) uses CuO–ZnO–based catalysts and operates at lower temperatures (180 °C–300 °C) to displace the equilibrium, since WGS reaction is exothermic. The whole cycle has the big disadvantage to be accompanied by large emissions of CO_2. The use of this three to four stages of reaction purification can be replaced by a single stage, the MR, in which reaction and separation occur in the same vessel and conversion significantly higher than traditional system can be reached (Figure 6.7).

Figure 6.8 compares CO conversion as a function of the temperature obtained for the MR and the traditional process operating at the same inlet conditions, that is, same GHSV (20,000 h^{-1}) and temperature for the MR and the traditional process (first stage). The CO conversion achieved by MR is around 10% higher than the overall one of the traditional process; it also exceeds significantly (ca. 25%–30%) the traditional reactor equilibrium conversion (TREC). The hydrogen removal from the reaction side owed to the permeation shifts in the reaction toward further conversion. This effect is well operated in this MR because a reaction pressure of 15 bar promotes the hydrogen permeation well. This gain is more clearer considering that the MR conversion is ca. 33% higher than that achieved by the first stage of the traditional process (HT–WGS).

Most of the hydrogen produced is recovered as pure gas in the permeate stream (Figure 6.9). In addition, the retentate is compressed and concentrated in CO_2 (65% molar conentration) and, thus, CO_2 can be more easily captured, resulting a relevant/important reduction in successive separations. On the contrary, the H_2 exiting from the traditional process (at more or less 60% molar concentration) is still mixed with other gases (Figure 6.9) and, in particular, with ca. 5.5% CO that must be drastically reduced and, thus, requires a further separation/purification stage before further use. Moreover, the CO_2 concentration of the residual stream can be close to 70%, only if the whole H_2 present in the stream is separated. Its concentration does not exceed 60% considering the actual separation efficiency of industrial PSA, where the H_2 recovery does not exceed 80%–90% [23, 24].

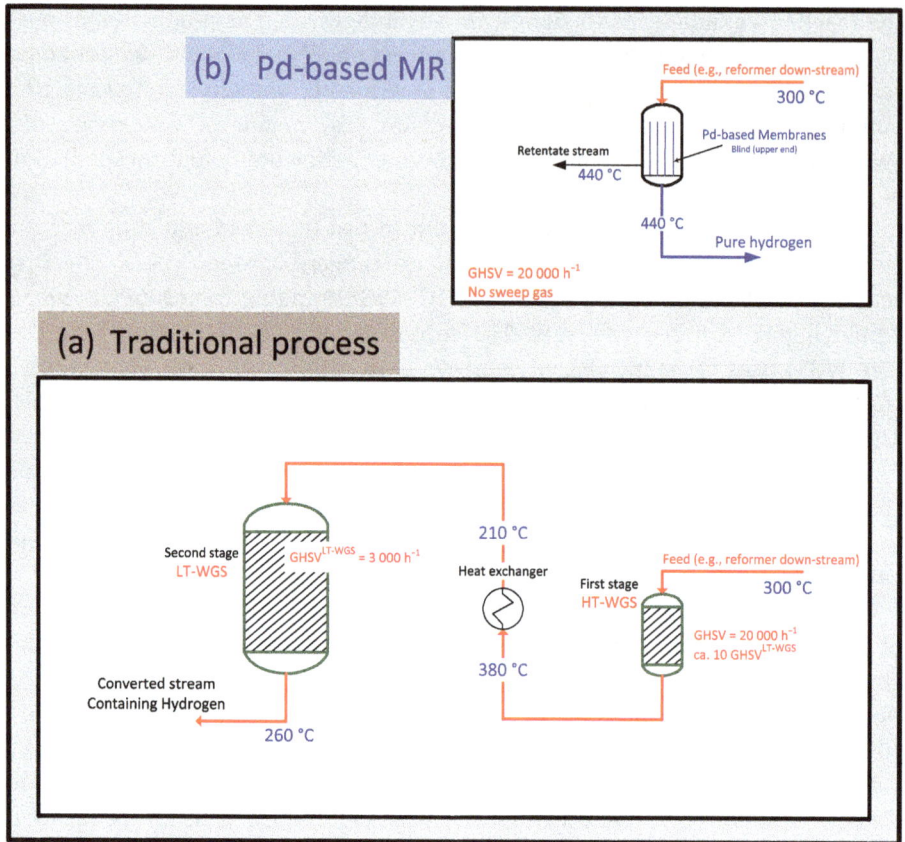

Figure 6.7: Schemes of "Pd-based MR" and "traditional process" for WGS reaction. The temperature values reported are indicative of a typical operation. (Reproduced from Barbieri et al. [18] with the permission from The Royal Society of Chemistry.)

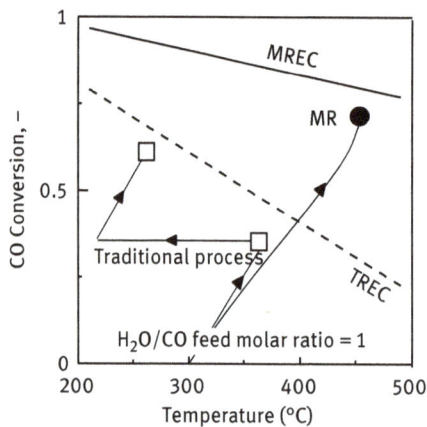

Figure 6.8: CO conversion as a function of temperature for MR and traditional process. (Reproduced from Barbieri et al. [18] with the permission from The Royal Society of Chemistry.)

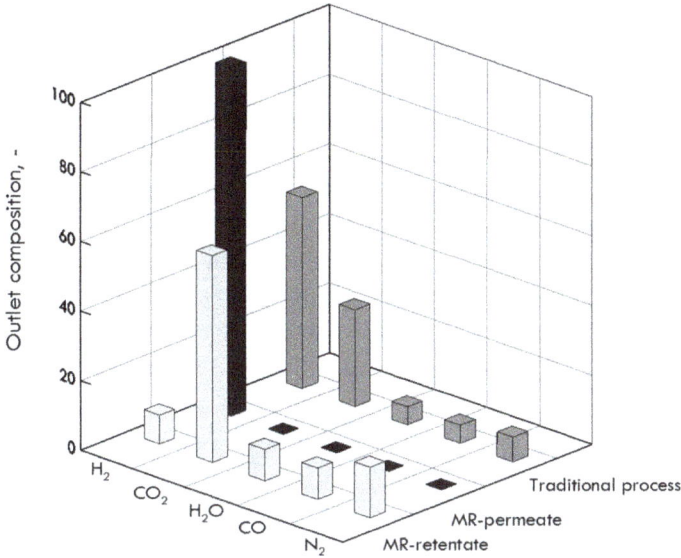

Figure 6.9: Outlet stream composition of the MR and the traditional process. (Reproduced from Barbieri et al. [18] with permission from The Royal Society of Chemistry.)

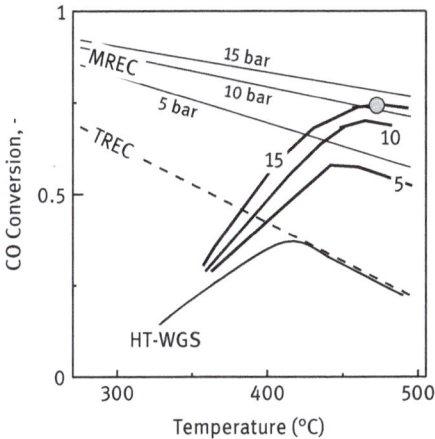

Figure 6.10: Outlet conversion versus outlet temperature of MR and the traditional process for three feed pressures. (Reproduced from Barbieri et al. [18] with the permission from The Royal Society of Chemistry.)

The dependence of the MR performance on the temperature and pressure was investigated as shown in Figure 6.10 and compared with that achieved by the HT−WGS (the first stage of traditional process) in the same operating conditions to understand the differences between the two reacting systems. Any point in the four curves is the outlet conversion and temperature of the MR. In the whole temperature range investigated, CO conversion achieved in MR is significantly higher than that achieved in HT−WGS and exceeds the TREC for temperatures higher than 370 °C and pressure exceeding 5 bar. CO conversion curves, simulated for the three feed pressures, initially follow an increasing trend with the temperature reaching up to a maximum, followed by a slight decrease.

The gas hourly space velocity (GHSV) is a variable generally used to indicate the reverse of the residence time of reactants over a catalytic bed. A low GHSV indicates a high residence time and, thus, favors the conversion, whereas the contrary happens for a high GHSV. However, a high GHSV is highly desirable because this means that the amount of low catalyst convertedwith a high feed flow rate and, in addition, a low reactor volume is required. Figure 6.11 shows the MR CO conversion and the corresponding H_2 recovery index as a function of GHSV for three feed pressures. CO conversion of the HT–WGS is also reported for comparison (dashed line). An increase in GHSV corresponds to a decrease in CO conversion. This trend is much more emphasized with the much higher feed pressure of the MR, because the conversion at the lowest GHSV considered is significantly higher and thus there is much more room for its reduction. However, in all cases, the MR CO conversion is higher than the HT–WGS one (five times higher at 15 bar and 20,000h[-1]) and also exceeds the TREC at a lower GHSV and above 10 bar. As a consequence, the hydrogen recovery has the same decreasing dependence on the GHSV. A high CO conversion means, in fact, a high H_2 production or, rather, a high H_2 partial pressure on the reaction side. This is traduced into a high permeation driving force and, thus, in more H_2 recovery in the permeate. In particular, H_2 recovery is always higher at $20,000\,h^{-1}$ and 15 bar, reaching a value of 92%. The MR was also simulated for a 30 bar feed pressure, because this value has a higher industrial interest even if no such resistant self-supported membrane is available on the market yet. The results are very interesting particularly at $40,000\,h^{-1}$; the highest GHSV is considered, where the conversion of 80% is much higher (ca. four times) than the one achieved at 15 bar by the same MR. The higher pressure significantly favors the permeation of hydrogen, in fact, the stage cut is 55% instead of 30% at 15 bar.

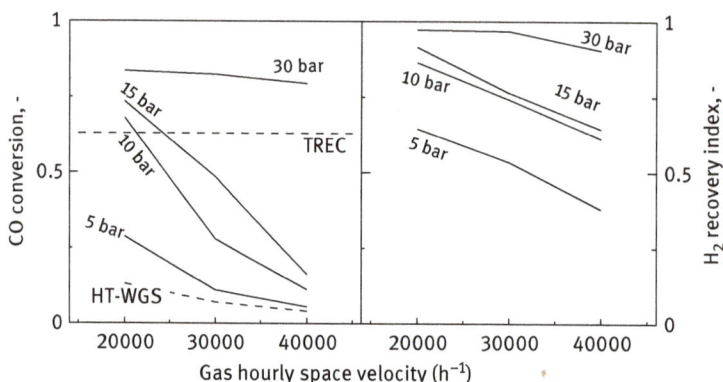

Figure 6.11: CO conversion in the MR and in the first stage of the traditional process (HT–WGS) and hydrogen recovery index as a function of GHSV. (Reproduced from Barbieri et al. [18] with the permission from The Royal Society of Chemistry.)

In the last five years, significant advances were made in the evaluation of phenomena such as polarization and inhibition that influence the permeation flux through a Pd–Ag membrane and whose estimation is an important aspect to take into account in the MR design. In most cases, these two phenomena are combined and, as aforementioned, their effect can be estimated by means of the novel overall permeance reduction coefficient (*PRC*), which takes into account the permeance decrease with respect to intrinsic value [25].

In particular, Caravella et al. [26] quantified these phenomena in so-called concentration polarization maps and inhibition maps for hydrogen-separation membrane systems assisted with a Pd–Ag membrane. These maps provide the value of appropriate factors – that is, the concentration polarization coefficient (CPC) and the inhibition coefficient (IC) – as functions of the external operating conditions of temperature, pressures and hydrogen composition, using the Sieverts' law as the reference driving force for evaluating the reduction in driving force. The analysis showed that the polarization effect can be relevant (CPC higher than 20%) specifically when using very thin membranes (1–5 μm).

Most recently, concentration polarization distribution in Pd-based MRs was investigated for the WGS by a simulation approach coupling (i) an improved characterization of a 3.6 μm thick membrane using permeation data from the literature, (ii) CFD simulations of particle beds and (iii) a complex model of an MR [27]. The simulation results indicate that the maximum concentration polarization in the reactor is ca. 20%. This high value, present at the reactor end, is caused by the low hydrogen concentration, which implies a larger resistance to mass transport owing to the nonpermeating species. However, the weight of this reactor section on the overall CPC is not so high. In fact, the average CPC is ca. 10.5%, which is significantly lower than the maximum value.

6.2.5 Process intensification metrics for membrane reactors

In the last decade, many efforts were made to transform the traditional industrial growth into a sustainable growth. The process intensification strategy, as new design philosophy recently introduced for bringing drastic improvements in manufacturing and processing, aims to pursue this growth in a competitive but sustainable way, reducing the energy consumption, better exploitation of raw materials, minimizing the wastes, increasing the plan efficiency, reducing the plant size and capital costs, increasing the safety, improving remote control and so on [28–32].

A deep understanding of the process intensification principles places the membrane technology and the membrane engineering in a crucial role for the implementation of this strategy [33]. Among other new unit operations involving membranes, MRs are expected to play a decisive role in the scenario of the sustainable growth.

Currently, they represent a solution for several processes involving petrochemical industry [34, 35], energy conversion [36, 37], hydrogen production [38–43] and well fulfil the requirements of process intensification, offering better performance, lower energy consumption and lower volume occupied with respect to the conventional operations. The synergic effects offered by MRs through combining reaction and separation in the same unit, their simplicity and the possibility of advanced levels of automation and control, offer an attractive opportunity to redesign industrial processes [44–46]. However, to make the use of a new technology more attractive, it is fundamental to define a new way of analyzing its performance and highlighting its potentialities with respect to the well-consolidated traditional technologies. Hand-in-hand with the redesign of new processes comes, thus, the identification of new indexes, so-called metrics that together with the traditional parameters are usually used to analyze a process that can supply additional and important information to support decision-making process on the type of operation and the identification of the operating condition windows that make a process more profitable. Up to now, many efforts are being made to define indicators of the industrial processes [47, 48] and most of them are calculated in the form of appropriate ratios that can provide a measure of impact independent of the scale of the operation, or to weigh costs against benefits and, in some cases, they can allow the comparison between different operations [49]. The use of these new indexes can lead to an innovation in the analysis of the performance of the unit operations and, in the case of the membrane technology, can clearly and easily show the advantages and drawbacks that the choice of that specific technology can provide in comparison with the traditional units. On the light of the above-mentioned considerations, the upgrading of syngas via WGS by means of an MR is considered as a case study for introducing a non-conventional analysis of the performance of an alternative unit operation. In particularly referring to the evaluation of the MR's performances, the following indexes were defined [12, 50, 51]:

- Volume index, defined as the ratio of the catalytic volume of an MR and a traditional reactor (TR) for reaching a set CO conversion

$$\text{Volume index} = \frac{\text{Volume}^{MR}}{\text{Volume}^{TR}}\bigg|_{\text{Conversion}} \tag{6.16}$$

- Conversion index, the ratio between the conversion of an MR and a TR, for a set reaction volume.

$$\text{Conversion index} = \frac{\text{Conversion}^{MR}}{\text{Conversion}^{TR}}\bigg|_{\text{Catalyst}} \tag{6.17}$$

- Mass intensity, defined as the ratio between the total H_2 fed to the MR and produced by the reaction and total mass entering the reactor

$$\text{Mass intensity, (MI)} = \frac{\text{Total } H_2 \text{ fed and produced by reaction}}{\text{Total inlet mass}}, \frac{kg_{H_2}/s}{kg/s} \tag{6.18}$$

$$= \frac{F_{Total}^{Feed}\left(X_{H_2}^{Feed} + X_{CO}^{Feed} \cdot X_{CO}^{Actual}\right)M_{H_2}}{Mass_{Total}^{Feed}} \tag{6.19}$$

$$MI^{TREC\,or\,MREC} = \frac{F_{Total}^{Feed}\left(X_{H_2}^{Feed} + X_{CO}^{Feed} \cdot X_{CO}^{TREC\,or\,MREC}\right)M_{H_2}}{Mass_{Total}^{Feed}} \tag{6.20}$$

– Energy intensity, defined as the ratio between the total energy involved in the reactor and, as for mass intensity, the total H_2 fed to the MR and produced by the reaction, that is, the whole hydrogen exiting the system

$$\text{Energy intensity, (EI)} =$$

$$= \frac{\text{Total energy produced (or consumed) by reaction within the reactor}}{\text{Total } H_2 \text{ fed and produced by reaction}}, \frac{J/s}{kg_{H_2}/s} \tag{6.21}$$

$$= \frac{F_{Total}^{Feed} X_{CO}^{Feed} X_{CO}^{Actual} \Delta H^{Reaction}}{F_{Total}^{Feed}\left(X_{H_2}^{Feed} + X_{CO}^{Feed} \cdot X_{CO}^{Actual}\right)M_{H_2}} \tag{6.22}$$

$$EI^{TREC\,or\,MREC} = \frac{F_{Total}^{Feed} X_{CO}^{Feed} X_{CO}^{TREC\,or\,MREC} \Delta H^{Reaction}}{F_{Total}^{Feed}\left(X_{H_2}^{Feed} + X_{CO}^{Feed} \cdot X_{CO}^{TREC\,or\,MREC}\right)M_{H_2}} \tag{6.23}$$

The volume index is an important parameter in installing new plants that must be characterized by low size and high productivities. The volume index is an indicator of the productivity of an MR and it compares the MR reaction volume with that of a TR, necessary to achieve the same conversion. The volume index ranges from 0 to 1. A low volume index means that the reaction volume required by an MR to reach a set CO conversion is much lower than that necessary for a TR. As a consequence, the catalyst weight necessary in MR is significantly reduced.

Considering the WGS reaction as an example, it can be seen that volume index is a decreasing function of the feed pressure, owing to the positive effect that the latter has in an MR on CO conversion. MR reaction volume is 75% of the TR one at 600 kPa and reduces down to 25% at 1,500 kPa, when an equimolecular mixture is fed and a final conversion of ~80% (corresponding to 90% of the traditional reactor equilibrium conversion) is considered. This means a reduction in plant size (Figure 6.12) and hence related costs. Feeding a typical syngas stream also containing hydrogen $(CO:H_2O:H_2O:CO_2=20:20:50:10)$ to the Pd–Ag MR, the volume index is further lower owing to the low value of the equilibrium conversion. Therefore, the amount of catalyst necessary to reach a suitable conversion is drastically reduced with also a clear gain in terms of plant size reduction.

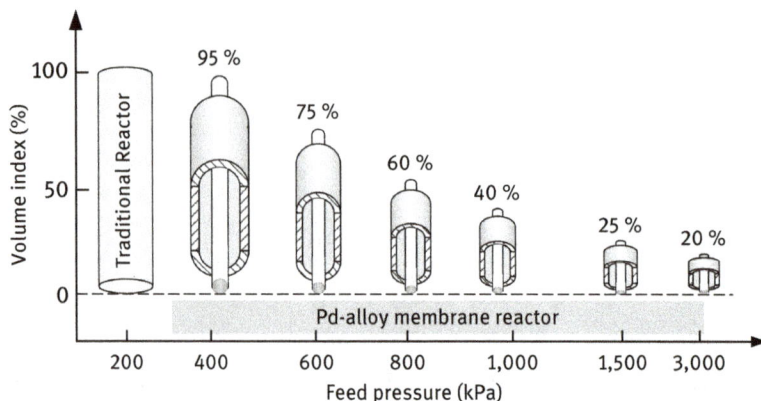

Figure 6.12: Volume index as a function of feed pressure feeding an equimolecular mixture. Furnace temperature=280 °C. Set CO Conversion 90% of the TREC. Reprinted from Journal of Membrane Science, 306, Brunetti A.; Caravella C.; Barbieri G.; Drioli E. Simulation study of water-gas shift in a membrane reactor, Pages 329–340. Copyright 2007, with permission from Elsevier.

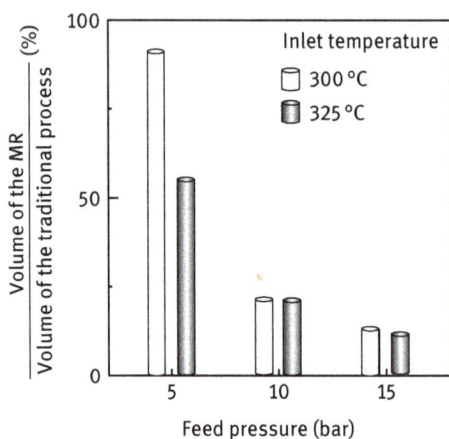

Figure 6.13: Ratio between the MR volume and the volume of the traditional process as a function of feed pressure for an inlet temperature of 300 °C and 325 °C. (Reproduced from Barbieri et al. [18] with the permission from The Royal Society of Chemistry.)

Figure 6.13 shows the volume index calculated as the ratio between the reaction volume required by an MR with respect to that necessary for the whole traditional process (high temperature and low temperature reactors), for achieving the same conversion, as a function of the feed pressure for inlet temperatures of 300 °C and 325 °C. The evaluation was done, first, calculating CO conversion achieved in the traditional process (with a defined reaction volume for each high and low temperature WGS reactors) for its suitable operating conditions, and then evaluating the reaction volume required by the MR for obtaining the same CO conversion. The huge difference between the two reaction systems mainly depends on the low temperature WGS requiring a significant higher volume because it operates at a temperature of 220 °C–300 °C and at a low GHSV (3,000 h^{-1}) owing to the slow kinetics of the CuO–ZnO catalyst. This means a very bigger amount of catalyst used for converting a relatively small feed flow rate and it counts for a lot in the determination of the reaction volume of the whole traditional process. As expected, the reaction volume required by MR always results in lower than that of the whole traditional process and it is much lower as much higher is the feed pressure. At 5 bar, for an inlet temperature of 300 °C, the MR reaction volume is around 90% of traditional process, owing to a limited H_2 permeation making the MR not really work better than the TR. This value drastically reduces at a higher feed pressure, becoming ca. 13% at 15 bar. Furthermore, at a temperature higher than 325 °C, it is still reduced by passing from 55% at 5 bar to ca. 10% at 15 bar. A high feed pressure and a high temperature, in fact, imply that more H_2 permeates through the membrane and, thus, less catalyst is required for achieving a set conversion. A higher pressure (30 bar) produces a further reduction of the reaction volume of the MR, as shown for a GHSV of 40,000 h^{-1}.

The capability of reaching a conversion higher than a TR, exceeding the TR equilibrium limits is a typical property of an MR. The *conversion index*, defined as the ratio between the conversion achieved in an MR and that of a TR, for a set reaction volume, gives an evaluation of the gain in terms of conversion and its use is particularly indicated when the feed mixture also contains reaction products. A high conversion index implies a relevant gain in terms of conversion achieved in an MR with respect to the conventional reactor one, with the same reaction volume, meaning better raw material exploitation and lower waste. The MRs are pressure-driven systems. Therefore, the conversion index is an increasing function of the feed pressure as shown in Figure 6.14. In particular, a *conversion index* of ca. 2 was achieved at 200 kPa, whereas one of ca. 6 was reached at 1,500 kPa feeding the reformate stream ($CO:H_2O:H_2O:CO_2$=20:20:50:10). However, already at 500 kPa, a conversion index is equal to 4. When an equimolecular feed containing only reactants is fed to MR, conversion index ranges from 1.5 to 1 because the TR conversion is already high. However, a conversion index equal to 1.5 indicates around 95% of CO conversion, implying not only a pure H_2 stream in the permeate side but also a concentrated CO_2 retentate stream that is easy to recover.

The mass intensity is defined as the ratio between the total H_2 fed to the MR and produced by the reaction and total mass entering the reactor. The higher is its value the more intensified is the process. In any case, it cannot be higher than 1 when pure hydrogen is fed to the system; this is a singular case where neither the reaction nor the separation occurs. The value of this index depends on the conversion and on the composition of the feed stream. In this example, the nominator of mass intensity consists of the H_2 fed to the reactor plus the hydrogen given by the reactor since the reaction stoichiometry says that one mole of H_2 is produced by one mole of CO converted by WGS reaction. The maximum or ideal value of mass intensity is the one at the reactor equilibrium conversion; in the case of the TR it will be referred as TREC, whereas for the MR as MREC [52].

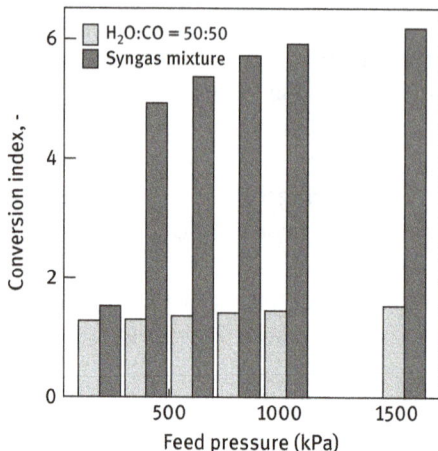

Figure 6.14: Conversion index as a function of the feed pressure for different feeds. Furnace temperature = 280 °C. Reprinted from Comprehensive Membrane Science and Engineering – Volume 3: Catalytic membranes and catalytic membrane reactors, Edited by E. Drioli and L. Giorno; Chapter 3.03, Barbieri, G., Scura F., Brunetti A., Modeling and simulation of membrane reactors and catalytic membrane reactors, Pages 57–79. Copyright 2010, with permission from Elsevier.

The energy intensity is defined as the ratio between the total energy involved in the reactor and, as for mass intensity, the total H_2 fed to the MR and produced by the reaction, that is, the whole hydrogen exiting the system. For this index, the higher is its value the more intensified is the process. Also the value of this index depends on the conversion and on the composition of the feed stream and the ideal energy intensity is achieved at the equilibrium conditions. The highest energy intensity (considering the absolute value when an exothermic reaction as the WGS is considered) means more energy developed by the system and, thus, the best performance for the reactor.

In the comparison between TR and MR, the latter always yields more material and is energy intensive than a traditional reactor, particularly at a high feed pressure indicating that MR requires less material as feed and makes available more energy in producing the same amount of H_2. For instance, looking at the Figure 6.15, for a GHSV of 40,000 h^{-1}, the temperature range of 350 °C–380 °C appears the most suitable implying the achievement of a more intensified process, since both mass and energy indexes for TR and MR show the highest values. In particular, at 350 °C and 1,500 kPa, the MR achieves mass intensity = 0.031 and energy intensity = −12.6 kJ/g_{H2}, whereas the value of mass intensity and energy intensity for the TR are only 0.023 and −9.00 kJ/g_{H2}. The result is interesting since it can be also observed from a different point of view. To get the same values of the indexes achieved by TR at 350 °C, it would be sufficient for the MR to operate at 320 °C and 5 bar or at 300 °C and 10 bar. This means milder temperature conditions with indirect gains also in terms of catalyst lifetime and so on.

The advantage offered by an MR was also quantified in terms of ratios referred to the equilibrium condition of the traditional reactor. The MR always resulted in more intensity than a traditional reactor operated in similar conditions and it also exceeded the ideal performance that was achievable by a traditional reactor at a temperature higher than 350 °C (Figure 6.15).

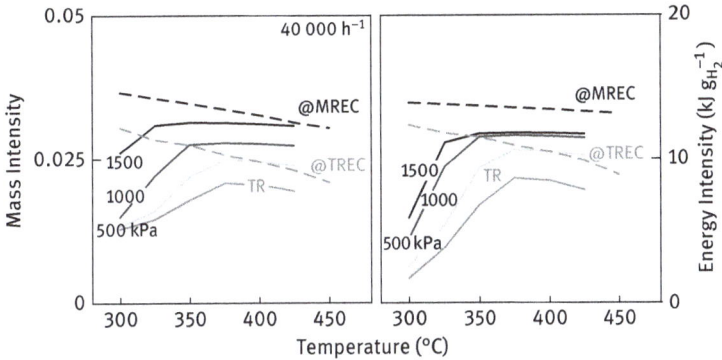

Figure 6.15: Mass intensity and energy intensity as a function of the temperature for different values of reaction pressure. Dashed lines: values calculated at TREC or MREC (at 1,500 kPa). Black continuous curves referring to MR.

The advantage offered by an MR with respect to the traditional reaction unit usually used is clearly highlighted in Figure 6.16, where the ratios between the actual indexes of the MR and the correspondent ideal (calculated at equilibrium, TREC) indexes of TR are shown.

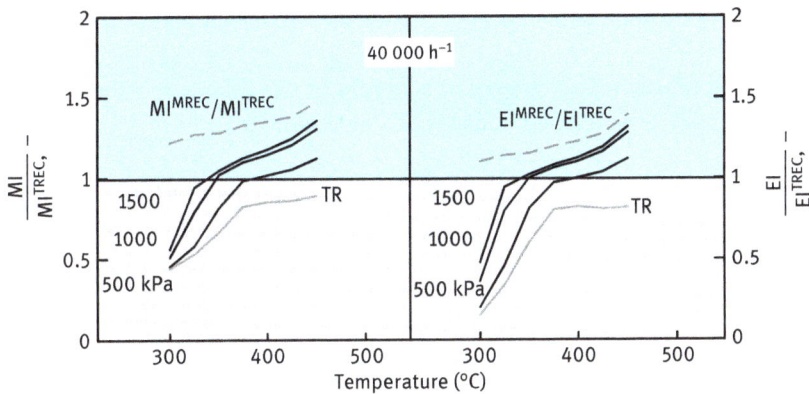

Figure 6.16: Mass intensity and energy intensity ratio referred to the TREC as a function of the temperature for different values of reaction pressure. Black continuous curves are referring to MR.

In this figure, the curves that are also calculated for actual values of TR are plotted for comparison. In the graphs, two zones can be identified: the first relative to values higher than 1 and the second for values lesser than 1. A ratio equal to 1 means that the MR allows the attainment of the best/ideal performance achievable by a TR in equilibrium under the same conditions. Values greater than 1 indicate that the process carried out with an MR results in more intensity; this condition can never be achieved by a TR. In general, the higher the ratio, the more intensified the process. The MR is always more intensified than TR operated in actual conditions and exceeds the ideal performance of a TR, at a temperature higher than 350 °C. This temperature can be lower as much higher is the feed pressure, since it promotes the conversion. Mass and energy intensities demonstrated, in line with the process intensification strategy, the assets of the MR technology also in terms of better exploitation of raw materials (reduction up to 40%) and higher energy efficiency (up to 35%). Figure 6.16 shows an overview of the most interesting results where a phase diagram of the mass intensity and energy intensity ratios referred to the TREC is plotted. The process is much more intensified as much higher are the mass and energy intensity ratios. The point on the graph of coordinates (1, 1) is the ideal case for a TR when it is under the equilibrium state.

A horizontal and a vertical line drawn through this point identify four regions:
1. In the first one, that is in the yellow area, both parameters have the most desired values: mass intensity (MI) and energy intensity (EI) ratios higher than 1
2. The second one in gray. The variables, mass intensity and energy intensity ratios, lower than 1
3. The other two areas are in white. Mass intensity and energy intensity ratios are lower and higher than 1, respectively in one of these areas, and vice versa in the other

The yellow area is achievable only by the MR, since mass intensity and energy intensity ratios are higher than that at TREC; when the variables fall in this area, the MR is more intensified of any TR that is also in its equilibrium state (Figure 6.17).
1. The gray area represents all the possible values for traditional reactor and the best values 1 and 1 are obtained under TREC condition. This area is also common to the MR
2. All the points of the ratios lie on the same curve crossing the point (1,1) independently of the operating conditions highlighting that as the mass intensity is better the energy intensity also becomes better; thus, the white regions are not accessible for any TR and MR.
3. As the temperature and the pressure increase, the values tend to be at the intensified zone. For instance, the MR operates in a profitable way exceeding also the performance of a TR in equilibrium state under a temperature higher than 350 °C, a feed pressure greater than 500 kPa and GHSV not exceeding 30,000 h^{-1}.

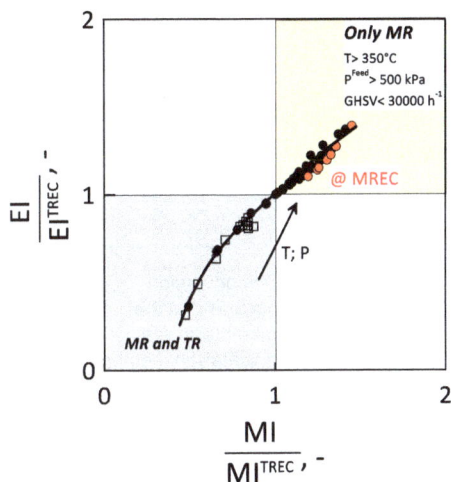

Figure 6.17: Energy intensity ratio referred to TREC as a function of mass intensity ratio is referred to the TREC for all the operating conditions. Reprinted from Fuel Processing Technology, 118, Brunetti A.; Drioli E; Barbieri G. Energy and mass intensities in hydrogen upgrading by a membrane reactor, Pages 278–286. Copyright 2014, with permission from Elsevier.

6.3 Membrane bioreactors: Definitions and configurations

Membrane bioreactors (MBR) are systems where a chemical reaction is catalyzed by a catalyst of biological origin is combined with a membrane that compartmentalizes the reaction environment and governs the mass transport from/to the reaction environment itself [53–57]. Figure 6.18 illustrates the most common MBR configurations. In general, MBRs operate in liquid phase, including monophasic aqueous [58–60] or organic phase [61–63], biphasic (two-separate phase) aqueous/organic [64, 65] or multiphasic mixture/suspension [66]. Solid-gas bioreactions can be also combined with and implemented by membranes.

In the most general case, the two processes, that is, the bioreactor and the membrane operation, are performed in individuals but in connected systems so that the reaction mixture is circulated along the membrane module and recycled back to the reactor tank, thus creating a common overall well-mixed environment (Figure 6.18a). This configuration is most commonly used at the productive scale. The membrane may work as a separation unit, such as micro- or ultrafiltration, that is capable of retaining reagents and catalyst while allowing the passage of the product along with the solvent. In such cases, to keep the volume constant, new solvent is added to the system to balance the permeated one, so that, overall, the MBR works as

(a)   (b)

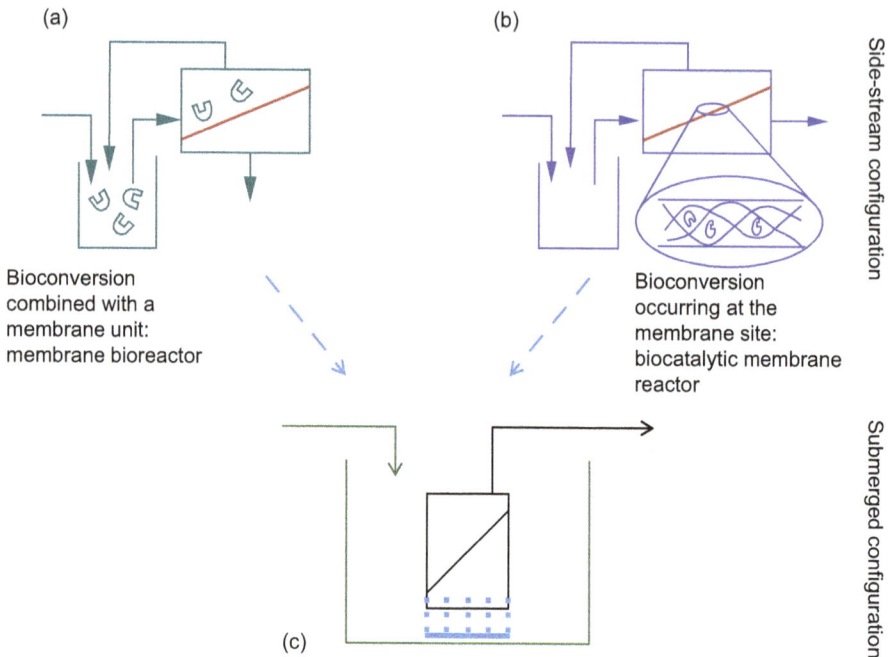

Bioconversion
combined with a
membrane unit:
membrane bioreactor

Bioconversion
occurring at the
membrane site:
biocatalytic membrane
reactor

(c)

Side-stream configuration

Submerged configuration

Figure 6.18: General scheme of membrane reactors using catalyst of biological origin, that is, MBRs: (a) a bioreactor and a membrane unit are combined into one system; (b) the membrane itself works as the reaction site and promotes mass transfer at the same time. (c) Both types of bioreactors can be obtained using the membrane module either outside on inside the bulk phase, that is, side-stream or submerged, respectively.

a continuous stirred tank reactor (CSTR) (see Figure 6.19 for a brief summary of the CSTR concept and mass balance).

Another common case is the one where the biocatalyst is located at the membrane level on the surface within the porous membrane matrix (Figure 6.18b). This is a specific type of MBR and it is more appropriately named biocatalytic membrane reactor (BMR), meaning that the membrane is involved in the catalytic process. In fact, both reaction and mass transport from/to the reaction environment take place at the membrane level and are both governed by the membrane properties and fluid dynamics.

Here the reactor volume is represented by the membrane pore void volume where the biocatalyst is located (Figure 6.20). Each pore works as a microreactor, the overall membrane matrix represents a high throughput of microreactors in parallel [67].

When convective flow through the membrane is promoted and it can be assured uniform concentration and density throughout the pores, each microreactor can be approximated to a CSTR and related equations for mass balance and reaction rate can be assumed for the overall balance region, that is, the overall void volume of the BMR [59, 68].

(a) Continuous bioreactor concepts

<u>Characteristics:</u>
✓ Time invariant conditions at steady state
✓ Continuous production

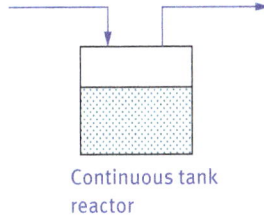

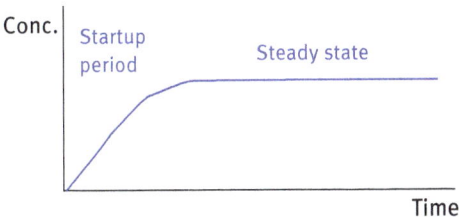

Continuous tubular reactor

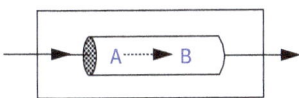

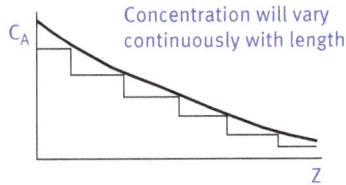

Continuous tank
reactor

Conc.

Startup period   Steady state

Time

<u>Operating variables</u>
• Inlet medium composition
• Temperature, pressure, pH
• Liquid flow rate (residence time)
• Mixing, aeration (with microrganism)

(b) Mass balance

♦ Stirred tank

Total mass $= \rho V$
Mass of $A = C_A V$

ρ V
CA

—Balance region

C_A

Well mixed => concentration
and density uniform throughout

Z

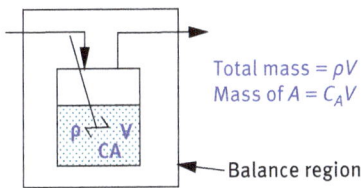

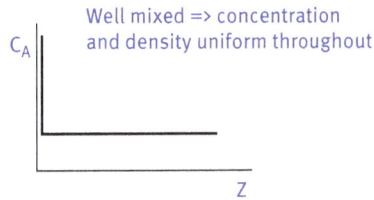

♦ Tubular reactor

A ·······► B

C_A

Concentration will vary
continuously with length

Z

This can be approximated by choosing the balance region small enough that C_A in each region
is uniform. In this case, many uniform property subsystems (well stirred tanks) form the reactor

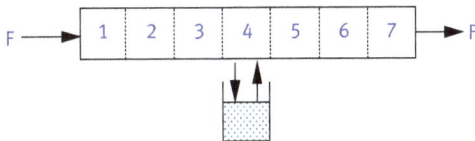

F ——► | 1 | 2 | 3 | 4 | 5 | 6 | 7 | ——► F

Figure 6.19: Summary of (a) continuous stirred tank reactor concept, (b) mass balance and
(c) reaction rate.

(c) Reaction rate

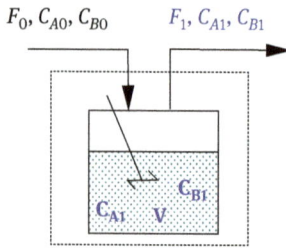

$$A \longrightarrow 2B$$

$$r_A = -kC_A$$

$$r_B = -2r_A = +2kC_A$$

$$\frac{d(VC_{A1})}{dt} = F_0 C_{A0} - F_1 C_{A1} + r_{A1} V$$

$$\frac{d(VC_{B1})}{dt} = F_0 C_{B0} - F_1 C_{B1} + r_{B1} V$$

For constant V:

$$\frac{dC_{A1}}{dt} = \frac{F}{V}(C_{A0} - C_{A1}) + r_A \qquad\qquad (F_0 = F_1 = F)$$

$$\frac{dC_{B1}}{dt} = \frac{F}{V}(C_{B0} - C_{B1}) + r_B$$

The solution of these equations gives C_{A1} and C_{B1} as a function of time

Note that, for a constant volume tank, the previous equation can be written also as:

$$\frac{dC_{A1}}{dt} = \frac{(C_{A0} - C_{A1})}{\tau} + r_A$$

Where $\tau = \dfrac{F}{V}$ is the mean residence time of the fluid in the tank reactor

For ACC = 0 the reaction rate can be evaluated by: $r_A = \dfrac{F(C_{A0} - C_{A1})}{V}$

Figure 6.19 (continued)

6.3.1 Membranes and membrane operations used in membrane bioreactors

In addition to previously mentioned pressure-driven membrane operations, other membrane processes can be combined with a bioreactor. Table 6.5 illustrates the most common combinations.

The membrane operation may not only remove the product, but also supply the reagent [69, 70] and the type of membrane process applied mainly depends on the properties of the reaction mixture, including the biocatalyst, the reagent(s) – or substrate(s) – and the product(s).

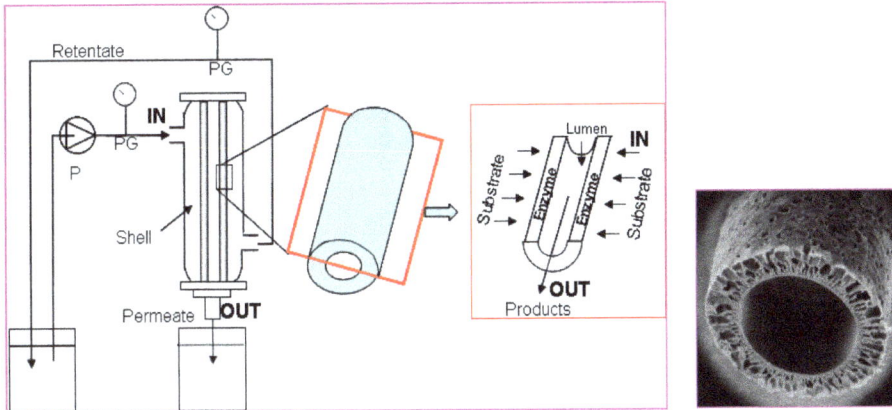

Figure 6.20: Schematic of a biocatalytic membrane reactor, where the biocatalyst is located within the porous membrane matrix, and the reactor volume is represented by the membrane void volume (From [68]).

Figure 6.21 summarizes the role of the membrane in MBR and BMRs.

Depending on the type of bioconversion and application, the membrane module can be located externally to the bulk phase (so-called side-stream configuration) or it can be immersed/submerged into it (Figure 6.18c) [71].

The MBR configuration that combines a bioreactor with a membrane operation is definitively the most applied compared to the configuration where the biocatalyst is compartmentalized within the membrane space. In particular, the side-stream module is more frequent in agro food, beverages and biorefineries [72–74]; the submerged MBRs are more popular in municipal and industrial wastewater treatment [71]; the BMRs are widely investigated in pharmaceutical, nutraceutical, biotechnology, diagnostic and air decontamination [75–79]. In biomedical applications, the biocatalysts, mainly represented by cells, are adhered to the membrane surface and compartmentalized in the outside space between the shell and the membrane or in the lumen of capillary membranes.

The membrane materials, modules and operations used in various MBRs are illustrated in Table 6.6. The types of membrane material, structure, operation and driving force applied govern the interaction between bulk components and membrane, transport phenomena, selectivity, flux, fouling, stability, cleaning and operation and maintenance.

As far as the transport through the membrane is concerned, it can be investigated by using the models already discussed for each type of membrane process in other chapters [80, 81].

In brief, the mass transfer through a membrane is the result of a driving force acting on the individual components in the feed through the membrane matrix and in the permeate. The flux equation depends on the mass transfer mechanism. For liquid phase, it mainly concerns convective viscous and diffusive flow through porous, mesoporous and microporous membranes or solution diffusion through

Table 6.5: Membrane operations combined with biocatalytic reactions.

Process	Concept	Driving force	Mode of transport	Species passed	Species retained
Microfiltration (MF)	Feed → Retentate, Solvent — Microporous membranes	Pressure difference 100–500 kPa	Size exclusion convection	Solvent (water) and dissolved solutes	Suspended solids, fine particulates, some colloids
Ultrafiltration (UF)	Feed → Retentate, Solvent — UF membranes	Pressure difference 100–800 kPa	Size exclusion convection	Solvent (water) and low molecular weight solutes (<1,000 Da)	Macrosolutes and colloids
Nanofiltration (NF)	Feed → Retentate, Solvent — NF membranes	Pressure difference 0.3–3 MPa	Size exclusion solution diffusion Donnan exclusion	Solvent (water), low molecular weight solutes, monovalent ions	Molecular weight compounds > 200 Da multivalent ions
Reverse osmosis (RO)	Feed → Retentate, Solvent — RO membranes	Pressure difference 1–10 MPa	Solution diffusion mechanism	Solvent (water)	Dissolved and suspended solids
Forward osmosis (FO)	Feed → Retentate, Solvent — FO membrane	Concentration difference	Solution diffusion mechanism	Solvent (water)	Dissolved and suspended solids

Process	Driving force	Mechanism	Permeate species	Retained species
Gas separation (GS)	Pressure difference 0.1–10 MPa	Solution diffusion mechanism	Gas molecules having low molecular weight or high solubility–diffusivity	Gas molecules having high molecular weight or low solubility–diffusivity
Pervaporation (PV)	Chemical potential or concentration difference	Solution diffusion mechanism	Highly permeable solutes or solvents	Less permeable solutes or solvents
Electrodialysis (ED)	Electrical potential difference 1–2 V/cell pair	Donnan exclusion	Solutes (ions) Small quantity of solvent	Nonionic and macromolecular species
Dialysis (D)	Concentration difference	Diffusion	Solute (ions and low MW organics) Small solvent quantity	Dissolved and suspended solids with MW>1,000 Da
Membrane contactors (MC)	Chemical potential, concentration difference, temperature difference	Diffusion	Compounds soluble in the extraction solvent; volatiles	Compounds nonsoluble in the extraction solvent; non-volatiles

(continued)

Table 6.5 (continued)

Process	Concept	Driving force	Mode of transport	Species passed	Species retained
Membrane-based solvent extraction (MBSX)	Feed / Retentate; Permeate / Sweep; Porous membrane	Chemical potential or concentration difference	Diffusion partition	Compounds soluble in the extraction solvent	Compounds nonsoluble in the extraction solvent
Membrane emulsification (ME)	Dispersed Phase / Cont. Phase; Emulsion; Porous membrane	Pressure difference	Convection	—	—
Membrane distillation (MD)	Warm feed / Warm concentrate; Liquid – vapour – Liquid; Cool Distillate / Cool stream; Porous membrane	Temperature difference	Diffusion	Volatiles	Non-volatiles
Supported liquid membranes (SLM)	Product A / Feed / Product B; Supported liquid membrane	Concentration difference	Diffusion	Ions, low MW organics	Ions, less permeable organics

Membrane

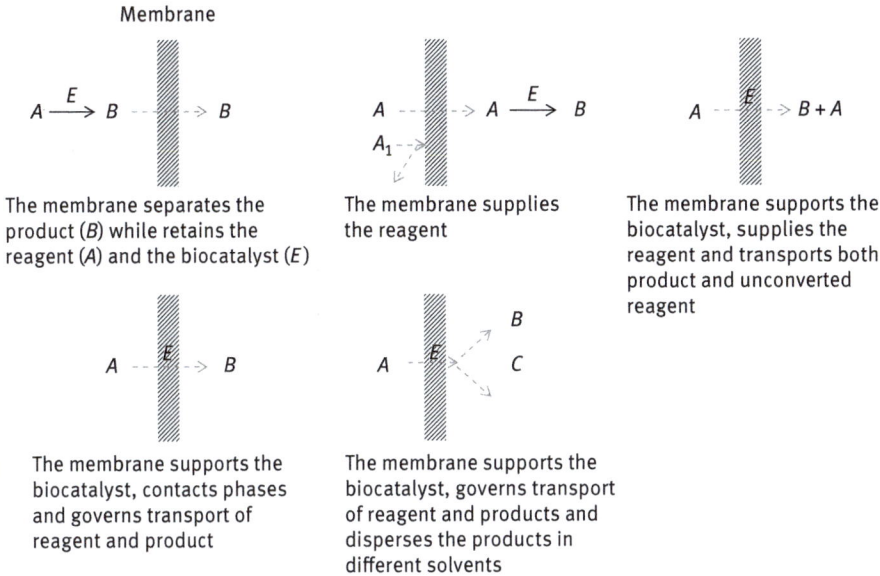

The membrane separates the product (B) while retains the reagent (A) and the biocatalyst (E)

The membrane supplies the reagent

The membrane supports the biocatalyst, supplies the reagent and transports both product and unconverted reagent

The membrane supports the biocatalyst, contacts phases and governs transport of reagent and product

The membrane supports the biocatalyst, governs transport of reagent and products and disperses the products in different solvents

Figure 6.21: Role of the membrane in MBRs.

Table 6.6: Common types of membranes and operations in membrane bioreactors.

MATERIAL	OPERATION					
	MF	UF	NF/RO	GS	PV	MD
Cellulose acetate	X	X	X	X	X	
Cellulose triacetate	X	X	X			
Blend CA/triacetate			X			
Cellulose esters	X					
Cellulose nitrate	X					
Blend CA/CN	X					
Polyvinyl alcohol	X					
Polyacrylonitrile		X			X	
Polyvinyl chloride	X					
PVC copolymer	X	X				
Acrylic copolymer	X					
Aromatic polyamide	X	X	X			
Aliphatic polyamide	X	X				
Polyimide	X	X	X	X		
Polysulfone	X	X				
Sulfonated polysulfone		X	X	X		
Polyetheretherketone (PEEK)	X	X		X		

(continued)

Table 6.6 (continued)

MATERIAL	OPERATION					
	MF	UF	NF/RO	GS	PV	MD
Polycarbonate	X					
Polyester	X					
Polypropylene	X				X	X
Polyethylene	X				X	X
Polytetrafluoroethylene (PTFE)	X	X			X	
Polyvinylidene difluoride (PVDF)	X	X			X	X
Collagen					X	
Chitosan					X	
Zeolites				X	X	
Polyorganophosphazene				X	X	
Polydimethylsiloxane (PDMS)				X	X	

dense membranes. In the first case, a selectivity based on size, shape and charge exclusion is obtained, whereas in the latter case the selectivity depends on the solubility of a component in the membrane matrix as well as on its diffusivity through it.

A brief summary of the flux equations for porous and dense membranes using liquids, which represent the most common fluid in MBR, is reported.

In many cases, the rate of passage through porous membrane is proportional to the driving force, for example, the flux–force relationship can be described by a linear phenomenological equation.

$$J = -A\frac{dX}{dx} \tag{6.24}$$

where A is the phenomenological coefficient and (dX/dx) is the driving force, expressed as the gradient of X (pressure, concentration and temperature) along a coordinate x perpendicular to the membrane transport barrier. Phenomenological equations for mass, volume, charge and so on are summarized in Figure 6.22.

If pores are assimilated to cylinders with constant diameter and perpendicular to the membrane surface, the flux of a Newtonian fluid through the membrane is given by the Poiseuille's law:

$$J = \frac{\varepsilon r^2 \Delta P}{8\mu\tau\Delta x} \tag{6.25}$$

Where J is the flux, Lt^{-1}, r is the mean pore radius, ΔP is the effective transmembrane pressure, μ is the viscosity, Δx is the length of the channel, ε is the surface porosity of membrane, and τ is the tortuosity.

Mass flux:	$J_m = -D\,dc/dx$	(Fick)
Vol. flux	$J_v = -Lp\,dP/dx$	(Darcy)
Heat flux	$J_h = -a\,dT/dx$	(Fourier)
Momentum flux	$J_n = -v\,dv/dx$	(Newton)
Electrical flux	$J_i = -1/R\,dE/dx$	(Ohm)

Figure 6.22: Phenomenological equations describing linear relationship between driving force and transport rate.

In the previous equation, $\Delta P = \Delta P_t - \Delta\pi$, where $\Delta P_t = P_f - P_p$ (ΔP_t is the hydrostatic transmembrane pressure (TMP); P_f and P_p are the feed and permeate pressure, respectively).

$\Delta\pi = \pi_f - \pi_p$ (difference in the osmotic pressure between feed, π_f, and permeate, π_p.

An equation used to describe the permeate flux through the membrane is the Merten's equation:

$$J = L_p(\Delta P_t - \Delta\pi) \tag{6.26}$$

where L_p is the permeability coefficient defined as follows:

$$L_p = \frac{\varepsilon r^2}{8\mu\tau} \tag{6.27}$$

In most MF and UF processes, the osmotic pressure of the solute is considered negligible and $\Delta P = \Delta P_t$ is normally used:

$$J = L_p\Delta P \tag{6.28}$$

In nature and in most real systems, various driving forces (such as pressure and concentration) take place simultaneously. The simple phenomenological equations cannot describe coupling phenomena, which are better discussed in terms of nonequilibrium thermodynamics.

Most transport processes take place because of a difference in chemical potential $\Delta\mu$. Under isothermal conditions (T constant), pressure and concentration contribute to the chemical potential of component i according to the following equation:

$$\mu_i = \mu_i^0 + RT\ \ln a_i + V_i P \tag{6.29}$$

The concentration or composition is given in terms of activity (a_i) in order to express nonideality

$$a_i = \gamma_i\,x_i \tag{6.30}$$

where y_i is the activity coefficient and x_i the mole fraction.

The difference in chemical potential is expressed as follows:

$$\Delta\mu_i = \mu_i^0 + RT\,\Delta\ln a_i + V_i\Delta P \qquad (6.31)$$

Not far from the equilibrium, it can be assumed that each force is linearly related to the fluxes, or each flux is linearly related to the forces.

The general flux equation that also includes an electrical potential and describes the transport through dense membranes is known as the extended Nernst–Planck equation:

$$J_i = D_i \underbrace{(\nabla C_i + C_i \nabla \ln y_i}_{} + \underbrace{C_i \tilde{V}\frac{\nabla P}{\tilde{R}T}}_{} + \underbrace{Z_i C_i F_k \frac{\nabla \psi}{\tilde{R}T})}_{}$$

$$\qquad\qquad \uparrow \qquad\qquad \uparrow \qquad\qquad\qquad \uparrow \qquad\qquad\qquad \uparrow \qquad\qquad (6.32)$$

| Flux of i | Electrical driving force | Pressure driving force | Activity (Conc.) driving force |

In MBR, the evaluation of transport is important to match rate of transport of molecules from/to the bioreactor with the reaction kinetics so that to allow the system to work in reaction limited regime, that is, the transport should not limit the productivity of the MBR.

6.3.2 Biocatalysts used in membrane bioreactors

Table 6.7 illustrates the most common biocatalysts used in membrane bioreactors.

When combining a membrane operation with a bioconversion, for example, with the membrane working as an extractor separating the product (Figure 6.23); besides the membrane properties and transport phenomena, it is also important to consider the biocatalyst kinetics to tune the separation process as appropriate. For example, if enzymes or whole cells are used, it can make a significant difference and appropriate operating conditions that are needed to guarantee high biocatalyst performance and productivity.

Enzymes follow the Michaelis-Menten reaction model (Figure 6.24). The initial enzyme concentration can be assumed as a constant during the operation time (besides deactivation/loss of enzyme, there is no change in biocatalyst concentration).

Microorganisms, such as bacteria, have a kinetic growth profile as described in the Monod equation (Figure 6.25). The concentration of bacteria increases as a function of time; therefore, the combination with the membrane separation process

Table 6.7: Common biocatalysts used in membrane bioreactors.

Biocatalyst	Status	Commonmembrane role	Properties of reaction components
Enzymes	Compartmentalized	Enzyme recycle, product separation and/or reagent supply	Enzymes need cofactors, substrates are large polymers, reaction mixture is very viscous
	Immobilized on membrane surface	Support for the catalyst	The size of the substrate is too big to enter the membrane matrix, products can pass through the membrane
	Immobilized within the membrane matrix	Support for the catalyst, reagent supply, product separation	The size of the substrate and products is suitable to be transported through the membrane
Bacterial cells	Compartmentalized	Cell recycle, products and/or purified compo- nents separation and/or reagent supply	Cells operate the transformation of interest during the growth phase of the fermentation
	Immobilized	Support for the catalyst, reagent supply, product separation	Cells can operate the transformation of interest during a phase different than the growing one
Fungi	Compartmentalized and/or attached on the membrane surface	Biocatalyst recycle, components separation	Biocatalyst grows in the bulk phase and/or it needs to attach on a surface to form a biofilm
Yeast Virus Algae/ microalgae		Biocatalyst growth	
Mammalian cells	Compartmentalized	Cell recycle metabolites supply, catabolites removal	Cells such as in blood need to be kept alive in bulk phase
	Attached	Supply for cell growth, metabolites supply, catabolites removal	Cell's anchorage dependents such as hepatocytes need to adhere to surface for the differentiation and biotransformation

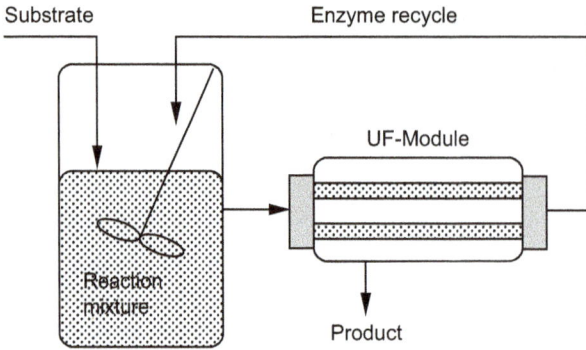

Figure 6.23: Membrane bioreactor using enzyme as catalyst, capillary/hollow fiber ultrafiltration membranes to remove the reaction product form the reaction compartment.

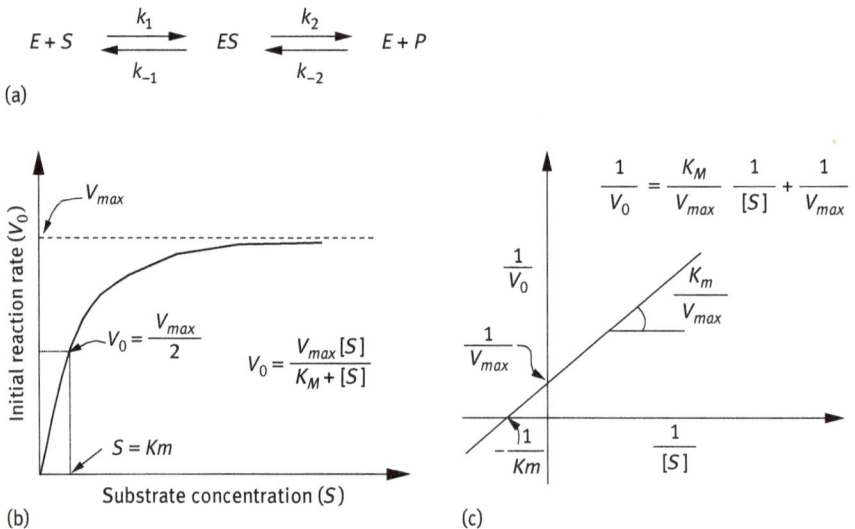

$$E + S \underset{k_{-1}}{\overset{k_1}{\rightleftarrows}} ES \underset{k_{-2}}{\overset{k_2}{\rightleftarrows}} E + P$$

(a)

$$V_0 = \frac{V_{max}}{2}$$

$$V_0 = \frac{V_{max}[S]}{K_M + [S]}$$

$$\frac{1}{V_0} = \frac{K_M}{V_{max}} \frac{1}{[S]} + \frac{1}{V_{max}}$$

(b)

(c)

Figure 6.24: Enzyme kinetics models: (a) enzyme reaction equation; (b) Michaelis-Menten equation and curve; and (c) Lineweaver-Burk linear equation and plot for graphical estimation of kinetic parameters.

might be started when constant biocatalyst concentration has been achieved before the availability of nutrients becomes limiting and excess of products have inhibition effects.

Mass transfer and kinetics properties might be optimized in order that the overall system may operate in reaction limited conditions.

The reason why the configuration where the bioreaction and the mass transfer occur in individual units connected by pumps that recirculate the bulk phase along them (Figure 6.18a) is more popular in productive systems compared to the case

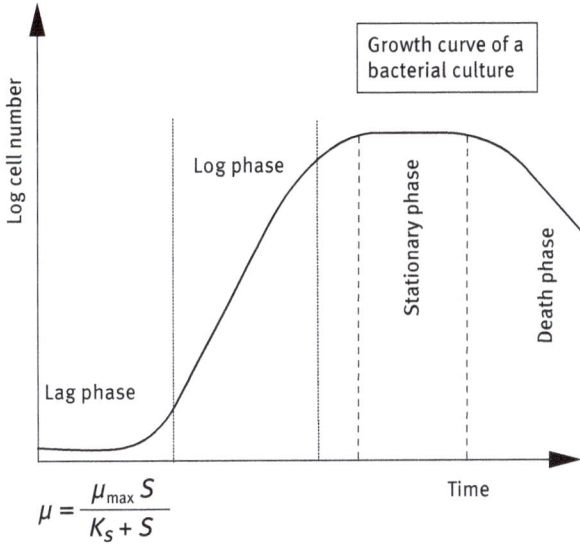

$$\mu = \frac{\mu_{max}\, S}{K_S + S}$$

Figure 6.25: Kinetic model and curve growth of bacterial culture.

where the biocatalyst is located within the membrane (Figure 6.18b); it is in fact that in the first case, process parameters and conditions can be tuned and controlled individually, so that, for example, membrane cleaning and maintenance to control fouling can be operated without negatively affecting the biocatalyst stability.

The continuous membrane bioreactor has some advantages compared to the batch system. These include continuous operation, reuse of biocatalyst, low product inhibition, high cell density, high productivity, product in clear solution, reduced waste, low labour costs. Despite the common observation that immobilized enzyme in turn decrease their catalytic activity when heterogenized, it has been demonstrated in various case studies that this is not a general rule and that suitable conditions permit to have high catalytic activity and stability [59, 68, 82, 83]. Even though predictive approaches are not available as yet to guide the selection of such suitable conditions, an increase in research efforts might soon change this scenario.

Table 6.8 illustrates the common enzyme categories and catalyzed reactions.

The unique feature of enzymes compared to inorganic catalysts is their specificity; they bring two specific molecules together letting their atoms form bonds while themselves remaining unchanged, such as inorganic catalysts, but they link only two particular molecules, not others, assuring the overall specificity. In light of the seek for precise, clean and low energy input technologies, BMRs using immobilized enzyme will play an important role in future redesigning of intensified production processes.

Table 6.8: Common enzyme class of interest.

Enzymes class	Reaction
Oxidoreductases	$A^- + B \rightarrow A + B^-$ Transfer of electrons (hydride ions or H atoms)
Transferases	$A\text{–}X + B \rightarrow A + B\text{–}X$ Group transfer reactions
Hydrolases	$A\text{–}B + H_2O \rightarrow A\text{–}OH + B\text{–}H$ Hydrolysis reactions (transfer of functional groups to water)
Lyases	$ATP \rightarrow cAMP + PP_i$ Addition of groups to double bonds, or formation of double bonds by removal of groups
Isomerases	$A \rightarrow B$ Transfer of groups within molecules to yield isomeric forms
Ligases	$A + B \rightarrow A\text{–}B$ Formation of COC, COS, COO and CON bonds by condensation reactions coupled to ATP cleavage

6.3.3 Biocatalytic membrane reactors

BMRs are distinguished by the fact that the biocatalyst is immobilized in membranes. Membranes represent a suitable microenvironment for enzyme immobilization; credit goes to the fact that they can mimic confinement that usually such catalytic proteins experience in biological systems [84].

Methods for loading an enzyme on a polymeric or inorganic membrane fall under the two major categories: (1) physical entrapment and (2) chemical binding (Figure 6.26). Although one may tend to consider carrying out the immobilization by one of the illustrated mechanisms, a combination of them may often occur, unless nonspecific interactions between membrane material and protein macromolecules can be avoided.

The physical entrapment can be obtained either during the formation of the membrane itself (by adding the enzyme into the casting solution) or the enzyme can be loaded on an already prepared membrane. The case study of an enzyme entrapped in an asymmetric hollow fiber by ultrafiltration is illustrated in Figure 6.27.

Here the enzyme remains entrapped within the membrane because of the fact that the protein size is larger than the selective layer of the membrane and cannot pass through it. In addition, depending on the relative physical and chemical properties of enzymes and membranes, van der Waals interactions, ionic binding and so on can also occur. In other words, the process takes advantage of fouling mechanism, that is, the tendency of membranes to get fouled by proteins. The macromolecules that have loose interactions with the membrane material will be washed out during

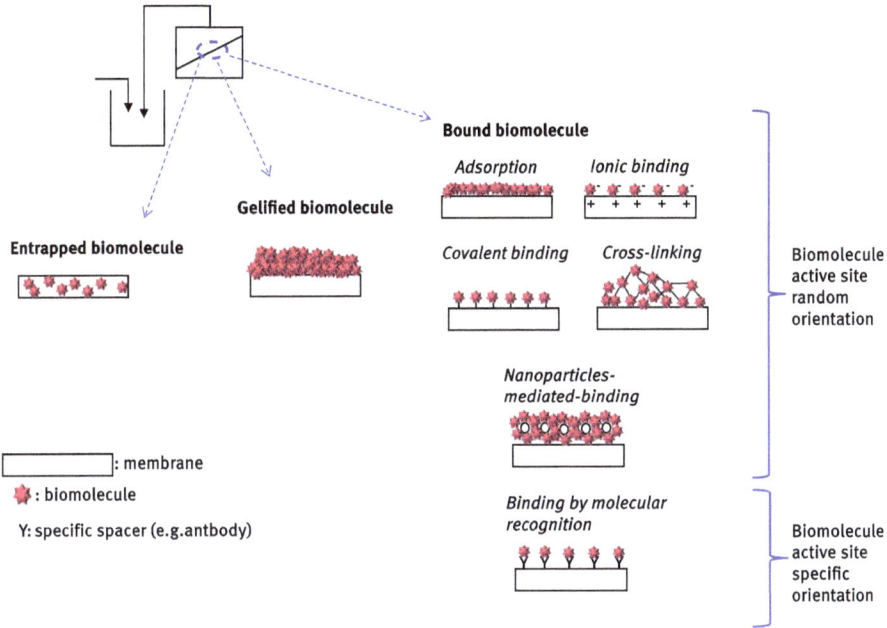

Figure 6.26: Major immobilization methods to load enzymes on membranes.

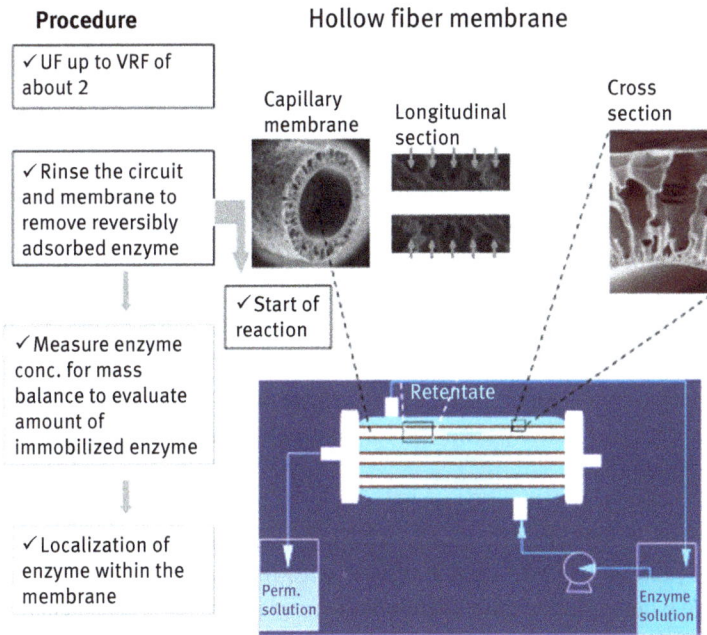

Figure 6.27: Enzyme immobilization by entrapment in an asymmetric hollow fiber by ultrafiltration.

the rinsing steps. Since all these mechanisms of interactions randomly attract the enzyme macromolecules to the membrane, they do not significantly affect the overall reactor performance [85].

A different situation is when the immobilization is obtained by molecular recognition [86] or site-specific [87], which permits the enzyme to immobilize in a specific orientation, that is, with the active site exposed toward the bulk phase to maximize interactions with the reagent. In this case, any random interaction that may obstruct the active site should be avoided.

The covalent binding is usually obtained by a multistep procedure that aims at creating reactive groups on the membrane that then reacts with a spacer and/or a bifunctional molecule that can promote a covalent bond with the enzyme in mild conditions [88].

Table 6.9 illustrates common agents used in enzyme immobilization via covalent bond.

Table 6.9: Common agents used in enzyme immobilization via covalent bond.

Glutaraldehyde (GA)	
Dimethylaminopropane (DAMP)	
(3-aminopropyl) triethoxysilane (APTES)	

One of the most important parameter that needs to be considered when operating with BMRs is the amount of immobilized enzyme. Enzymes are affected by crowding phenomena; they are inhibited by their own high concentration because of aggregation and interactions that occur between their active site (this is a kind of feedback regulation in biological systems that avoids uncontrolled bioconversions). Furthermore, high enzyme mass and protein aggregates also represent a barrier to the transport, so that molecules that are in internal layer or inside the aggregate do not meet the substrate. Very often, the lower enzyme activity observed for immobilized enzyme is an overall result of these phenomena.

The behavior of catalytic activity and specific activity as a function of immobilized enzyme on membranes is illustrated in Figure 6.28. In general, the catalytic activity (i.e., the amount of product per unit time) increases with th amount of immobilized enzyme in the range where no crowding phenomena or resistance to mass transfer is promoted and the enzyme is still saturated by the substrate. In the same range, the specific activity is constant as additional formed products per unit time normalized by

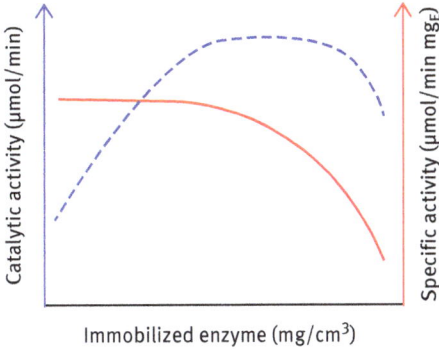

Figure 6.28: Behavior of catalytic activity and specific activity as a function of immobilized enzyme.

the additional enzyme available that maintains the overall ratio constant. In the range where additional immobilized enzyme does not increase the amount of biocatalyst that can contribute to the bioconversion, the catalytic activity is constant (i.e., enzyme may distribute in multilayers that obstruct the one located behind, so that the working amount of enzyme is constant even though the overall immobilized enzyme increases). In this range, the specific activity decreases, as the same amount of product per unit time is normalized by a higher enzyme amount. In the range where additional enzyme provokes inactivation, for example, by crowding phenomena, the catalytic activity decreases and the slope of the decrease in specific activity becomes steeper.

For what concerns the kinetic parameters of immobilized enzyme can be calculated with the same model of the free enzyme taken into the account; in this case, observed kinetic parameters can be obtained rather than intrinsic ones and as long as the system works in reaction limited regime. In fact, the reaction equation for immobilized enzymes is modified as illustrated in Figure 6.29a and the appearance of the product in the bulk (Pb) as a function of time shows a latency (Figure 6.29b) because of the need for the substrate to transfer from the bulk to the enzyme-loaded membrane and for the product to transfer from the membrane reaction site into the bulk where it can be sampled and analyzed [53].

Enzymes can be immobilized on the membrane surface or within the membrane matrix (or both).

For biocatalytic membranes where the enzyme is present on the membrane surface at a steady-state and at the interface, the mass transfer of the substrate must be counterbalanced by the consumption rate of the substrate:

$$\text{Transport rate} = \text{Reaction rate}$$

$$k_S(S_b - S_0) = \frac{V_{max}[S]}{K_M + [S]} \tag{6.33}$$

where S_b and S_0 are the substrate concentration n the bulk and at the membrane level, respectively.

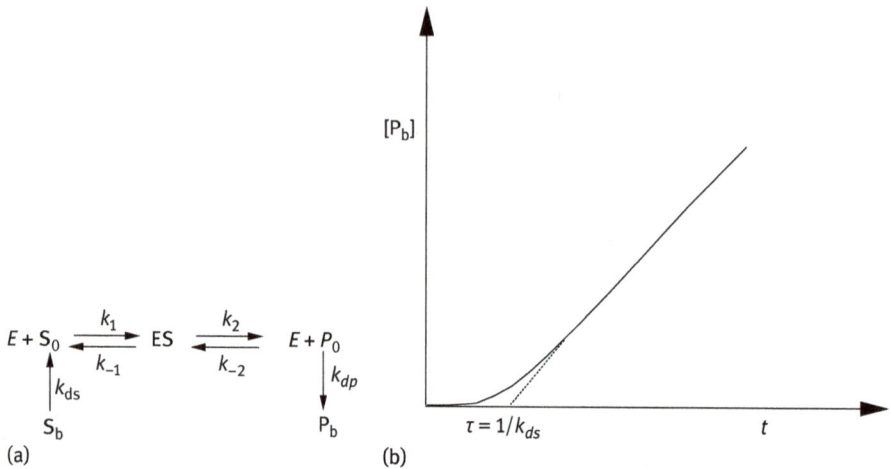

Figure 6.29: (a) Reaction equation for immobilized enzyme and (b) behavior of product in the bulk as a function of time for the reactor configuration where the product accumulates into the reactor tank.

The Damköhler number is a ratio between the maximum reaction rate and maximum mass transfer rate:

$$Da = \frac{V_{max}}{k_s \, S_b} \tag{6.34}$$

The Damköhler number has an important physical meaning and its estimation can give information on the regime in which the system works. When $Da \ll 1$, it means that the denominator, that is, the maximum transfer rate is much larger than the maximum reaction rate. Therefore, the system works at low mass transfer resistance, which is the case when the system works in reaction-limited regime, that is, the system is kinetically controlled and the Michaelis-Menten equation can be assumed valid and the substrate concentration in the bulk can be used to calculate observed kinetic parameters.

The effectiveness factor, η, is usually used to evaluate the influence of mass transfer on the overall reaction:

$$\eta = \frac{\text{Observed reaction rate}}{\text{Reaction rate in absence of transfer effect}} \tag{6.35}$$

For $\eta \leq 1$, the mass transfer resistance is large and as an effect a reduction of observed catalytic activity is obtained.

For Da approaching to zero (maximum transfer rate is very high compared to the maximum reaction rate), η approaches to 1, which means that for systems working in reaction limited regime, the observed kinetics can be approximated to the intrinsic kinetics.

For BMRs that contain the enzyme within the internal surface of the porous support, the concentration profile of substrate within the membrane diffusion layer must be considered to calculate the observed substrate conversion.

This is given by the effective diffusion coefficient:

$$D_{\text{eff}} = D_{S_0} \frac{\varepsilon_p}{\tau} \frac{K_p}{K_r} \tag{6.36}$$

At steady state, the mass transfer is balanced by the conversion rate.

Also for these systems, a dimensionless parameter (Thiele modulus) that has the physical meaning of a ratio between the *max reaction rate/max diffusion rate*, can be used to evaluate the regime in which the system works:

$$\phi = L \left(\frac{V_{\max}}{D_{\text{eff}} K_M} \right)^{1/2} \tag{6.37}$$

For $\phi \leq 1$, the reaction is essentially controlled by kinetics and the mass transfer limitation is negligible.

For BMRs, it is crucial to establish the reaction rate equation that applies for the given system. Table 6.10 summarizes the component balances for tank reactors that might be used when the microporous reactor can be approximated to one of them.

Table 6.10: Component balances for various tank reactor systems.

Reactor type	Terms in Eq.	Final form
Batch reactor	Flow term = 0	Acc = Production
Fed-batch or semi-continuous reactor	Flow term out = 0	Acc = In + Production
Steady-state CSTR	Acc = 0	In − Out + Prod = 0
Component Ci does not react	Prod = 0	Acc = In − Out
Constant volume CSTR	Simplifies to $V(ds/dt) = F(S0 - S1) + v_r V$	Acc = In − Out + Prod

When a BMR works under continuous stirred tank reactor conditions and no accumulation is obtained in the pores, the component balance reduces to the following:

$$\text{IN} - \text{OUT} + \text{PROD} = 0$$

$$FC_{\text{IN}} - FC_{\text{OUT}} + v_r V = 0 \tag{6.38}$$

$$v_r = \frac{F(C_{\text{IN}} - C_{\text{OUT}})}{V} \tag{6.39}$$

This volumetric reaction rate is expressed in terms of measurable variables, which can give observed kinetics. It is worth to recall that V is the reactor volume (L^3) and corresponds to the void volume of the membrane pore. F is the flow rate (L^3/t), C_{IN} and C_{OUT} are the concentration (mol/L^3)of substrate or product that enters and exits the reactor, respectively.

Table 6.11: Common examples of biocatalytic membrane reactors application.

Biocatalyst	Status	Application
Lactase	Immobilized	Hydrolysis of beta-D-galactosidic linkage of lactose milk (Industrial scale)(2)
Glucose isomerase	Immobilized	Conversion of D-glucose to D-fructose (Industrial scale)
Acylase	Immobilized	Production of L-aminoacids (Industrial scale)
E. coli	Immobilized	Production of L-aspartic acid (Industrial scale)
Pseudomonas dacunahe	Immobilized	Production of L-alanine (Industrial scale)
Aminoacilase, and dehydrogenase	Free and immobilized	Production of L-amino acids
Brevibacterium ammoniagenes	Immobilized	Production of L-malic acid (Industrial scale)
Pectic enzymes	Free or immobilized	Hydrolysis of pectins to improve processability (industrial scale)
Thermolysin	Immobilized	Production of aspartame (Industrial scale)

Table 6.11 illustrates common examples of application of BMRs.

Although some examples of BMRs using immobilized enzymes are applied at the industrial scale, the process is still underexploited as compared to its potentialities. More research efforts are needed to proof the robustness of the technology and to overcome the trial and error approach moving to a more predictive one.

6.3.4 Side-stream membrane bioreactors and fermentors

This type of MBR is usually obtained by the combination of a bioreactor (working as a stirred tank reactor [STR]) with membrane operations. Table 6.12 summarizes common examples of STR with an ultrafiltration that forms an overall continuous stirred ultrafiltration membrane reactor.

As a case study, a continuous membrane fermentor will be discussed [89]. This type of system is also named cell-recycle membrane fermentor (Figure 6.30). The production of lactic acid from glucose by means of *Lactobacillus bulgaricus* will be used as a model reaction. The membrane operation combined with the fermentation process is represented by ultrafiltration using polysulfone material.

Table 6.12: Common examples of membrane bioreactors applications formed by CSTR + UF.

Reaction	Membrane bioreactor	Purpose
Hydrolysis of starch to maltose (α-, β-amylase, pullulanase)	CSTR with UF membrane	Production of syrups 42 DE and HFCS
Fermentation of all fermentable sugars (yeast)	CSTR with UF membrane	Brewing industry
Anaerobic fermentation (*S. cerevisiae*)	CSTR with UF membrane	Production of alcohol
Hydrolysis of pectins (pectinase)	CSTR with UF membrane	Production of bitterness and clarification of fruit juice and wine
Fermentation of *L. bulgaricus*	CSTR with UF membrane	Continuous fermentation for production of carboxylic acids
Removal of limonene and naringin (β-cyclodextrin)	CSTR with UF membrane	Production of bitterness and clarification of fruit juice
Hydrolysis of K-casein (endopeptidase)	CSTR with UF membrane	Milk coagulation for dairy products
Hydrolysis of collagen and muscle proteins (protease, papain)	CSTR with UF membrane	Tenderization of meat, particularly beef
Dehydrogenation reactions (NAD(P)H dependent enzyme systems)	CSTR with UF-charged membrane	Production of enantiomeric amino acids
Hydrolysis of triglycerides to fatty acids and glycerol (lipase)	UF capillary membrane reactor	Production of food, cosmetics, emulsificants
Hydrolysis of cellulose to cellobiose and glucose (cellulose/β-glucosidase)	Asymmetric hollow fiber reactor	Production of ethanol and protein
Hydrolysis of raffinose (α-galactosidase and invertase)	Hollow fiber reactor with segregated enzyme	Production of monomer sugars

Initially, information on the individual performance of the biocatalyst (bacteria) and the behavior of the ultrafiltration with the fermentation broth will be discussed. Then the behavior of continuous systems as a function of operating parameters will be presented.

A typical profile of bacterial growth, decrease in substrate and product formation as a function of time is reported in Figure 6.31. From this figure, it is possible to see that at the beginning of the process, bacteria are in the latent phase; the consumption of glucose and the production of lactic acid are marginal. At this stage, it is not convenient to start ultrafiltration, since only loss of substrate would occur (UF cannot retain glucose). Later on, when the concentration of bacteria increases and the glucose concentration is reduced, the continuous process can be started.

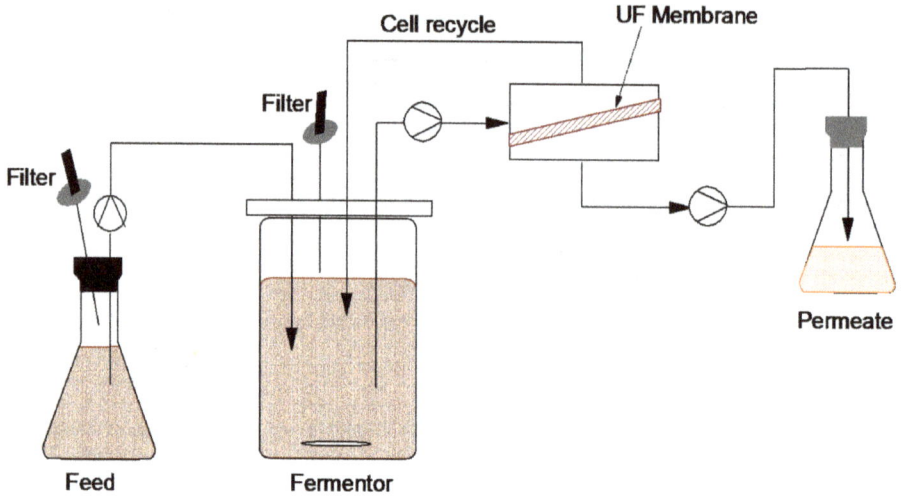

Figure 6.30: Membrane bioreactor as a continuous or cell-recycle membrane fermentor combined with an ultrafiltration step. Adapted from [55, 89].

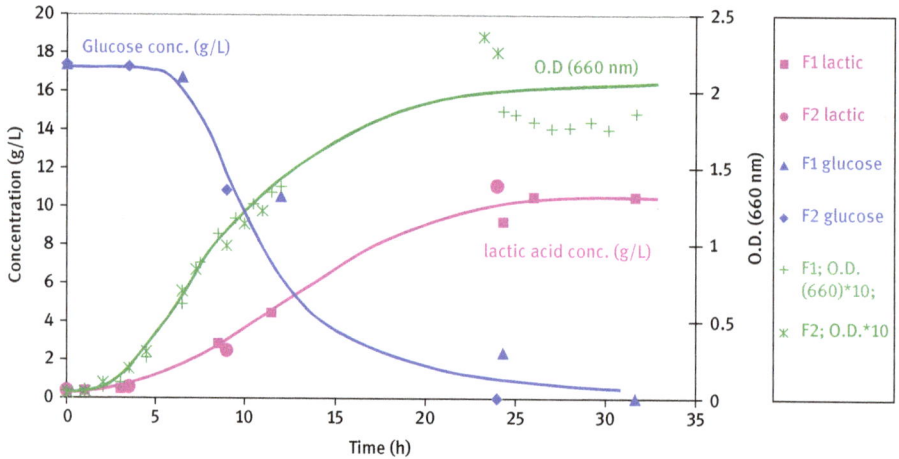

Figure 6.31: Behavior of batch fermentation. Adapted from [89].

By proper tuning of the residence time (or dilution rate), the performance of the system can be increased as shown in Figure 6.32. However, the product inhibition must also be taken into account. For example, increasing the residence time of the product in the reactor over a certain range may decrease the bioconversion capability of the bacteria.

Overall, the operating parameters must be tuned so that at steady state, the reaction rate matches the flux rate and the system performs at its maximum. For this

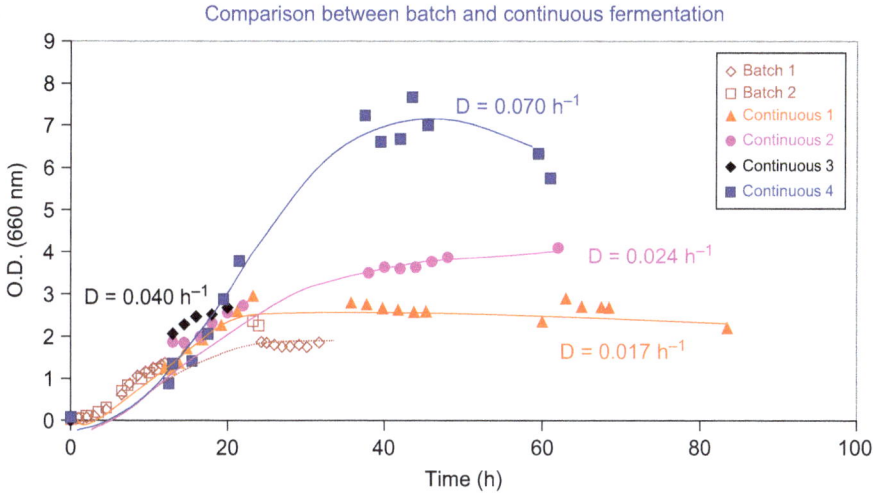

Figure 6.32: Increase of fermentation performance with dilution (residence) time in continuous membrane fermentor. Adapted from [89].

purpose, as mentioned earlier, both reaction and transport for the continuous system at steady state need to be well understood.

The continuous MBR has some advantages as compared to the batch system. These include continuous operation, reuse of biocatalyst, low product inhibition, high cell density, high productivity, product in clear solution, reduced waste and low labor costs. As a drawback, it can be accounted for the low concentration of the product in the permeate.

The batch bioreactor presents a high concentration of the product at the end of the process. On the other hand, it is present in a very complex matrix that makes its downstream separation difficult. Furthermore, the batch bioreactor suffers from start-up and shut-down procedures, high labor costs, low biocatalyst concentration, product inhibition and low productivity.

6.3.5 Submerged membrane bioreactors

Although some example of submerged BMR is reported [90], submerged MBRs mainly refer to the combination of a biodegradation with microfiltration or ultrafiltration process and are mostly applied for waste water treatment [71, 91]. The concept of membranes immersed in a sludge was published in 1989 by Yamamoto et al. from the University of Tokyo [92]. Compared to other MBR configurations and applications, they have been growing very fast, moving from laboratory investigation to practical use in less than 15 years. This extraordinary success was mainly because of the many research efforts that are devoted by both public and private organizations under the pressure of water scarcity stimulated by governmental regulation that prohibit the dispersion of waste water into the environment.

The key benefit of MBR in wastewater treatment is that they are able to obtain high quality of treated water that is suitable for being discharged in surface water flows or for urban irrigation. It is also suitable for being further being easily purified and sanitized for potable use.

Submerged MBRs work at low hydrostatic pressure with low flux. This implies higher membrane surface area;however, there is easier control of fouling by air scouring and longer membrane lifetime. In the external system, the permeate flux generally varies between 50 and 120 $l/m^2 \cdot h$ and the TMP is in the range of 1 to 4 bar. In the submerged configuration, the permeate flux varies from 15 to 50 $l/m^2 \cdot h$ and the TMP is about 0.5 bar. Furthermore, they do not require a recirculation pump; they use a suction pump and can take advantage of the hydrostatic pressure of the liquid in which they are immersed; therefore, they can work with low energy input. The energy requirement for SMBR can be two order of magnitude less compared to sidestream configuration. The replacement of membrane modules or their movement that takes place out of the tank for intervening when membrane clogging occurs is more complex compared to external sidestream modules and requires special procedure with highly specialized procedures.

SMBR have small footprint (<50%) compared to traditional activated sludge (purifying bacteria) plants. In fact, they can replace various separate and independent steps with a single MBR unit (Figure 6.33). For example, while the purification stage (aeration tank) and the separation of the biomass from the purified waste water (settling tank) are carried out separately and independently of each other in activated sludge processes, they are combined in one tank in SMBRs. S MBRs can be easily combined in existing

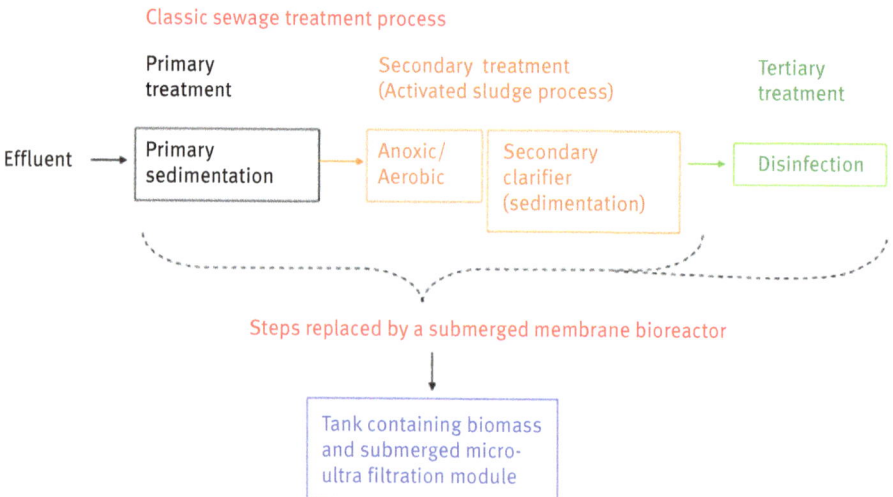

Figure 6.33: Schematic of steps in activated sludge and submerged MBR units. Dashed lines highlight the activated sludge steps replaced by SMBR.

plants with relatively simple retrofitting. The membrane separates the biomass and produces purified water with better quality. In fact, the use of microfiltration membranes with pore sizes usually between 0.1 and 0.4 µm ensures the complete retention of suspended matter and leads to a considerable reduction in the amount of bacteria in the outflow of the plant. Considering the capability of the membrane to retain bacteria, SMBRs show high biomass concentration, short hydraulic retention time and less sludge (< 40%) production compared to conventional activated sludge.

The type of membranes used in SMBR are either hollow fibers aligned vertically or horizontally or flat plates aligned vertically (Figure 6.34). In all cases, the permeate is removed by suction to avoid pressurizing the bioreactor; bubbling is used as a fluid mechanical method to control fouling and clogging. Air is introduced at the base of the membrane module and distributed in order to promote appropriate air scouring action along the membrane surface. The membrane cost decreased quite linearly in the course of years with increasing research efforts that promoted improvements in process design, operation and maintenance procedures, membrane life, element standardization and so on. SMBR costs are prioritized as equipment > energy > chemicals > membranes. The most common membrane characteristics used in SMBR are illustrated in Table 6.13. Other studied membranes include polypropylene (PP), cellulose acetate (CA), polysulfhone (PS), polyacrylonitile (PAN) or inorganic membranes, such as stainless steel flat membrane and tubular alumina ceramic membranes. Nowadays, the trend is to develop robust naturally based polymeric membranes that may not represent a critical issue in terms of disposal after their use.

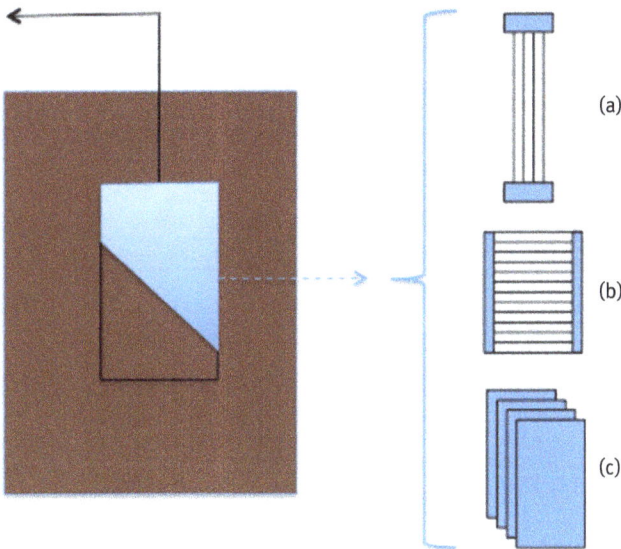

Figure 6.34: SMBR with (a) vertically aligned hollow fibers; (b) horizontally aligned hollow fibers; and (c) vertically aligned flat-sheet membranes.

Table 6.13: Characteristics of membranes used in submerged membrane bioreactors.

Membrane configuration	Membrane material	Membrane pore size (μm)
Hollow fiber	PVDF	0.04
Hollow fiber	PVDF	0.1
Hollow fiber	PES	0.05
Hollow fiber	PE	0.4
Flat sheet	PES	0.08
Flat sheet	PVDF	0.08
Flat Sheet	PE	0.4

Membrane fouling reduces productivity, shortens membrane lifespan and increases operation costs. Therefore, many studies focus on this research area. Most of them focus on the characteristics, causes and modeling of the membrane fouling. Some effective methods to prevent or reduce membrane fouling have been developed.

In SMBRs gas/liquid two-phase flow is used to control fouling. Aeration scours the membrane surface, provides oxygen to the biomass and maintains solids in suspension. Therefore, this is a key parameter to guarantee both flux and bioconversion processes.

Currently, SMBRs operate with lower hydraulic retention time (HRT between 10–20 days) and mixed liquor suspended solids (MLSS 10–15 g/L) compared to first generation process (HRT 100 days and MLSS up to 30 g/L, respectively). These conditions allow better operation of the process with increased life time and lower costs. In fact, in general, it is observed that while the reduction of COD is increased with increased biomass concentration (up to 99%), this levels off at concentration over 15 g/L. This can be because of the lower oxygen transport and active sites accessibility when density of the liquor becomes too high.

A summary of the major parameters affecting the MBR performance is reported in Table 6.14.

Table 6.14: Factors affecting membrane bioreactors performance.

Biochemical properties	Physical properties	Fluid dynamics properties	Membrane	Operation
SMLL concentration	Rheology of the mixture	Flow rates of inlet/outlet/recycle	Material	Membrane relaxation
Composition/type of contaminants	Gas/liquid/solid density	Baffles/mixer	Configuration	Back-flushing
Type of Biomass	–	Residence time distribution	Packing density	Aeration
–	–	–	Orientation	–

Biomass can be suspended in the bioreactor or it can grow as a biofilm on the membrane surface. The membrane can also feed the oxygen to the biofilm obtaining the so-called submerged membrane aerated biofilm reactor. The membrane side containing the biofilm is flushed with the wastewater while the other side is in contact with oxygen. The membrane transports the oxygen to the biofilm. In this way, the addition of oxygen can be controlled independently and only the biofilm next to the membrane can have sufficient oxygen to grow. Oxygen and contaminants meet at the membrane level.

In other applications, where a specific contaminant is intended to be eliminated, the membrane may serve as a selective extractor that is able to capture the contaminant and transport it to the other side of the membrane where it is biodegraded by the biofilm located there.

Thanks to the continuous decrease of cost and reasonable performance, SMBR have been increasingly installed worldwide.

However, despite the higher energy requirement, sidestream MBRs have continued to be applied, especially for better applications. Thanks to the easy maintenance, easy module replacement, high flux and research continued to explore this configuration promoting advances that may operate the process with energy need of about $0.3 \, \text{kWh/m}^3$ of purified water.

6.4 Conclusions

Nowadays membrane reactors are a promising innovative technology in various sectors, spcifically in hydrogen production. Their use allow better performance to be achieved than conventional reactors in terms of high recovery of pure hydrogen streams, higher conversion and reduced catalyst amount. The traditional process can thus be redesigned as more compact and efficient thereby obtaining an intensified process with a reduced plant size and higher yield.

Membrane reactors have been demonstrated as multifunctional units able to significantly increase the conversion achievable (up to 5 times) with respect to the one of a traditional reactor, significantly reducing the reaction volume required (down to 15% of a traditional reactor). Moreover, the analysis of its performance in terms of mass and Energy intensities highlights a region where only membrane reactor has access demonstrating the assets of this technology also in terms of better exploitation of raw materials (reduction up to 40%) and higher energy efficiency (up to 35%).

MBRs are becoming very much in demand; credit goes to fact that (i) biocatalysts are extremely selective, work in mild conditions of temperature, pressure and pH, (ii) the membrane can implement simultaneous separation of reaction products (thus increasing the reaction yield on the basis of Le Chatelier principle) while retaining the biocatalyst that can be reused in continuous processes. This means not only energy saving, low cost, low environmental impact and high mass intensity, but also safer production systems. MBRs have been already explored in food, pharmaceutical, biomedical, cosmetic, biofuel and water treatment and play more important roles in biorefinery, green chemistry and bioremediation. The synthesis of artificial catalysts mimiking the enzyme activity with increased stability would definitively promote a breakthrough in the field.

6.5 Notation

A	Surface area, m^2
C	Concentration, $mol\ m^{-3}$
Cp	Specific heat, $J\ mol^{-1}\ K^{-1}$
D	Diffusivity, $m^2\ s^{-1}$
E	Activation energy, $J\ mol^{-1}$
F	Molar flow rate, $mol\ s^{-1}$
h	Enthalpy, $J\ mol^{-1}$
ID	Inner diameter, m
J	Permeating flux, $mol\ m^{-2}\ s^{-1}$
k	Kinetic constant, see related equation
$K_{equilibrium}$	Equilibrium constant, -z
K_p	Equilibrium constant in terms of partial pressures, –
k	Thermal conductivity, $W\ m^{-1}\ K^{-1}$
L	Length, m
m	Reactant feed molar ratio, ·
n	Number of mole, -
N	Molar flux, $mol\ m^{-2}\ s^{-1}$
OD	Outer diameter, m
P	Pressure, Pa
Permeability	$mol\ m^{-1}\ s^{-1}\ Pa$
Permeability	$mol\ m^{-1}\ s^{-1}\ Pa^{-0.5}$ (Sieverts)
Permeance	$mol\ m^{-2}\ s^{-1}\ Pa$
Permeance	$mol\ m^{-2}\ s^{-1}\ Pa^{-0.5}$ (Sieverts)
Permeating Flux	$mol\ m^{-2}\ s^{-1}$
Q	Volumetric flow rate, $m^3(STP)\ s^{-1}$
R	Gas law constant, $8.314\ J\ K^{-1}\ mol^{-1}$
RI	Recovery index
r	Radial coordinate, m
r_{ij}	j^{th} reaction rate referred to the ith species, $mol\ m^{-3}\ s^{-1}$
T	Temperature,°C or K
t	Time, s
U	Overall heat transfer coefficient, $W\ m^{-2}\ K^{-1}$
V	Volume, m^3
X	Conversion, –
z	Axial coordinate, m

Greek letters

φ	Enthalpy flux associated to hydrogen permeation, $W\ m^{-2}$
ψ	Heat generated by chemical reactions, $W\ m^{-2}$

δ	Membrane thickness "Shell" thickness for interfacial reaction, m
ε	Porosity, –
$V_{l,\varphi}$	Stoichiometric coefficient with respect to the reference component of ith Species in j^{th} reaction, -
ρ	Density, g m^{-3}
τ	Space time, s
τ	Tortuosity, -

Superscripts

Annulus	Annulus side in a luminal (tubular) MR
Exit	MR exit referred
Feed	Membrane module inlet stream referred
Input	MR input referred
Lumen	Lumen side in a tubular MR
Membrane	Membrane phase referred
Output	MR output referred
Permeate	Membrane module permeate stream referred
Permeating	Membrane module permeating stream referred
Permeation	Membrane module permeation stream referred
Reaction	Membrane module stream on the reaction volume referred
Retentate	Membrane module outlet stream on the reaction referred
Shell	Membrane module shell side referred
Sweep	Membrane module inlet stream on permeate side referred

Acronyms

B.C.	Boundary condition
CPC	Concentration polarization coefficient
CSTR	Continuous stirred tank reactor
GHSV	Gas hour space velocity, h^{-1}
I.C.	Initial condition
MR	Membrane reactor
MREC	Membrane reactor equilibrium conversion
MSR	Methane steam reforming
PDE	Partial differential equation
SEM	Scanning electron microscope
STP	Standard temperature (0 °C) and pressure (100 kPa)
TR	Traditional reactor
TREC	Traditional reactor equilibrium conversion
WGS	Water gas shift

References

[1] Marigliano, G., Barbieri, G., Drioli, E. Equilibrium conversion for a Pd-alloy membrane reactor. Dependence on the temperature and pressure. Chem. Eng. Process. 2003; 42(3): 231–236.
[2] Barbieri, G., Scura, F., Drioli, E. Equilibrium of a Pd-alloy membrane reactor. Desalination. 2006; 200(1–3): 679–680.

[3] Abo-Ghander, N.S., Grace, J.R., Elnashaie, S.S.E.H. and Lima,C.J. Modeling of a novel membrane reactor to integrate dehydrogenation of ethylbenzene to styrene with hydrogenation of nitro-benzene to aniline. Chem.Eng. Sci. 2008; 63(7): 1817–1826.

[4] Gobina, E., Hou, K. and Hughes, R. Ethane dehydrogenation in a catalytic membrane reactor coupled with a reactive sweep gas. Chem. Eng. Sci. 1995; 50: 2311–2319.

[5] Barbieri, G., Scura, F. and Brunetti, A. Series "Membrane Science and Technology", Volume 13 "Inorganic Membranes: Synthesis, Characterization and Applications"; Chapter 9 - Mathematical modelling of Pd-alloy membrane reactors. Elsevier B.V , Edited by R. Mallada and M. Menendez 2008. (ISBN 978 0 444 53070 7; ISSN 0927-5193, DOI: 10.1016/S0927-5193(07)13009-6).

[6] Barbieri, G. "Pd-Based Tubular Membrane Reactor"; In "Encyclopedia of Membranes" – (Live Reference ISBN 978-3-642-40872-4) edited by Prof. Enrico Drioli, Dr. Lidietta Giorno. Springer-Verlag GmbH Berlin Heidelberg. 2015, doi:10.1007/978-3-642-40872-4_439-1.

[7] Barbieri, G. "Continuous Stirred Tank Membrane Reactor (CST-MR)" In "Encyclopedia of Membranes" - (Live Reference ISBN 978-3-642-40872-4) edited by Prof. Enrico Drioli, Dr. Lidietta Giorno. Springer-Verlag GmbH Berlin Heidelberg 2015. doi:10.1007/978-3-642-40872-4_152-1.

[8] Barbieri, G. "Hydrogen Production by Membrane Reactors"; in "Encyclopedia of Membranes" – (Live Reference ISBN 978-3-642-40872-4) edited by Prof. Enrico Drioli, Dr. Lidietta Giorno. Springer-Verlag GmbH Berlin Heidelberg 2015, DOI: 10.1007/978-3-642-40872-4_708-1.

[9] Barbieri, G., Brunetti, A., Granato, T., Bernardo, P., Drioli, E. "Engineering evaluations of a catalytic membrane reactor for water gas shift reaction". Ind. Eng. Chem. Res. 2005; 44: 7676–7683.

[10] Brunetti, A., Barbieri, G., Drioli, E., Lee, K.-H., Sea, B., Lee, D.W. "WGS reaction in a membrane reactor using a porous stainless steel supported silica membrane". Chem. Eng. Process. 2007; 46: 119–126.

[11] Brunetti, A., Barbieri, G., Drioli, E., Granato, T., Lee, K.-H. "A porous stainless steel supported silica membrane for WGS reaction in a catalytic membrane reactor". Chem. Eng. Sci. 2007; 62: 5621–5626.

[12] Brunetti, A., Caravella, C., Barbieri, G., Drioli, E. "Simulation study of water gas shift in a membrane reactor". J. Membr. Sci. 2007; 306(1–2): 329–340.

[13] Brunetti, A., Barbieri, G., Drioli, E. "A PEM-FC and H2 membrane purification integrated plant". Chem. Eng. Process: Process Intensification. 2008; 47(7): 1081–1089. – special issue Euromembrane 2006.

[14] Barbieri, G., Brunetti, A., Tricoli, G., Drioli, E. "An innovative configuration of a Pd-based membrane reactor for the production of pure hydrogen. Experimental analysis of water gas shift". J. Power Sources. 2008; 182(1): 160–167.

[15] Brunetti, A., Barbieri, G., Drioli, E. "Upgrading of a syngas mixture for pure hydrogen production in a Pd-Ag membrane reactor". Chem. Eng. Sci. 2009; 64: 3448–3454.

[16] Brunetti, A., Barbieri, G., Drioli, E. "Pd-based membrane reactor for syngas upgrading". Energy and Fuel. 2009; 23: 5073–5076.

[17] Brunetti, A., Barbieri, G., Drioli, E. "Integrated membrane system for pure hydrogen production: a Pd-Ag Membrane Reactor and a PEMFC". Fuel Process. Technol. 2011; 92: 166–174.

[18] Barbieri, G., Brunetti, A., Caravella, A., Drioli, E. "Pd-based membrane reactors for one-stage process of water gas shift". RSC Adv. 2011; 1 (4): 651–661.

[19] Brunetti, A., Drioli, E., Barbieri, G. "Medium/high temperature Water Gas Shift reaction in a Pd-Ag membrane reactor: an experimental investigation". RSC Adv., 2012; 2(1): 226–233. DOI: 10.1039/C1RA00569C.

[20] Brunetti, A., Caravella, A. Drioli, E., Barbieri, G. "Process intensification by membrane reactors: high temperature water gas shift reaction as single stage for syngas upgrading". Chem. Eng. Technol. 2012; 35: 1238–1248

[21] Brunetti, A., Caravella, A., Fernandez, E., Pacheco Tanaka, D.A., Gallucci, F., Drioli, E., Curcio, E., Viviente, J.L., Barbieri, G. "Syngas upgrading in a membrane reactor with thin Pd-alloy supported membrane", Int. J. Hydrogen Energy, 2015; Vol. 40(34): 10883–10893.

[22] Barbieri, G. "Water Gas Shift (WGS)"; doi:10.1007/978-3-642-40872-4_598-1 in "Encyclopedia of Membranes" – (Live Reference ISBN 978-3-642-40872-4) edited by Prof. Enrico Drioli, Dr. Lidietta Giorno. Springer-Verlag GmbH, Berlin Heidelberg, 2015.

[23] Miller, G.Q., Stöcker, J. "Selection of a Hydrogen Separation Process", 1989 NPRA Annual Meeting held March 19–21, San Francisco, California (USA), 1989.

[24] Miller, G.Q., Stöcker, J. "Selection of a Hydrogen Separation Process" 4th European Technical Seminar on Hydrogen Plants, Lisbon (Portugal), Oct 2003.

[25] Caravella, A., Scura, F., Barbieri, G., Drioli, E. "Inhibition by CO and polarization in Pd-based membranes: a novel permeance reduction coefficient". J. Phys. Chem. B. 2010; 114: 12264–12276. DOI: 10.1021/jp104767q.

[26] Caravella, A., Barbieri, G., Drioli, E. Concentration polarization analysis in self-supported Pd-based membranes. Sep. Purif. Technol. 2009; 66: 613–624. DOI: 10.1016/j.seppur.2009.01.008.

[27] Caravella, A., Melone, L., Sun, Y., Brunetti, A., Drioli, E., Barbieri, G. "Concentration polarization distribution along Pd-based membrane reactors: A modelling approach applied to water-gas shift". Int. J. Hydrogen Energy. 2016; 41(4): 2660–2670.

[28] Van Gerven, T., Stanckiewicz, A. Structure, energy, synergy, time-the fundamentals of process intensification. Ind. Eng. Chem. Res. 2009; 48: 2465–2474.

[29] Stankiewicz, A., Moulijn, J.A. Process intensification. Ind. Chem. Eng. Res. 2002; 41: 1920–1924.

[30] Tsouris, C., Porcelli, J.V. Process intensification – Has its time finally come?. Chem. Eng. Progr. 2003; 99: 50–54.

[31] Dautzenberg, F.M., Mukherjee, M. Process intensification using multifunctional reactors. Chem Eng Sc. 2001; 56: 251–267.

[32] Lutze, P., Gani, R., Woodley, J. Process intensification: A perspective on process synthesis. Chem. Eng. Proc. 2010; 49: 547–558.

[33] Drioli, E., Brunetti, A., Di Profio, G., Barbieri, G. Process intensification strategies and membrane engineering. Green Chem. 2012; 14: 1561–1572.

[34] Bortolotto, L., Dittmeyer, R. Direct hydroxylation of benzene to phenol in a novel microstructured membrane reactor with distributed dosing of hydrogen and oxygen. Sep. Pur. Tech. 2010; 73: 51–58.

[35] Ye, S., Hamakawa, S., Tanaka, S., Sato, K., Esashi, M., Mizukami, F. A one-step conversion of benzene to phenol using MEMS-based Pd membrane microreactors. Chem. Eng. J. 2009; 155: 829–837.

[36] Luo, H., Tian, B., Wei, Y., Wang, H., Jiang, H., Caro, J. Oxygen permeability and structural stability of a novel tantalum-doped perovskite BaCo0.7Fe0.2Ta0.1O3-δ, AIChE J. 2009; 56: 604–610.

[37] Luo, H., Wei, Y., Jiang, H., Yuan, W., Lv, Y., Caro, J., Wang, H. Performance of a ceramic membrane reactor with high oxygen flux Ta-containing perovskite for the partial oxidation of methane to syngas. J. Mem. Sci. 2010; 350: 154–160.

[38] Barbieri, G., Brunetti, A., Caravella, A., Drioli, E. Pd-based membrane reactors for one-stage process of water gas shift. RSC Adv. 2011; 1: 651–661.

[39] Brunetti, A., Barbieri, G., Drioli, E. Upgrading of a syngas mixture for pure hydrogen production in a Pd-Ag membrane reactor. Chem. Eng. Sci. 2009; 64: 3448–3454.

[40] Brunetti, A., Barbieri, G., Drioli, E. Chapter 12 in "Membrane engineering for the treatment of gases", Volume 2 "Gas-separation Problems Combined with Membrane Reactors", pages 87–109, 2011, Editors E. Drioli and G. Barbieri, The Royal Society of Chemistry, Cambridge, The United Kingdom, ISBN 978-1-84973-239-0.

[41] Dittmeyer, R., Hollein, V., Daub, K. Membrane reactors for hydrogenation and dehydrogenation processes based on supported palladium. J. Mol. Cat. A: Chem. 2001; 173: 135–184.

[42] Liu, P.K.T., Sahimi, M., Tsotsis, T. Process intensification in hydrogen production from coal and biomass via the use of membrane-based reactive separations. Current Opinion in Chemical Engineering. 2012; 1: 342–351.

[43] Koc, R., Kazantzis, K., Ma, Y.H. Process safety aspects in water-gas-shift [WGS] membrane reactors used for pure hydrogen production. J. Loss Prev. Proc. Ind. 2011; 24: 852–869.

[44] Abashar, M.E.E., Alhumaizi, K.I., Adris, A.M. Investigation of methane-steam reforming in fluidized bed membrane reactors. Chem. Eng. Res. Des. 2003; 81(2): 251–258.

[45] Tsotsis, T.T., Champagnie, A.M., Vasileiadis, S.P., Ziaka, Z.D., Minet, R.G. Packed bed catalytic membrane reactors. Chem. Eng. Sci. 1992; 47: 2903–2908.

[46] Adris, A.M., Lim, C.J., Grace, J.R. The fluidized-bed membrane reactor for steam methane reforming: Model verification and parametric study. Chem. Eng. Sci. 1997; 52(10): 1609–1622.

[47] Criscuoli, A, Drioli, E. New metrics for evaluating the performance of membrane operations in the logic of process intensification. Ind. Chem. Eng. Res. 2007; 46: 2268–2271.

[48] Sikdar, S.K. Sustainable development and sustainability metrics. AIChE J. 2003; 49: 1928–1932.

[49] IChemE. Sustainable development progress metrics: Recommended For Use In The Process Industries; Institution of Chemical Engineers: Rugby, U.K., (2006), 1. http://www.icheme.org/sustainability/metrics.pdf

[50] Brunetti, A., Drioli, E., Barbieri, G. Energy and mass intensities in hydrogen upgrading by a membrane reactor. Fuel Process. Technol. 2014; 118: 278–286.

[51] Barbieri, G. "Volume Index"; doi:10.1007/978-3-642-40872-4_780-1 in "Encyclopedia of Membranes" - (Live Reference ISBN 978-3-642-40872-4) edited by Prof. Enrico Drioli, Dr. Lidietta Giorno. Springer-Verlag GmbH Berlin Heidelberg 2015.

[52] Marigliano, G., Perri, G., Drioli, E. Conversion-temperature diagram for a palladium membrane reactor. Analysis of an endothermic reaction: Methane steam reforming. Ind. Eng. Chem. Res. 2001; 40: 2017–2026.

[53] Drioli, E., Giorno, L. Biocatalytic Membrane Reactors: Application in Biotechnology and the Pharmaceutical Industry, Taylor & Francis Publisher, London, UK, 1999.

[54] Giorno, L., Drioli, E. Biocatalytic membrane reactors: Applications and perspectives. Trends in Biotechnol. 2000; 18: 339–348.

[55] Giorno, L., Mazzei, R., Drioli, E. Biological membranes and biomimetic artificial membranes (2010) Comprehensive Membrane Science and Engineering, 1, pp. 1–12. DOI: 10.1016/B978-0-08-093250-7.00055-4

[56] Mazzei, R., Drioli, E., Giorno, L. Biocatalytic membranes and membrane bioreactors Comprehensive Membrane Science and Engineering. 2010; 3: pp. 195–212. DOI: 10.1016/B978-0-08-093250-7.00058-X

[57] Giorno, L., De Bartolo, L., Drioli, E. Membrane Bioreactors. (2011) Comprehensive biotechnology, Second Edition, 2, pp. 263–288. DOI: 10.1016/B978-0-08-088504-9.00101-X

[58] Pastore, M. and Morisi, F. Lactose reduction of milk by fiberentrapped β-galactosidase. Methods Enzymol. 1976; 44: 822–830.

[59] Giorno, L., Drioli, E., Carvoli, G., Donato, L., Cassano, A. Study of an enzyme membrane reactor with immobilized fumarase for production of L-malic acid. Biotech& Bioeng. 2001; 72(1): 77–84.

[60] Ranieri, G., Mazzei, R., Poerio, T., Bazzarelli, F., Wu, Z., Li, K., Giorno, L. Biorefinery of olive leaves to produce dry oleuropein aglycone: Use of homemade ceramic capillary biocatalytic membranes in a multiphase system. Chem. Eng. Sci. 2018; 185: pp. 149–156. DOI: 10.1016/j.ces.2018.03.053

[61] Klibanov, A.M. and Zaks, A. Enzyme-catalysed processes in organic solvents. Proc. Natl. Acad. Sci. U. S. A. 1985; 82: 3192–3196.

[62] Klibanov, A.M. Asymmetric transformations catalysed by enzymes in organic solvents. Acc. Chem. Res. 1990; 23: 114–120.

[63] Giorno, L., Molinari, R., Natoli, M., Drioli, E. Hydrolysis and regioselective transesterification catalyzed immobilized lipases in membrane bioreactors. J. Membr. Sci. 1997; 125: 177–187.

[64] Giorno, L., Molinari, R., Drioli, E., Bianchi, D., Cesti, P. Performance of a biphasic organic/aqueous hollow fibre reactor using immobilized lipase. J. Chem. Technol. Biotechnol 1995; 64: 345–352.

[65] Lopez, J.L. and Matson, S.L. A multiphase/extractive enzyme membrane reactor for production of diltiazem chiral intermediate. J. Membr. Sci. 1997; 125: 189–211.

[66] Giorno, L., Piacentini, E., Mazzei, R., Drioli, E. Distribution of phase transfer biocatalyst at the oil/water interface by membrane emulsifier and evaluation of enantiocatalytic performance. Desalination. 2006; 199: 182–184.

[67] Mazzei, R., Drioli, E., Giorno, L. Enzyme membrane reactor with heterogenized beta-glucosidase to obtain phytotherapic compound: Optimization study. J. Membr. Sci. 2012; 390–391. 121–129, Doi: 10.1016/j.bios.2017.02.003

[68] Mazzei, R., Giorno, L., Piacentini, E., Mazzuca, S., Drioli, E. Kinetic study of a biocatalytic membrane reactor containing immobilized β-glucosidase for the hydrolysis of oleuropein. J. Membr. Sci. 2009; 339(1–2): pp. 215–223. DOI: 10.1016/j.memsci. 2009.04.053

[69] Livingston, A.G., Arcangeli, J.-P., Boam, A.T., Zhang, S., Maragon, M., Freita dos Santos, L.M. Extractive membrane bioreactors for detoxification of chemical industry wastes: process development. J Memb Sci. 1998; 151: 29–44.

[70] Almeida, J.S., Maria, M., Crespo, J.G. Development of extractive membrane bioreactors for environmental applications. Environ. Protect. Eng. 1999; 1–2(25): 111–121.

[71] Judd, S. The MBR Book: Principles and Applications of Membrane Bioreactors for Water and Wastewater Treatment, Elsevier, 2nd Edition. Burlington,USA, 2011.

[72] Bélafi-Bako, K., Eszterle, M., Kiss, K., Nemesto´thy, N., Gubicza, L. Hydrolysis of pectin by Aspergillus niger polygalacturonase in a membrane bioreactor. J. Food. Eng. 2007; 78: 438–442.

[73] Mazzei, R., Piacentini, E., Drioli, E., Giorno, L. Boodhoo, K and Harvey, A. Membrane bioreactors for green processing in a suitable production system in Process Intensification for Green Chemistry: Engineering Solutions for Sustainable Chemical Processing edited by , John Wiley & Sons, Ltd, Chichester, UK (2013), Chapter 8, p. 311–353.

[74] Giorno, L., Mazzei, R., Piacentini, E., Drioli, E. Food Applications of Membrane Bioreactors, In "Engineering Aspects of Membrane Separation and Application in Food Processing" et al. Eds. Taylor & Francis, 2016; pp. 299–350.

[75] Sakaki, K., Giorno, L., Drioli, E. Lipase-catalyzed optical resolution of racemic naproxen in biphasic enzyme membrane reactors, J. Membr. Sci. 2001; 184: 27–38. Doi: 10.1016/S0376-7388(00)00600-1

[76] Giorno, L., Mazzei, R., Piacentini E., Biocatalytic membrane reactors for the production of nutraceuticals. Integrated Membrane Operations in the Food Production, 2014; pp. 311–322. DOI: 10.1515/9783110285666.311.

[77] Gebreyohannes, A.Y., Bilad, M.R., Verbiest, T., Courtin, C.M., Dornez, E., Giorno, L., Curcio, E., Vankelecom, I.F.J. Nanoscale tuning of enzyme localization for enhanced reactor performance in a novel magnetic-responsive biocatalytic membrane reactor. J. Membr. Sci. 2015; 487: pp. 209–220. DOI: 10.1016/j.memsci.2015.03.069

[78] Gebreyohannes, A.Y., Mazzei, R., Yahia Marei Abdelrahim, M., Vitola, G., Porzio, E., Manco, G., Barboiu, M., Giorno L. Phosphotriesterase-magnetic nanoparticles bioconjugates with

improved enzyme activity in a biocatalytic membrane reactor. Bioconjugate Chem, 2018; 29(6): pp 2001–2008, DOI: 10.1021/acs.bioconjchem.8b00214

[79] Vitola, G., Mazzei, R., Fontananova, E., Porzio, E., Manco, G., Gaeta, S.N., Giorno, L. Polymeric biocatalytic membranes with immobilized thermostable phosphotriesterase (2016) J. Membr. Sci, 516, pp. 144–151. DOI: 10.1016/j.memsci.2016.06.020

[80] Baker, R.W. Membrane technology and applications, 2nd Edition, John Wiley & Sons, Ltd, Chichester, UK, 2004.

[81] Strathmann, H., L. Giorno, E. Drioli. An Introduction to Membrane Science and Technology, CNR Publisher, Roma, 2006, ISBN 88-8080-063-9

[82] Giorno, L., D'Amore, E., Mazzei, R., Piacentini, E., Zhang, J., Drioli, E., Cassano, R., Picci, N. An innovative approach to improve the performance of a two separate phase enzyme membrane reactor by immobilizing lipase in presence of emulsion. J. Membr. Sci. 2007; 295 95–101, Doi: 10.1016/j.memsci.2007.02.041

[83] Giorno, L., D'Amore, E., Drioli, E., Cassano, R., Picci, N. Influence of -OR ester group length on the catalytic activity and enantioselectivity of free lipase and immobilized in membrane used for the kinetic resolution of naproxen esters. J. Catal. 2007; 247, 194–200, Doi: 10.1016/j.memsci.2007.07.016

[84] Giorno, L., Mazzei, R., Drioli, E. Biological membranes and biomimetic artificial membranes Comprehensive Membrane Science and Engineering. 2010; 1, pp.1–12. DOI: 10.1016/B978-0-08-093250-7.00055-4

[85] Mazzei, R., Drioli, E., Giorno, L. Biocatalytic membranes and membrane bioreactors . Comprehensive Membrane Science and Engineering. 2010; 3, pp. 195–212. DOI: 10.1016/B978-0-08-093250-7.00058-X

[86] Militano, F., Poerio, T., Mazzei, R., Salerno, S., De Bartolo, L., Giorno, L. Development of biohybrid immuno-selective membranes for target antigen recognition, Biosensors and Bioelectronics. 2017; 92, pp. 54–60. DOI: 10.1016/j.bios.2017.02.003

[87] Butterfield, D. A., Bhattacharyya, D., Dauner, S., Bachas, L. Catalytic biofunctional membranes contining site-specifically immobilized enzyme arrays: A review. J. Membr. Sci. 2001; 181: 29–37.

[88] Giorno, L., Vitola, G., Ranieri, G., Militano, F. Biocatalytic membrane reactors with chemically bound enzyme. Encyclopedia of Membr. 2016; pp. 194–200, Doi: 10.1007/978-3-662-44324-8_1338.

[89] Giorno, L., Chojnacka, K., Donato, L., Drioli, E. Study of a Cell-Recycle membrane fermentor for the production of Lactic acid by Lactobacillus bulgaricus. Ind. Eng. Chem. Res. 2002;41, 433–440.

[90] Chakraborty, S., Drioli, E., Giorno, L. Development of a two-separate phase submerged biocatalytic membrane reactor for the production of fatty acids and glycerol from residual vegetable oil streams. Biomass and Bioenergy. 2012; 46: 574–583.

[91] Judd, S., Judd, C. Industrial MBRs: Membrane Bioreactors for Industrial Wastewater Treatment, Bedfordshire, United Kingdom IWA Publishing, 2014.

[92] Yamamoto, K., Hiasa, H., Talat, M., and Matsuo, T. Direct solid liquid separation using hollow fiber membranes in activated sludge aeration tank. Water. Sci. Technol. 1989; 21, 43–54.

Francesca Macedonio, Jeong F. Kim, Enrico Drioli

7 Preparation of synthetic (polymeric and inorganic) membranes and their characterization

7.1 Introduction

The most important part of any membrane-based process is the membrane itself. The core definition of a membrane is a selective or nonselective barrier that separates and/or contacts two adjacent phases and allows or promotes the exchange of matter, energy and information between the phases in a specific or nonspecific manner. Membranes are different in terms of their material, structure, function(s), transport property and transport mechanism. Hence, it is important to understand all of these factors simultaneously in order to fully exploit the advantages of membrane processes (Figure 7.1).

The membrane fabrication technology has improved significantly over the past 50 years, and there exists a wide array of literature on different fabrication methods [1–3, 7]. By carefully tuning the fabrication parameters, it is now possible to fabricate various types of membranes with the desired performance. A competent membranologist, however, must understand the complex relationship between the key parameters in order to fabricate membranes with appropriate characteristics.

An appropriate fabrication method(s) must be selected depending on the material, processability and desired application. For instance, polymeric membranes are typically fabricated via phase inversion method, and ceramic membranes are generally fabricated by sol–gel and sintering techniques. Hence, a membranologist must understand the basics of material science as well as the thermodynamic fundamentals of membrane fabrication techniques.

A topic closely related to the membrane preparation is its characterization. In addition to the determination of the membrane transport properties, membranes are also characterized by various methods in terms of their morphology and their mechanical, chemical, and electrochemical properties. Which of the different membrane characterization techniques are used depends on the membrane type and its possible application.

Francesca Macedonio, Enrico Drioli, National Research Council of Italy, Institute for Membrane Technology (ITM-CNR), University of Calabria, Rende, Italy
Jeong F. Kim, Membrane Research Center, Korea Research Institute of Chemical Technology (KRICT), Daejeon, South Korea

https://doi.org/10.1515/9783110281392-007

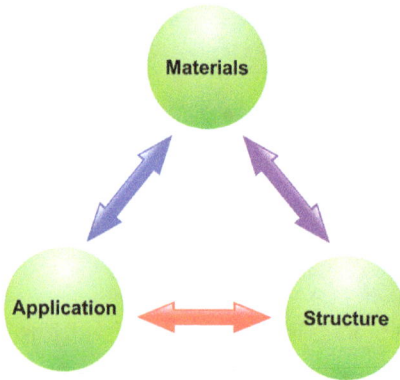

Figure 7.1: Importance of material, structure, and respective application. All three components must be simultaneously considered when developing membrane-based processes.

In this chapter, the basic principles of membrane preparation methods will be discussed. In addition, state-of-the-art characterization methods will also be discussed.

7.2 Membrane preparation methods

7.2.1 Effect of material properties

Membranes can be prepared by different materials and different functional groups can be incorporated to give them specific features. There are roughly five different categories of membrane materials: polymers, ceramics, metals, liquids, and biological materials. For specific applications, a mixture of two different types of materials can be used to fabricate membranes with desired characteristics. For example, metals can be incorporated into polymer matrix to produce the membranes with improved properties [4].

Among the available membrane types, polymers are most widely employed in membrane engineering, not only due to the lower cost compared to other materials, but also due to the unparalleled versatility and processability.

7.2.2 Structure–property relationship in polymers

Some of the key structural parameters such as the type of monomeric unit, type of sequence, polymer molecular weight, crystallinity and glass transition temperature, all have effects on the thermal, chemical and mechanical properties of the polymers, which in turn affect the membrane formation and performance. Understandably, choosing the right type of polymer is important, and the very basics of polymer chemistry is covered in this section.

Polymers, by definition, are long chain molecules composed of repetitive linkage of small molecules known as monomers (Figure 7.2). If a polymer is made up of a single repeating unit, such as in polyethylene, it is called a homopolymer. If a polymer is made up of two or more repeating units, it is a copolymer. Depending on the order in which the monomers are arranged in a polymer, it can be assumed whether it is a block copolymer, random copolymer, or graft copolymer sequence. Different chains can also cross-link, chemically and/or sometimes physically, to form more rigid and chemically stable structures.

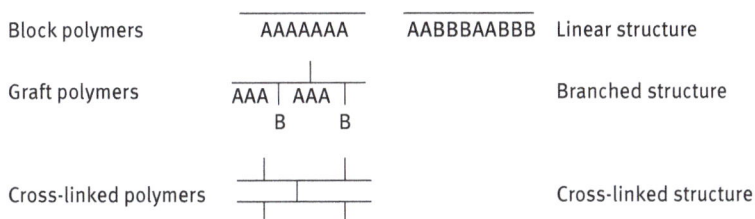

Block polymers	AAAAAAA	AABBBAABBB	Linear structure
Graft polymers	AAA | AAA | B B		Branched structure
Cross-linked polymers			Cross-linked structure

Figure 7.2: Different polymer molecular structures.

The properties of polymers are strongly dependent on their structure. For example, crystalline polymers are more brittle compared to amorphous polymers, and cross-linked polymers have generally better mechanical, thermal and chemical resistance properties.

Isotactic structure gives rise to crystalline structures due to their capability to fold in a more compact form (Figure 7.3). The folding is more restricted in the other two configurations (syndiotactic and atactic), mainly due to the steric hindrance.

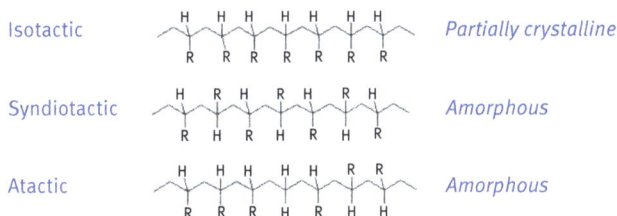

Isotactic		Partially crystalline
Syndiotactic		Amorphous
Atactic		Amorphous

Figure 7.3: Structure–property relationship in polymers.

The property of a polymer also depends significantly on the chemical functionality of the backbone and the side chain, which consequently determines the chain flexibility. Some of the common types of polymers used for membranes are summarized in Table 7.1.

As the polymerization reaction proceeds, the length of each chain grows to different size, resulting in a distribution of molecular weight. It is important to finely control the molecular weight of a polymer as it has significant effect on the inherent

Table 7.1: Effect of side groups of vinyl polymers on glass transition temperature.

Side Group	Polymer	Glass transition temperature
−H	Polyethylene	−120
−CH3	Polypropylene	−15
−C6H5	Polystyrene	100
−C1	Polyvinyl Chloride	87
−CN	Polyacrylonitrile	120
−F	PTFE	126

film forming ability. Two of the commonly employed ways to represent the molecular weight distribution is by the number average (M_n) and weight average (M_w) molecular weight, as defined in eqs (7.1) and (7.2).

$$M_n = \frac{\sum N_i M_i}{\sum N_i} \tag{7.1}$$

$$M_w = \frac{\sum N_i M_i^2}{\sum N_i M_i} \tag{7.2}$$

$$\text{PDI} = \frac{M_w}{M_n} \tag{7.3}$$

where N_i represents the number of molecules with a molecular weight M_i.

The ratio of M_w to M_n is referred to as polydispersity (PDI). If PDI approaches 1, the molecular weight distribution becomes narrow; at PDI of 1, the polymer is considered to be monodisperse.

Within the solid-state phase of polymers, the polymers may be in a rubbery or glassy state, with drastically different properties. The temperature at which transition occurs from the glassy to the rubbery state is defined as the glass transition temperature (T_g). Understanding the concept of T_g and the polymer state is critical as it has a significant effect on the performance of the resulting membranes, particularly for gas separation membranes. For instance, T_g affects the permselectivity and permeability of the membranes under various temperature conditions. Semicrystalline polymers can exhibit both crystallinity and amorphous structure. When the temperature is raised further, for semi-crystalline polymers, there exists a temperature at which the polymer melts (T_m) and loses its crystallinity (Figure 7.4).

The structural factors (molecular weight, chain flexibility, interaction) of a polymer influence its thermal, chemical, and mechanical properties. For instance, a polymer backbone composed of flexible single-bond carbon chains shows higher chain flexibility compared to a polymer backbone with double- or triple-bond carbon chains.

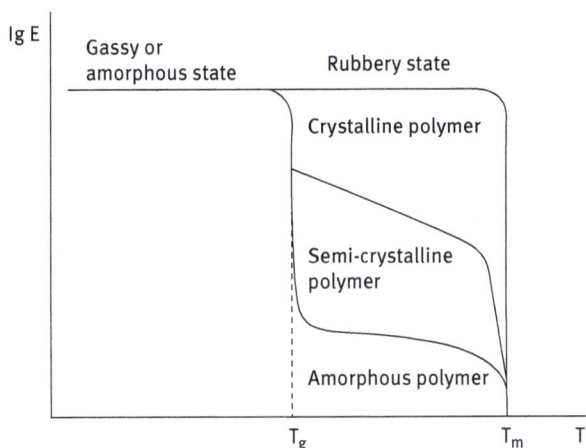

Figure 7.4: Modulus of crystalline, semicrystalline and amorphous polymers as function of temperature.

Similarly, the nature and size of the side chains also induce significant effect on the overall chain flexibility of the polymer.

High chain rigidity, presence of aromatic rings and strong chain interactions all lead to high thermal and chemical stability. For example, incorporating heterocycles and aromatic groups improves the chemical and thermal stability at the expense of chain flexibility. On the other hand, such characteristics lead to poor processability, making it difficult to fabricate it into membranes of desired shape.

7.3 Preparation of porous and nonporous membranes via phase inversion

7.3.1 General overview

There are many methods available to prepare membranes such as phase inversion, sintering technique, track etching, dip coating, etc. (Table 7.2). Understandably, the material influences the choice of the preparation techniques and the application, and hence it is important to tailor the membrane material and the structure to maximize the membrane performance (i.e., permselectivity) to the desired application.

Among the developed methods, phase inversion technique is by far the most versatile preparation method. The phase inversion method can be employed to prepare membranes with different morphology (porous or dense), structures (asymmetric or symmetric) and functions.

Table 7.2: Commonly employed preparation techniques of porous membranes and respective applications.

Membrane type	Membrane material	Pore size (micron)	Preparation process	Application
Symmetric porous structures	Ceramic, metal, polymer, graphite	0.1–20	Powder pressing and sintering	Microfiltration, gas separation
Symmetric porous structures	Polymer of partial crystallinity	0.2–10	Extruding and stretching of films	Microfiltration, battery separator
Symmetric porous structures	Polymer, mica	0.05–5	Irradiation and etching of films	Microfiltration, point-of-use filter
Symmetric porous structures	Polymer, metal, ceramic	0.5–20	Template leaching of films	Microfiltration
Symmetric porous structures	Polymer	0.5–10	Temperature-induced phase inversion	Microfiltration
Asymmetric porous structures	Polymer	<0.01	Diffusion–induced phase inversion	Ultrafiltration
Asymmetric porous structures	Ceramic	<0.01	Composite membrane sol–gel process	Ultrafiltration

In the phase inversion process, a homogeneous polymer solution consisting of a polymer dissolved in an appropriate solvent, in a single phase (liquid), is transformed into a two-phase system. During the phase separation, a polymer-rich phase eventually solidifies into the matrix of the membrane, and the polymer lean phase becomes the membrane pores. In other words, a thermodynamically stable polymer solution is exposed to an environment where the solution is no more stable (i.e., at thermodynamic nonequilibrium), thereby the solution spontaneously phase separates into a polymer-rich phase (membrane matrix) and a polymer-lean phase (membrane pores). It is critical to control the thermodynamic state of the dope solution as well as the kinetics of the phase separation, in order to fully gain control over the membrane morphology and performance. Therefore, the relevant parameters for the preparation of membranes through phase inversion are: polymer, solvent, additives and nonsolvent.

There are mainly four different types of phase inversion method: nonsolvent induced- *or* diffusion-induced phase separation (NIPS/DIPS), thermally-induced phase separation (TIPS), vapor-induced phase separation (VIPS), and evaporation-induced phase separation (EIPS). The only thermodynamic presumption for all procedures is that the system must have a miscibility gap over a defined concentration/temperature range. The principle behind these four methods is essentially the same, but each method has its own unique advantages with different process parameters to control. Among the four aforementioned methods, the NIPS method is the most widely employed technique closely followed by the TIPS method. Each method is discussed in detail.

7.3.2 Nonsolvent induced phase separation

Among the phase inversion method, the NIPS method offers a versatile way to prepare variety of morphologies including porous membranes for MF/UF and nonporous dense membranes for gas separation and pervaporation. It is a well-established technique and it should be noted that in the literature, the NIPS method is referred by several other names such as Loeb-Sourirajan method (after its inventors), immersion precipitation, and DIPS.

There are mainly three components in NIPS (Table 7.3): polymer, solvent, and nonsolvent. Although in most cases NIPS method employs additives as the fourth component, it will not be discussed for the sake of simplicity. In a typical NIPS process, a polymer dope solution in an appropriate solvent is cast onto a glass plate or a nonwoven support (thickness between 100 and 300 μm), and then immersed into a nonsolvent bath to initiate phase inversion.

Table 7.3: Commonly employed polymers and solvents in NIPS process.

Typical polymers	Typical solvents	Nonsolvents
Cellulose acetate, polysulfone, polyethersulfone, polyvinylidene fluoride, polyacrylonitrile, polyimide, polyether imide	DMF, NMP, DMAc, DMSO, THF, acetone, dioxane, GBL	Water (mostly), alcohol

Upon immersion, the solvent diffuses out into the nonsolvent bath and the nonsolvent simultaneously diffuses into the polymer solution. As the solvent–nonsolvent mass exchange proceeds, the composition of the dope solution changes and becomes thermodynamically unstable, at which point the solution begins to phase separate, or demixes, into a polymer-rich phase and a polymer-lean phase. The polymer-rich phase then solidifies into the membrane and the polymer-lean phase becomes the pores. In the NIPS process, two mechanisms need to be delicately controlled: (i) the rate of solvent–nonsolvent mass exchange, and (ii) the rate of phase separation & solidification. These two mechanisms ultimately determine the final membrane morphology and performance. It should be stressed that by simply controlling these parameters, using the same polymer, it is possible to tune the membrane morphology from porous to dense structures.

A typical ternary NIPS phase diagram is illustrated in Figure 7.5. Each corner of the ternary diagram represents the pure component (polymer, solvent, and nonsolvent), and any point within the diagram represents a mixture of the three components. The curve that separates a metastable region from an unstable region in the coexistence region of a binary fluid is referred as "spinodal curve." The spinodal is the limit of stability of a solution denoting the boundary of absolute instability of a

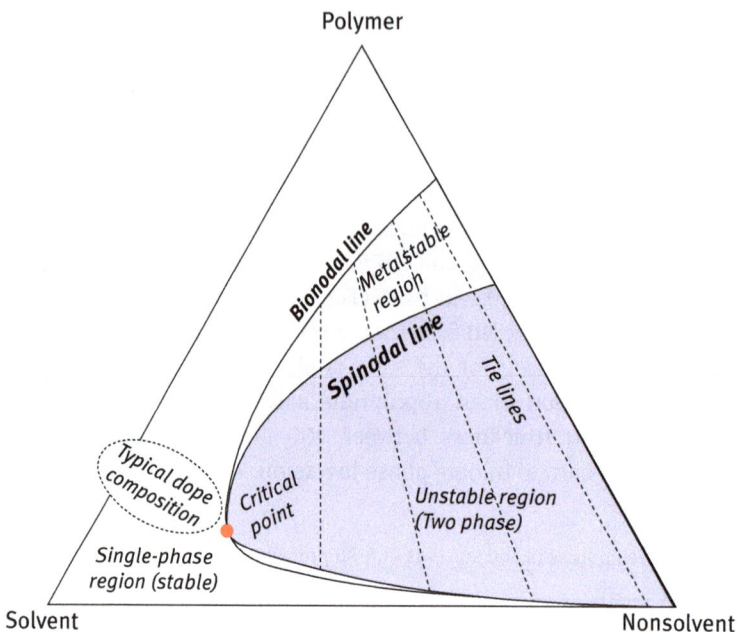

Figure 7.5: Typical NIPS phase diagram.

solution to decomposition into multiple phases. The initial dope composition lies within the homogeneous single-phase region where the solution is thermodynamically stable. As the NIPS proceeds, the composition of the solution changes and enters the binodal and spinodal line, where demixing begins to take place.

By definition, binodal (or coexistence curve) denotes the condition in which two distinct phases may coexist. Equivalently, it is the boundary between the set of conditions in which it is thermodynamically favorable for the system to be fully mixed and the set of conditions in which it is thermodynamically favorable for it to phase separate.

In general, the binodal is defined by the condition in which the chemical potential of all solution components is equal in each phase. The extremum of a binodal curve in temperature coincides with the spinodal curve and is known as a critical point.

Within the binodal region (outside the spinodal region), the solution is metastable and the demixing only proceeds when a stable polymer-poor nucleus forms. On the other hand, within the spinodal region, the solution is thermodynamically unstable and demixing occurs spontaneously into a polymer-rich and a polymer-lean phase. Within the unstable region, the phase separation follows the tie-line (shown in Figure 7.5) and the compositions of the polymer-rich and polymer-lean phase lie at each end of the tie-line (binodal line).

Importantly, the *path* and the *rate* at which the solution follows along the phase diagram has significant influence on the final membrane morphology. Particularly, when passing the binodal line, it is important to cross above the critical point in order to initiate polymer-rich phase that forms the matrix of the membranes. If the binodal line is crossed below the critical point, polymer-rich phase nucleates and grows into individual droplets, resulting in a membrane with very low or no mechanical integrity. The shape and size of the binodal and spinodal lines depend on solution thermodynamics such as the polymer–solvent–nonsolvent compatibility.

There are mainly two types of phase separation, or demixing, processes: instantaneous demixing and delayed demixing (Figure 7.6). Generally, instantaneous demixing yields a highly porous substructure (with macrovoids) and a finely porous skin layer, while delayed demixing route induces the formation of membranes with a porous (often closed-cell, macrovoid free) substructure and a dense, relatively thick skin layer. Hence, one needs to promote an appropriate demixing type depending on the desired applications. Which specific process dominates is mainly determined by the solvent/nonsolvent affinity and the solvent concentration in the coagulation bath.

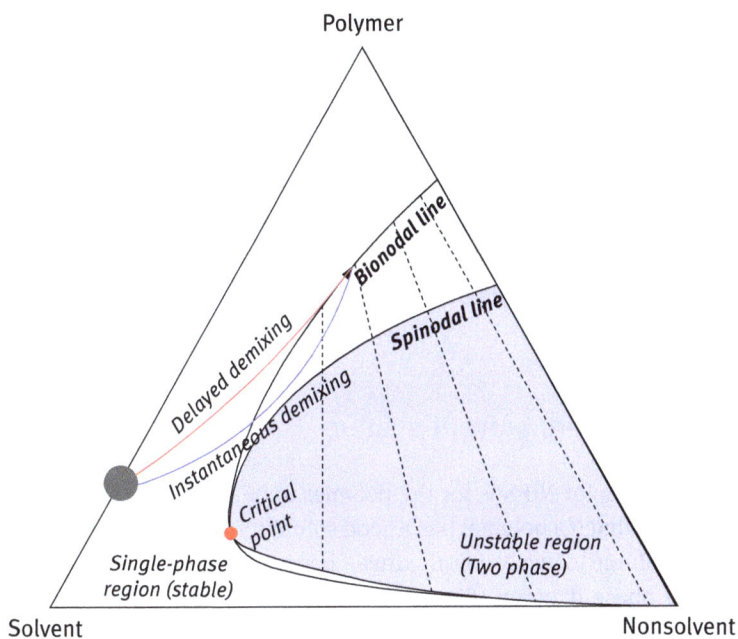

Figure 7.6: Two different demixing routes: instantaneous demixing and delayed demixing.

In particular, in the case of good solvent/nonsolvent miscibility, the latter can easily penetrate into the casting solution and create a porous structure. In this light, the good solubility of polymers in solvents having different polarities and miscibilities with nonsolvents thus allows one to obtain various membrane morphologies.

In simple terms, if the membrane forms (solidification) instantly upon immersion into the nonsolvent bath, it is referred to as instantaneous demixing. On the other hand, if it takes some time until the membrane forms, it is referred to as delayed demixing. The general rule of thumb is that above 10–20 s, it is considered to be delayed demixing. The demixing process can be determined visually, or using light transmission apparatus, by measuring the time at which the solution becomes opaque (nontransparent).

The type of demixing and respective rates can be controlled using several parameters. In large category, it can be split into thermodynamic parameters and kinetic parameters, and they are dependent on each another.

7.3.3 Determination of the polymer/solvent/nonsolvent phase diagram by cloud-point measurements

In order to determine the composition or temperature at which the solution is no longer thermodynamically stable, turbidity or cloud points must be determined. Ternary phase diagrams can be determined by visual observation of the cloud points. Cloud points are defined as the moment when the solution changes from clear to turbid. They can be determined by titration: pure nonsolvent (typically, water) or a solvent/nonsolvent solution is added slowly to a stirred solution of the polymer and solvent. The turbidity point can determined visually. Upon addition of a nonsolvent, instantaneous precipitation of the polymer can be observed locally. Stirring of the mixture should be continued until the precipitated polymer redissolves or until the solution becomes homogeneously opaque. In the latter case, the cloud point is reached.

7.3.4 Polymer/solvent/nonsolvent system

The first important criterion for NIPS is for the polymer to be soluble in the system solvent. One can imagine that if a polymer has a good solubility in a certain solvent, it would have a wide soluble concentration range, hence smaller binodal region (unstable area) in the phase diagram (Figure 7.7). In addition, the system solvent needs to be miscible with the nonsolvent (typically water) for the NIPS to proceed. Naturally, the difference in affinity of the solvent toward the polymer and the nonsolvent determines the demixing process. For instance, if the solvent has a poor miscibility or affinity toward the nonsolvent, a delayed demixing would result. On the other hand, if the solvent has a much higher affinity toward the nonsolvent relative to that of polymer, an instantaneous demixing would occur.

The polymer-solvent compatibility can be semi-qualitatively estimated using the solubility parameter theory [6]. However, solubility behavior can be better

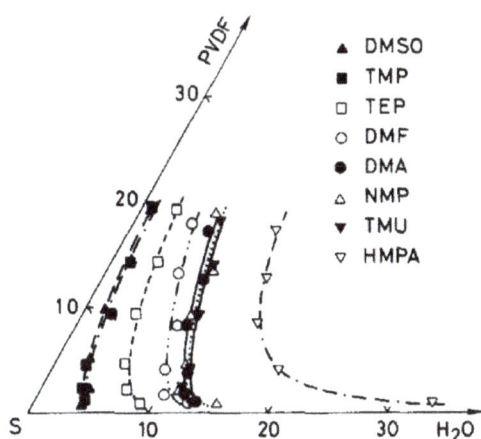

Figure 7.7: Ternary phase diagram for PVDF-solvent-water system at 20°C (From [5]). The shape and size of the binodal line change with the system solvent. Reprinted with Permission.

described by changes in the free enthalpy of mixing than via the solubility parameter approach [7].

What is the influence of the choice of solvent/nonsolvent system on membrane morphology? As described in the previous sections, the two different mechanisms for membrane formation lead to different structures, and the difference between the two mechanisms being characterized by the instant at which the onset of liquid-liquid demixing occurs.

From the observations depicted in Figure 7.7, it is to be expected that PVDF (polyvinylidenedifluoride) with HMPA or TMU as the solvent and water as the non-solvent result in a dense membrane (delayed demixing). When DMSO is used as solvent and water as nonsolvent, a porous type of membrane will be obtained (instantaneous demixing).

Another example is reported in Figure 7.8: open pore structures, such as observed for DMF, are formed by nucleation and growth of the polymer lean phase in the metastable region (between the binodal and the spinodal region). Given the position of the binodal demixing curve in the phase diagram (Figure 7.9), the nodular structure of DMA (dimethylacetamide) cannot be explained by the nucleation of the polymer lean phase. A possible explanation for the formation of a nodular structure could be that spinodal demixing occurs. Based on the phase diagrams, the demixing should be more instantaneous in the case of DMF than in the case of DMA. Instantaneous demixing is often closely related to the formation of macrovoids. The fact that with the present systems the tendency to form macrovoids is much stronger in the case of DMA than in the case of DMF is a clear indication that, besides thermodynamics, kinetic factors (e.g., rate of solvent-nonsolvent diffusion) also play an important role in the morphology development.

DMF   **DMA**   **THF**

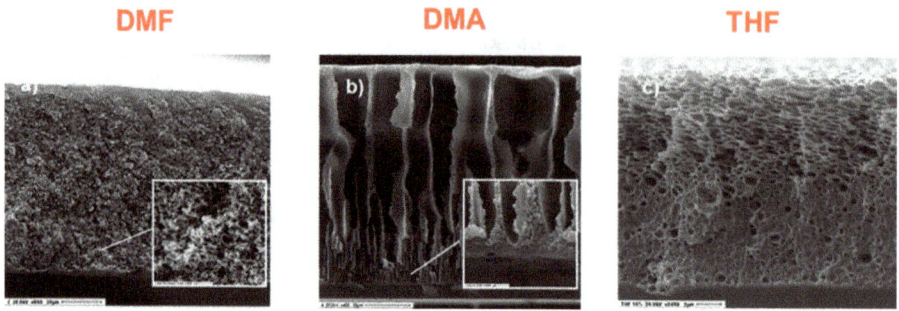

Figure 7.8: Effect of solvent in membrane morphology (PEEK-WC (poly-(etheretherketone) with Cardo) as polymer, water as nonsolvent). From Buonomenna et al. [8]. Reprinted with Permission.

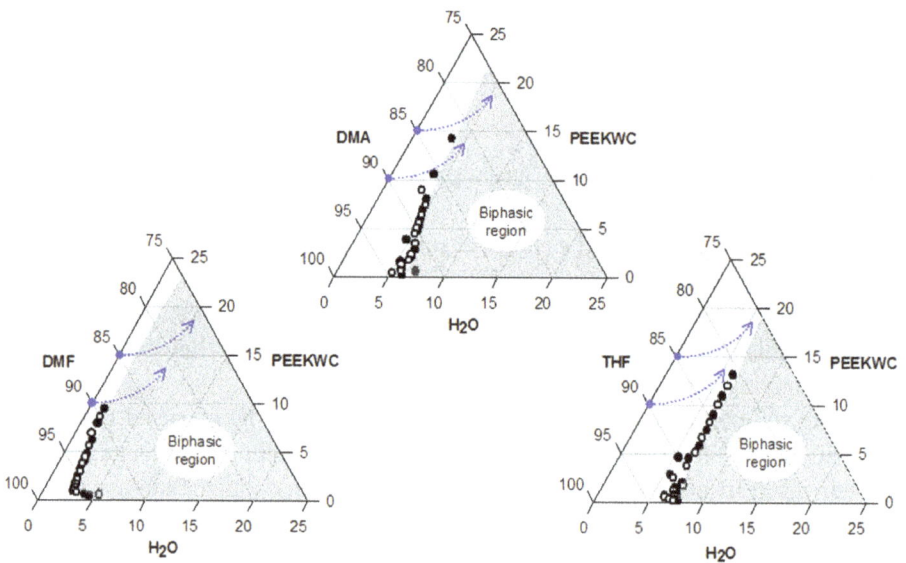

Figure 7.9: Different Solvents: the PEEK-WC/(DMF, DMA or THF)/H_2O phase diagram.

7.3.5 Dope Composition (polymer concentration and nonsolvent content)

Once the type of polymer, solvent, and nonsolvent are selected, the dope composition needs to be optimized to meet the requirements of the desired application: flux, selectivity, and mechanical properties.

First of all, increasing the initial polymer concentration in the casting solution leads to combination of effects. Changing the polymer concentration generally does not affect the nature of demixing (instantaneous or delayed demixing). Increasing the initial polymer concentration in the casting solution leads to a much higher polymer concentration at the interface. This implies that a lower porosity is obtained. In the case of porous membranes (e.g., for MF & UF membranes), with a higher initial polymer concentration in the casting solution, a higher polymer concentration at the film interface is obtained that results in a less porous top layer and a lower flux. For dense membranes, the thickness of the dense layer increases with increasing polymer concentration, again leading to lower flux.

Second, a third component can be included in the dope solution to add extra dimension in membrane morphology control. For instance, if a nonsolvent (i.e., water) is added to the dope solution, one can expect the solution thermodynamic stability to decrease, bringing the solution closer to the binodal line (Figure 7.10).

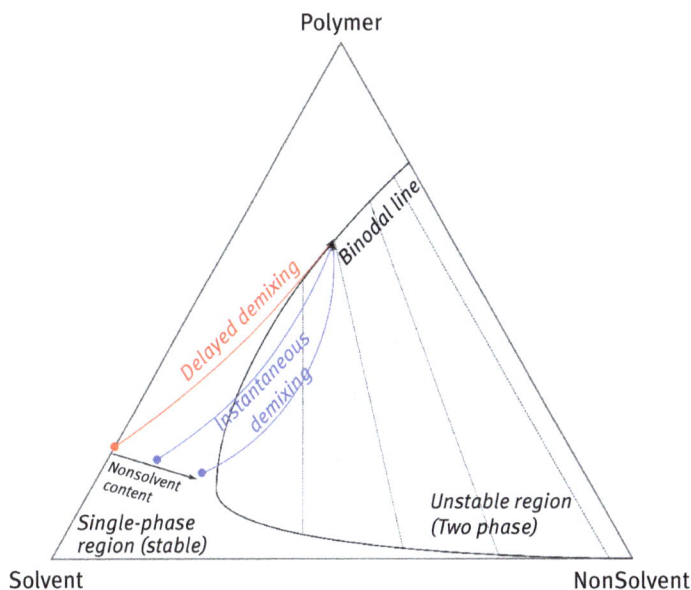

Figure 7.10: Effect of nonsolvent content in the dope solution in phase demixing.

Hence, by simply adding a nonsolvent in the dope solution, the type of demixing can be changed from delayed demixing to instantaneous demixing. In practice, compounds other than the nonsolvent is used and the nature of the third compound, or additive, has significant effect on the membrane morphology.

7.3.6 Coagulation bath composition and type

Controlling the coagulation bath composition is another way to manipulate the phase demixing phenomenon and, therefore, the type of membrane structure formed. The addition of solvent to the coagulation bath results in a delayed onset of liquid–liquid demixing. Indeed, it is even possible to change from porous to nonporous membranes by adding a solvent to the coagulation bath. By including the solvent in the coagulation bath, the rate of mutual diffusion slows, inducing delayed demixing (Figure 7.11). In addition, adding the solvent in the coagulation bath brings another interesting effect, where the polymer concentration at the surface in contact with the bath decreases. Hence, although the solvent in the bath induces delayed demixing, it simultaneously decreases the surface polymer concentration to promote a porous surface.

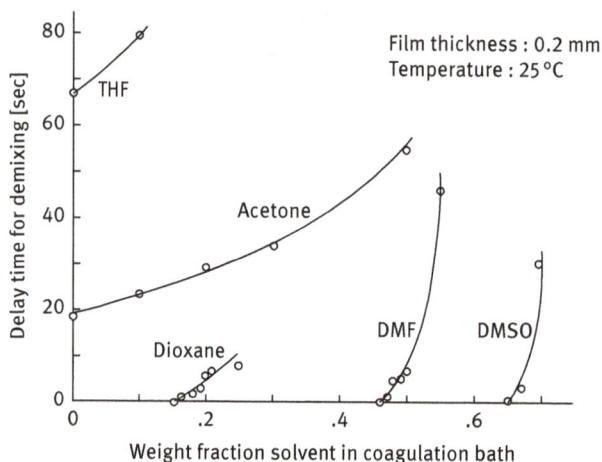

Figure 7.11: Effect of solvent and solvent concentration in water bath on phase separation delay. From Reuvers [9] (figure freely accessible online at https://research.utwente.nl/en/publications/membrane-formation-diffusion-induced-demixing-processes-in-ternar. Last access on 07/May/2018). (Cellulose acetate as polymer, water as nonsolvent.)

The higher alcohols give membranes with a much thicker dense top layer (Figure 7.12). In general, the top layer thickness of the membrane increases with increasing molar volume of the external nonsolvent species, due to their lower diffusion rate, which leads to delayed liquid–liquid phase demixing. In the present system, the

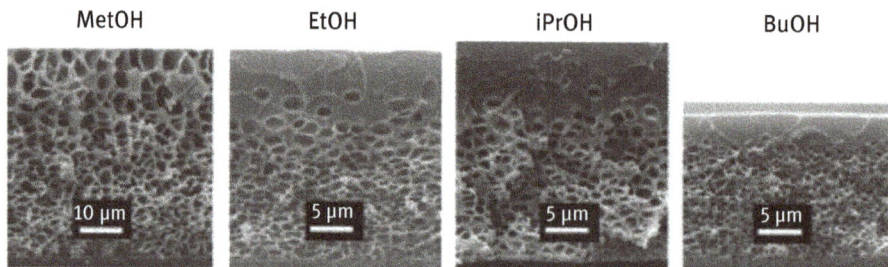

Figure 7.12: Effect of nonsolvent bath in membrane morphology (PEEK-WC as polymer, chloroform as solvent). From Jansen et al. [10]. Reprinted with permission.

thermodynamic and kinetic factors are synergistic. While the small dimensions of methanol favor its rapid diffusion into the cast film, it is also the most efficient nonsolvent. Being a strong nonsolvent, methanol causes phase separation into a concentrated polymer-rich phase and a relatively high volume of polymer-lean phase, responsible for the higher void fraction in the final membrane.

7.3.7 System temperature

For further optimization of the membrane performance, the kinetics and thermo-dynamics of the phase inversion process can be varied by the external conditions (Figure 7.13) One can expect that higher temperature enhances the rate of solvent/nonsolvent exchange (kinetic factor), but simultaneously increases the thermo-dynamic stability of the solution (thermodynamic factor). Hence, the effect of coagulation temperature varies for polymer–solvent–nonsolvent systems and is usually determined experimentally.

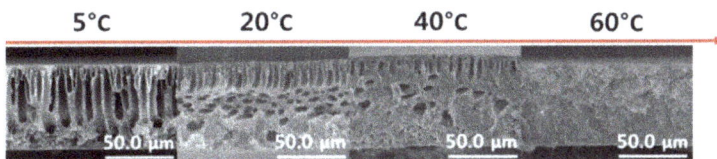

Figure 7.13: Effect of coagulation bath temperature on membrane morphology (PVDF/PolarClean system). From Jung [11]. Reprinted with permission.

The effect of the various parameters on membrane structure is summarized in Table 7.4.

Table 7.4: Effect of NIPS parameters.

Parameter	Porous membrane	Nonporous membrane
Polymer–solvent–nonsolvent	– Low polymer content – High solvent–nonsolvent affinity and mutual diffusion	– High polymer content – Low solvent–nonsolvent affinity and slow mutual diffusion (delayed demixing)
Dope composition	– Addition of nonsolvent to the polymer solution	– Addition of volatile co-solvent
Coagulation bath	– Addition of solvent to lower the surface polymer concentration – Higher temperature	– Addition of third compound to reduce the solvent–nonsolvent mutual diffusion

7.3.8 Macrovoid formation

Asymmetric membranes consist of a thin top layer supported by a porous sublayer and quite often macrovoids can be observed in the porous sublayer. The presence of macrovoids is not generally favorable, because they may lead to a weak spot in the membrane which is to be avoided especially when high pressures are applied, such as in gas separation.

Macrovoids form through two phases: initiation and propagation. As described above, the macrovoid formation is due to the combination of many factors but is mostly initiated during the liquid–liquid demixing process just beneath the top layer. The nuclei of the polymer-poor phase are also those responsible for macrovoid formation. Once initiated, the nucleus grows downward due to the diffusional flow of solvent from the surrounding polymer solution, forming a semi-stable liquid–liquid interface, closely chased by the solidification front. A nucleus can only grow if a stable composition is induced in front of it by diffusion. Growth will cease if a new stable nucleus is formed in front of the first formed nucleus, or if the solidification front takes over.

The membranes that exhibit macrovoids are usually the ones fabricated via the instantaneous demixing route. Therefore, the parameters which favor the formation of porous membranes may also favor the formation of macrovoids. The parameters that influence the onset of liquid–liquid demixing also determine the occurrence of macrovoids in systems that show instantaneous demixing. The main parameter involved is the choice of solvent/nonsolvent pair. Other parameters (such as the addition either of solvent to the coagulation bath or of nonsolvent to the casting solution, and the polymer concentration) can be varied to prevent macrovoid formation. Another method to prevent macrovoid formation is the addition of additives to the casting solution. Prevention of macrovoid formation in microfiltration/ultrafiltration membranes by encouraging delayed onset of liquid–liquid demixing also results in the densification of the top layer, which is unwanted.

7.4 Thermally-induced phase separation

Thermally-induced phase separation technique was first introduced and actively researched in the 1980s–1990s to fabricate microporous membranes, but it has not gained much attention since NIPS was deemed as a more convenient and versatile method to prepare polymeric membranes. Recently, however, with the advent of membrane contactors and membrane bioreactors, TIPS research is *re*-gaining its momentum due to many unique advantages such as process simplicity, high productivity, low tendency to form defects, high porosity, low tortuosity, and the ability to form interesting microstructures with narrow pore size distribution. In addition, the possibility to control the polymer polymorphism using solvent and process parameters is being highlighted as an irreplaceable feature.

The basic procedure for TIPS method (Figure 7.14) is composed of the following steps:
1. Dissolve a polymer of interest in a high-boiling, low MW solvent at an elevated temperature, typically near or higher than the melting point of the polymer to form a homogeneous melt-blend.
2. Cast the dope solution into the desired shape, for example, flat-sheet or hollow fiber.
3. Cool the cast solution in a controlled manner to induce phase separation and crystallization of the polymer.
4. Extract the diluent, often via solvent extraction, to yield a membrane.

Figure 7.14: Basic procedure of TIPS.

As described, a membrane is formed from a homogeneous dope solution by removing the thermal energy to induce phase separation. Hence, the phase inversion process is a delicate balance between polymer–solvent interaction, cooling rate, cooling media, and thermal gradient. As pointed out by Lloyd et al. [12], one of the distinct advantages of TIPS method is its ability to fabricate membranes from semi-crystalline polymers that are not usually soluble in solvents at ambient temperatures. In addition, TIPS process is usually a binary system, as compared to the ternary NIPS system, rendering the TIPS process inherently simpler than the NIPS with fewer variables to be controlled. The general concept of TIPS is very similar to NIPS. The phase diagram is typically drawn as a function of temperature (Figure 7.15) Compared to the NIPS phase diagram, there exists a solid–liquid phase boundary where polymer crystallizes straight out of the solution (solid–liquid phase separation), which results in spherulitic morphology.

The solvent must be thermally stable to be applicable for TIPS. Most of the employed TIPS solvents were phthalate-based chemicals that are environmentally unsustainable. However, recent developments toward environmental friendly alternatives [13, 14] give additional advantage for TIPS method.

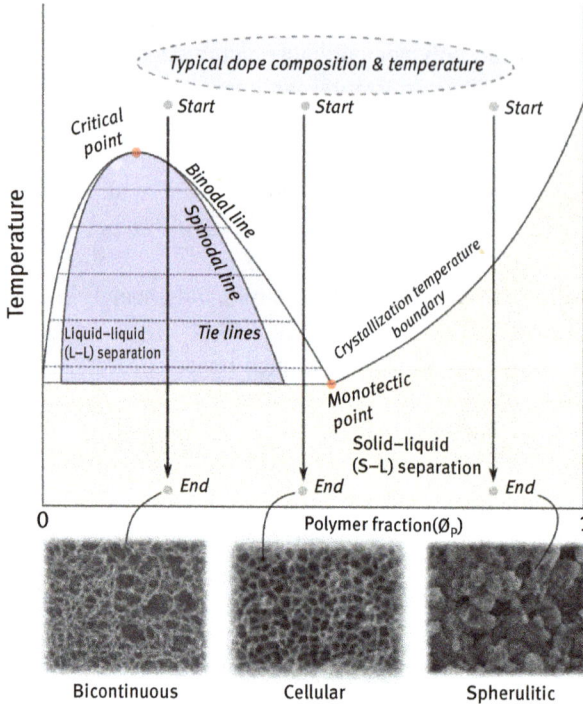

Figure 7.15: Typical TIPS phase diagram and corresponding morphology (adapted from Kim et al. [29]).

7.5 Vapor-induced phase separation and evaporation-induced phase separation

A simple technique for preparing phase inversion membranes is precipitation by solvent evaporation (Figure 7.16). In this method, a polymer is dissolved in a solvent and the polymer solution is cast on a suitable support. The solvent is allowed to evaporate in an inert atmosphere, in order to exclude water vapour, allowing a dense homogeneous membrane to be obtained.

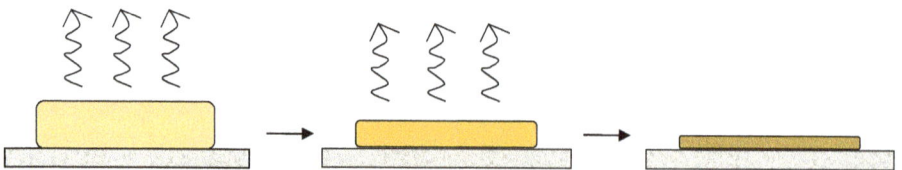

Figure 7.16: EIPS schematic.

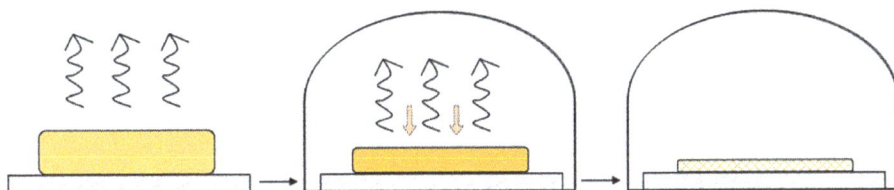

Figure 7.17: Schematic diagram of VIPS.

VIPS (Figure 7.17) occurs when the cast film (polymer and solvent) is placed in a vapor atmosphere where the vapour phase consists of a nonsolvent saturated with the same solvent. Membrane formation occurs because of the diffusion of nonsolvent into the cast film. This leads to a porous membrane without the toplayer (Figure 7.18).

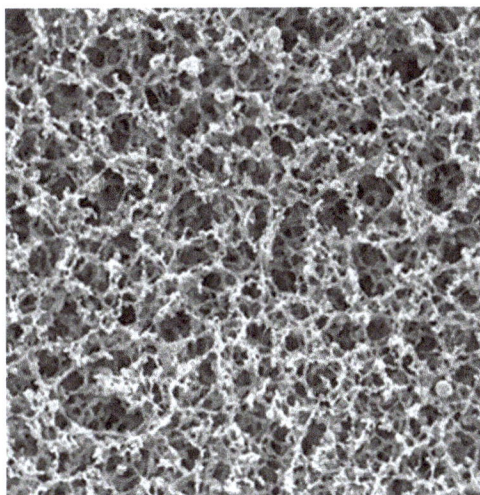

Figure 7.18: Cellulose nitrate membrane obtained by water vapor-induced phase separation.

7.6 Stretching method

Stretching method fabricates membranes by first extruding a thin film (or fiber) made from a semi-crystalline polymer material, and subsequently stretching the film perpendicular to the direction of extraction (Figure 7.19) so that the crystalline regions are located parallel to the extrusion direction. When a mechanical stress is applied, small ruptures occur and a porous structure is obtained (pore sizes from 0.1 to 3 µm – porosity up to 90%). An intermediate and final annealing step is generally required to control the polymer crystallinity and size of the lamellar nodes [15]. The key advantage of this method is that it is a solventless process, and no toxic solvent waste is generated. Polymers that do not melt in common solvents are applied for this method, mainly hydrophobic polymers such as polypropylene (PP), polyethylene (PE), and polytetrafluoroethylene (PTFE) (Figure 7.20). The stretched films have been applied to waterproof functional clothings (ePTFE), performance fabrics, and membrane distillation.

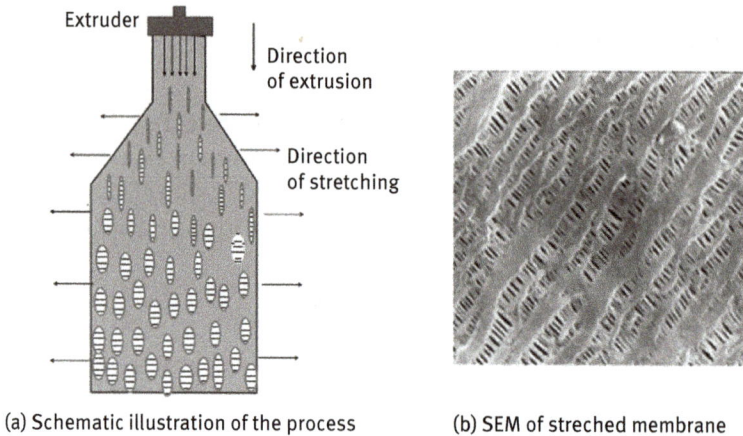

(a) Schematic illustration of the process   (b) SEM of streched membrane

Figure 7.19: Partially crystalline polymers can be stretched in parallel or transverse direction of extrusion.

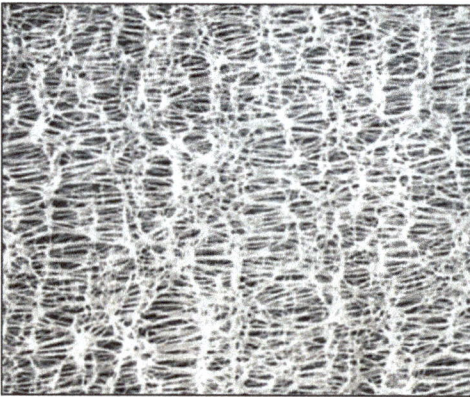

Figure 7.20: PTFE membrane obtained by stretching. Films are extruded at temperatures close to the Tm.

7.7 Preparation of composite membranes

The definition of composite membrane is a membrane composed of several different elements or layers. More generally, composite membranes refer to inorganic-embedded membranes or membranes reinforced with another material (polymeric and/or inorganic). Some of the well-known ones include zeolite-embedded membranes [16], metal–organic frameworks (MOFs) embedded membranes, and more recently, carbon nanotube (CNT) and graphene oxide (GO) embedded membranes.

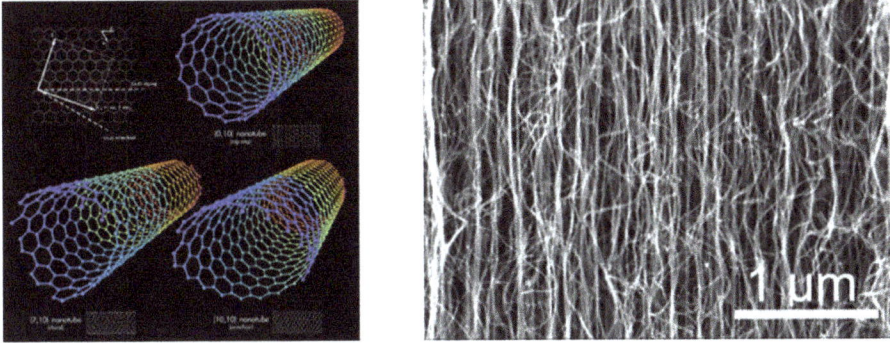

Figure 7.21: Schematic illustration of CNT and SEM images of CNT.

CNTs are nano-scale one-dimensional cylinders of rolled up graphene with inner diameters as small as 0.7 nm (Figure 7.21). CNTs are atomically smooth, molecule-sized channels in which the water behaves very differently when it is confined inside the tubes by adopting a unique structure that can be converted into a one-dimensional array of water molecules under certain conditions. It has been predicted that the water flow through CNTs can be several orders of magnitude higher than conventional channels.

The ideal structure, as shown in Figure 7.22, would be obtained when the space between the CNTs are filled with polymer and the closed ends of the CNTs are etched open.

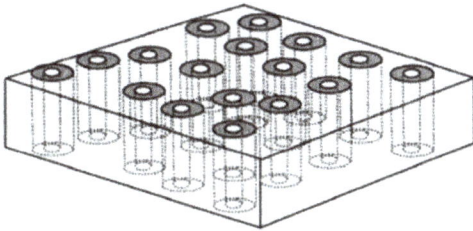

Figure 7.22: Targeted CNT-embedded membrane structure.

7.8 Preparation of thin film composite membranes

Thin film composite (TFC) membrane is an important class of membrane where a thin selective layer sits on top of a porous support (Figures 7.23 and 7.24). TFC

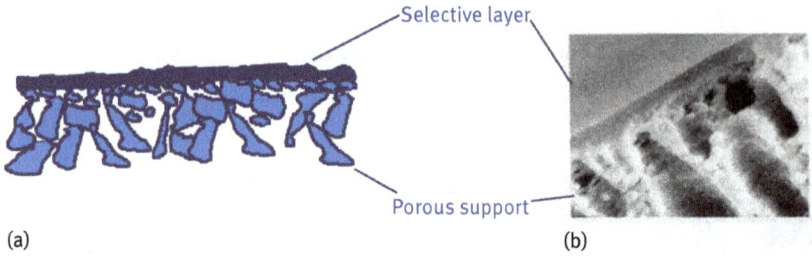

(a)

(b)

Figure 7.23: Composite membrane structure.

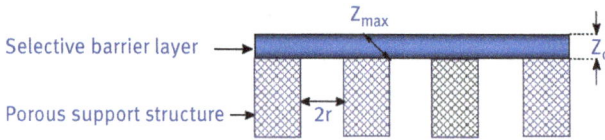

Z_{max}

Selective barrier layer →

Z_o

Porous support structure →

$2r$

Figure 7.24: Composite membrane schematic.

membranes are prepared in a two-step process. First, a porous support is fabricated, typically via NIPS method. Then, a thin selective layer is applied onto the support layer via coating or interfacial polymerization (IP) method.

Today, the most important technique for preparing composite membranes is via **IP** (Figure 7.25) using trimesoyl chloride (TMC) and m-phenylenediamine (MPD) chemistry [17]. Compared to asymmetric membranes, TFC membranes offer

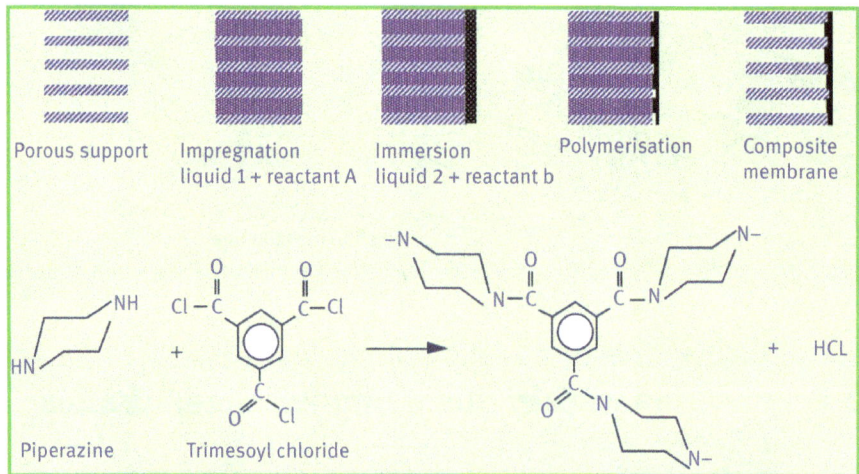

Figure 7.25: Formation of a composite membrane by IP of piperazine with TMC.

significantly higher flux and performance. Particularly, TFC membranes using TMC-MPD chemistry exhibit excellent water flux and salt rejection, and they currently dominate the reverse osmosis (RO) and water nanofiltration (NF) market.

The key advantage of TFC membrane is that both the selective and the support layer can be fine-tuned independently, allowing more versatile chemistry and materials to be applied. A majority of research is focused on improving the selective layer; however, the performance of a composite membrane is also significantly affected by the properties of the porous support.

7.9 Preparation of ceramic membranes

Ceramic membranes exhibit unparalleled chemical stability and resistance to swelling. However, their high manufacturing cost has been a major hurdle to compete against the polymeric membranes. Nevertheless, the fast growth of the membrane market combined with technology innovations in ceramic membranes have changed the market paradigm, and ceramic membrane is expected to take significant market portions in the future [18]. Ceramic membranes to be used in micro-, ultra- and nanofiltration as well as in pervaporation and gas separation are also prepared as multiple layer composite structures. Common materials include alumina, silica, titania, and zirconia membranes but several other materials also exist [19].

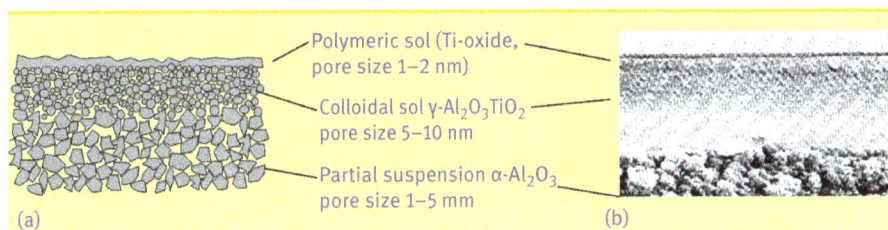

Polymeric sol (Ti-oxide, pore size 1–2 nm)
Colloidal sol γ-Al$_2$O$_3$TiO$_2$ pore size 5–10 nm
Partial suspension α-Al$_2$O$_3$ pore size 1–5 mm
(a)
(b)

Figure 7.26: (a) Schematic drawing illustrating the structure of a three-layer inorganic composite membrane. (b) Scanning electron micrograph of a three-layer ceramic membrane.

Ceramic membranes are generally composed of two or more layers (Figure 7.26). Taking alumina ceramic membrane as an example, it consists of the followiing:

- The first support layer is generally prepared from a suspension of ceramic powder (e.g., Al$_2$O$_3$) using an organic polymer as a binder and to increase the viscosity of the suspension. The suspension can be processed into the desired form (disk or tube type), which is dried and converted into the final membrane by sintering at ca. 1500°C. Such membranes give microfiltration range pore size.
- A second layer prepared by a so-called sol–gel process is deposited on the support layer by dip coating or filtration. The material used for the preparation

of the second layer is usually Al_2O_3 or TiO_2. Sol-coated membranes typically give ultrafiltration range pore size or lower.

– For the separation of lower molecular weight components in nanofiltration and pervaporation, a third layer (or more) is deposited.

7.10 Suspension coating and the sol–gel process

A suspension is a dispersion of solid particles in a liquid, a sol is the suspension of colloidal particles or polymers. The main difference between a suspension and sol is the size of the dispersed particles. Suspensions are prepared from a fine powder of aluminum oxide with an average particle diameter of 5–10 μm. The dispersion is mixed with an organic polymer such as polyvinyl alcohol or a cellulose derivative and poured in a mold or extruded in the desired shape as a flat sheet, tube or capillary, dried and sintered at 1500–1800°C. The porous inorganic membrane preparation scheme based on slip casting of a ceramic powder and sintering at 1500–1800 °C can be found in Figure 7.27.

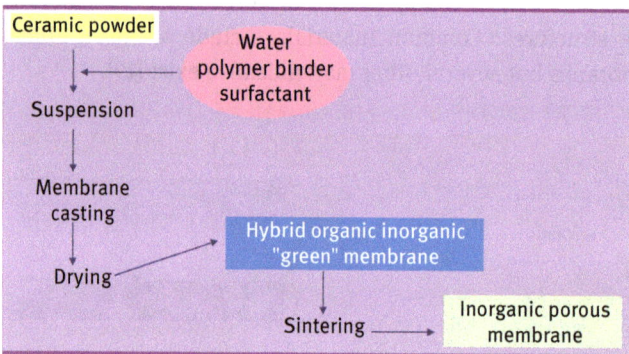

Figure 7.27: Preparation steps of inorganic composite membranes.

One surface of the support structure is then coated again with a suspension of finer particles, dried and sintered again. Sometimes, more than one coating is applied to obtain the desired pore size at the surface of the membrane. The process of preparing a multi-layer membrane by suspensions of different particle sizes is referred to as slip coating. Membrane with pore sizes less than 20 nm can be made with the slip-coating process. To obtain membranes with smaller pores, a sol–gel process is applied and a colloidal or polymeric gel is used which is prepared by controlled hydrolysis of a metal alkoxide to a hydroxide. As shown in Figure 7.28, the process can follow two paths: colloidal route and polymeric route.

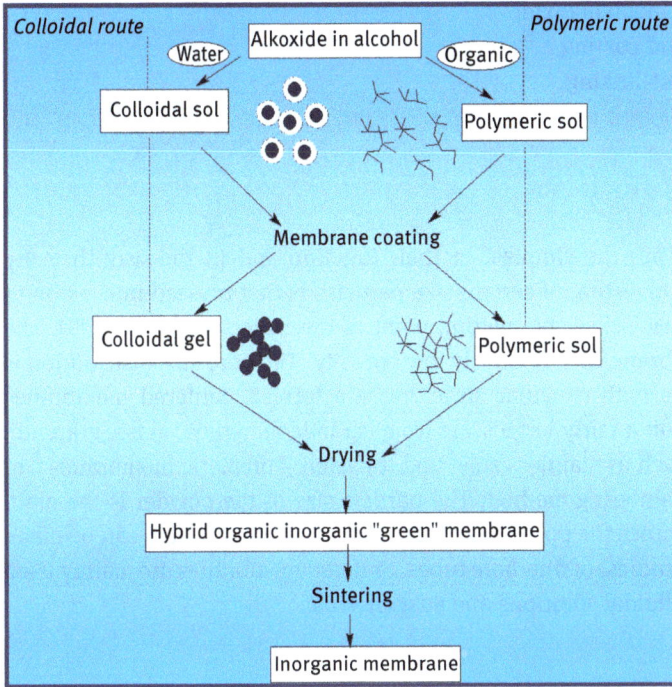

Figure 7.28: Sol–gel process.

In the colloidal process path, a metal oxide such as aluminum oxide is dissolved in alcohol and then hydrolyzed and precipitated by addition of excess water at elevated temperature to form a stable colloidal solution. The solution is cooled to form a sol, which is then coated on an appropriate support and sintered at ca. 800 °C. The average pore diameter obtained in the colloidal sol–gel process is in the range of 5–20 nm.

The overall process consists of three steps:
- Metal alkoxide formation
- Precipitation
- Coating and sintering

In the polymeric sol–gel process, the metal alkoxide is only partially hydrolyzed in an alcohol solution by a controlled addition of of water. The hydroxyl groups of the alkoxide react with each other and form an inorganic/organic polymer, which forms a clear sol that is coated on the support structure. It is dried and sintered at 500–800°C and further cross-linked to form a porous structure. For the polymer sol–gel process, often a titanium alkoxide is used which is converted into a cross-linked inorganic polymer in three steps:

- Hydrolysis of the metal alkoxide
- Polymerization and coating
- Sintering and cross-linking.

7.11 Sintering method

Sintered membranes are the simplest in their function and in the way they are prepared. A powder consisting of certain size particles is first pressed into a film or plate, and sintered just below the melting point of the material (Figure 7.29). The process yields a membrane with relatively low porosity. The pore size distribution is very broad and pores with irregular structures are formed. Sintered membranes (Table 7.5) are made on a fairly large scale from ceramic materials, glass, graphite, and metal powders such as stainless steel and tungsten. Polymeric membranes can also be prepared via sintering method. The particle size of the powder is the main parameter that determines the pore sizes of the final membrane, which can be made in the form of discs, candles, or fine-bore tubes. Sintered membranes are mainly used for the filtration of colloidal solutions and suspensions.

Figure 7.29: Schematic diagram of the sintering process. Typical pore sizes between 0.1 and 10 μm. Porosity 10–20% with polymer; 80% with metals.

Table 7.5: Materials used for the sintering method.

Powders of polymers	Polyethylene, PTFE, polypropylene
Powders of metals	Stanless steel, tungsten
Powders of ceramics	Aluminium oxide, zirconium oxide
Powders of graphite	Carbon
Powders of glass	Silicalite

7.12 Zeolite membranes

Zeolite is also an actively researched inorganic porous material (aluminosilicate) that allows some molecules to pass through and causes others to be either excluded or broken down. It is, to some extent, the inorganic equivalent of organic enzymes. The word "zeolite" comes from Greek and means "boiling stone," due to the fact that natural zeolites visibly lose water when heated.

In nature, zeolites are formed where volcanic rock of specific chemical composition is immersed in water so as to leak away some of the components. Lab-made zeolites in part mimicked natural zeolites, but many new ones have been developed, which are targeted toward specific purposes.

Zeolites are an interesting class of microporous materials with an ordered three-dimensional matrix. The basic building units (BBUs) are tetrahedral, where the central atom is typically Si or Al and the peripheral atoms are oxygen. The BBUs can be combined in larger composite building units (CBUs).

A zeolitic framework defines a regular system of voids and channels of discrete size that is accessible through pores of well-defined molecular dimensions.

Each confirmed zeolite framework type has a unique three-letter code (e.g., FAU for the faujasite framework, Figure 7.30) which is assigned by the Structure Commission of the International Zeolite Association (IZA).

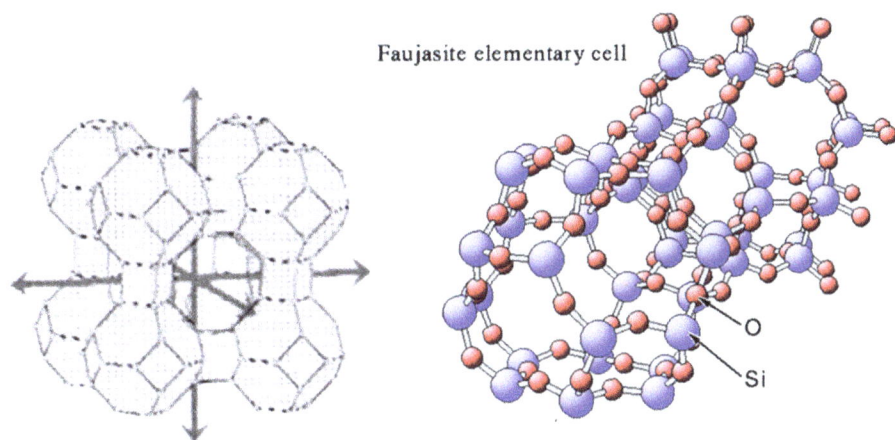

Figure 7.30: Zeolite FAU framework.

Techniques involved in the preparation of a zeolite membrane can be divided into four categories:
- Pretreatment of the supports, which can involve thermal and plastic treatment, chemical treatment, and mechanical treatment.
- Synthetic methods, including *in situ* synthesis on supports, seeding, nanosols (nanosized crystals), synthesis at interface between two phases, selective etching, seed-film, electro-trapping, pressurized sol–gel coating, binding, electrical orientation, microwave, and isomorphous substitution of framework atoms.

–   Impregnation of the support, which can be stable or temporary, vertical or horizontal.
–   Elimination of small defects, which can be achieved by selective cooking or by reaction with silicon alkoxide or other silylation agents.

Recently, there have been many interesting breakthroughs in fabricating zeolite membranes [20], and it is now possible to fabricate zeolite membranes quite reproducibly. And perhaps, scaling up high performance defect-free zeolite membranes (other than LTA zeolite) may be possible in near future.

7.13 Preparation of perovskite membranes

Perovskites are inorganic complex oxides with the empirical formula ABO3, where A-site cations are typically rare earth metals, while B-site cations are occupied by transition metals with mixed valence states. Some perovskites exhibit high mixed electronic and oxygen ionic conductivities, and for this reason are being widely studied for applications in solid oxide fuel cells (SOFCs), oxygen sensors and pumps, batteries, and oxygen-permeable membranes. Besides their oxygen semipermeability, mixed-conductive perovskite-type oxides also show catalytic activity for oxidation reactions [21]. Their catalytic properties are closely related to the nature of B-site cation. For example, in a study performed by Zeng et al. [22] on the catalytic properties of two perovskite oxides, La0.2Sr0.8CoO3 (LSC) and SrCo0.8Fe0.2O3 (SCF), for oxidative coupling of methane (OCM), LSC showed much better C2 selectivity than SCF due to the fact that the substitution of Co with Fe promoted the complete oxidation reactions [23]. This effect of B-site cation on the oxide catalytic properties was also found in propane oxidation [24]. Although A-site cations are in general catalytically inactive, they strongly affect the oxygen nonstoichiometry of the oxide and the mobility of oxygen ion in the bulk phase [25]. This, in turn, affects the ability of solid phase oxygen species to participate in the oxidation reactions. In general, high selectivity and yield for partial oxidation reactions could be achieved over those perovskite oxides with intermediate oxygen nonstoichiometry and oxygen ion mobility [21]. The type of the conductivity of the oxide is also found to play a certain role on its catalytic properties. For example, p-type or ion-conducting oxides in general are more selective for methane coupling reactions while n-type conductors are more selective for COx generating reactions [23, 26, 27]. Zeng et al. [22] evaluated four synthesis methods, which are hydrothermal synthesis, coprecipitation and calcination, spray-pyrolysis, and conventional ball milling and calcination, for the preparation of $La_{0.8}Sr_{0.2}Co_{0.6}Fe_{0.4}O_3$ (abbreviated as LSCF) powders.

In the coprecipitation method, the corresponding La, Sr, Co and Fe nitrates were dissolved into water and coprecipitated with potassium hydroxide. The coprecipitated gel was washed several times to remove the potassium salt impurities, and then dried and calcined at 800°C to form crystalline powder. Polyethylene glycol was added to the powder as a binder and green disk compacts were prepared by uniaxial pressing, followed by cold isostatic pressing at 50,000 psi. The final LSCF disks were sintered at 1250–1300 °C for 1–2 h.

In the hydrothermal process, a La–Co–Fe hydroxide gel can be prepared and reacted with strontium hydroxide under hydrothermal conditions, as detailed in the patent [28].

Spray-pyrolysis process involves the following steps:

1. Preparation of an aqueous nitrate solution of the desired metal cation stoichiometry;
2. Nebulization of the nitrate solution;
3. Pyrolysis in a heated chamber (pyrolysis temperature ~800°C) and
4. Collection of the resulting fine particle-size oxide powder.

LSCF powder can be also prepared through the conventional ball milling and calcination method (or solid-state reaction method), starting with La_2O_3, $SrCO_3$, $CoCO_3$ and Fe_2O_3 raw materials. First A-site cation-deficient LSCF powders have to be prepared. After ball milling and drying, the mixed LSCF powders are calcined at 1,000°C and single-phase perovskite powders are got.

7.14 Track-etching method

Track-etching method is a two-step process where a film is first subjected to high energy particle radiation followed by an immersion into an etching bath (Figure 7.31). The resultant membranes are symmetric with uniform and cylindrical pores (Figure 7.32). The pore density and diameter are controlled by the residence time in the irradiator and etching bath, respectively. In order to avoid combined pores (two pores merging), the membrane porosity has to be kept low (<5%). In general, polycarbonate material is used.

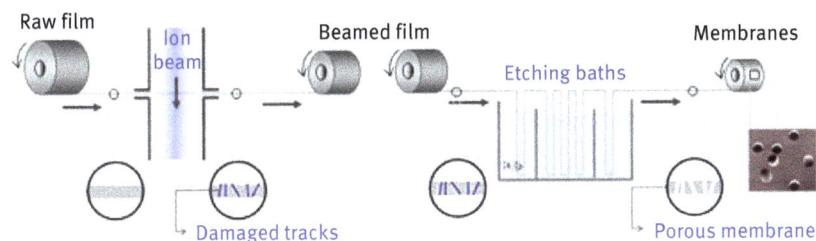

Figure 7.31: Schematic diagram of track-etching.

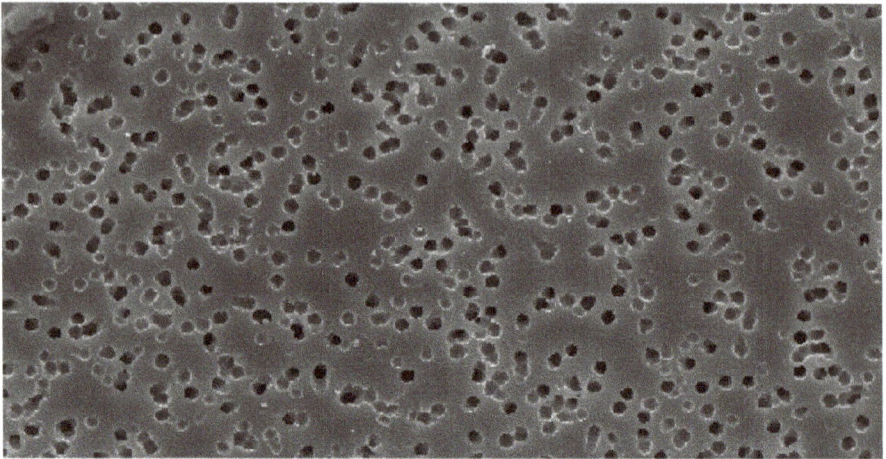

Figure 7.32: Track-etched membrane.

7.15 Template leaching

Template leaching (Figure 7.33) technique is suitable for polymers that do not dissolve into some solvents. First, a homogeneous film is prepared from a mixture of polymer and a leachable component. The leachable component is typically a soluble low molecular weight substance or even a macromolecular material such as poly(vinyl alcohol) (PVA) or poly(ethylene glycol) (PEG). The leachable component, after the film has been prepared, is removed by an appropriate chemical treatment.

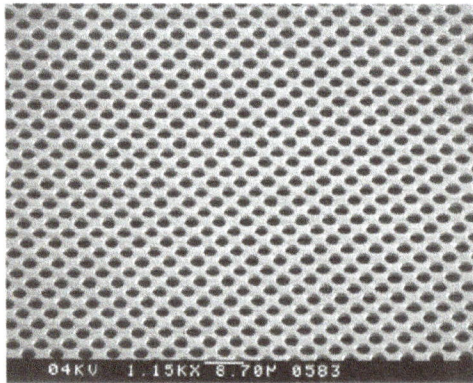

Figure 7.33: SEM image of membrane prepared through template leaching technique.

7.16 Electrospinning

Electrospinning can fabricate highly porous membranes with high porosity, excellent pore connectivity, and high surface area [29]. A typical apparatus consists of a high voltage supplier, a polymer solution feed system, and a collector (s). The choice of solvent is important, as it needs to respond to the electric field applied between the collector and the tip. Some of the drawbacks include low scalability, low productivity and pore size limitation (>100 nm). Various materials can be electrospun, as long as it exhibits solubility within highly dielectric solvent. Some of the common materials applied for membranes include PVDF, PI and PES.

7.17 Preparation of homogeneous solid membranes

The selective barriers of many composite membranes may be considered as homogeneous solid membranes. In homogeneous solid membranes the *entire* membrane consists of a dense, solid and pore-free structure. They are made from polymers as well as inorganic materials such as glass or metal. Because of their high selectivity for different chemical components, homogeneous membranes are used in various applications, which generally involve the separation of low molecular mass components with identical or nearly identical molecular dimensions. The most important applications are in gas separation.

One of the most important homogeneous solid metal membranes is the palladium or palladium alloy membranes used for the separation and purification of hydrogen.

In the last few decades, a great deal of attention has been attracted to the use of hydrogen as an energetic carrier to be employed for clean energy production by means of new technologies, such as polymer electrolyte fuel cells (PEMFCs). The global interest in the development of a "hydrogen economy" and an associated demand for hydrogen as a source of clean energy has led to the development of new materials and methods for hydrogen generation, storage, separation and sensing. Pd and Pd alloys are regarded as the most important materials for high quality hydrogen extraction from a mixture of gases.

7.18 Preparation of ion exchange membranes

There are two types of ion exchange membranes (IEM): cation exchange membrane (CEM) that permeates cations, and anion exchange membrane (AEM) that permeates anions. An ideal IEM should exhibit high permselectivity, low electrical resistance, good mechanical properties, and high chemical stability. The applications include desalination, fuel cells [30] and batteries.

Figure 7.34 shows schematically the matrix of a cation-exchange membrane with fixed anions and mobile cations, which are referred to as counter-ions. In contrast, the mobile anions, called co-ions, are more or less completely excluded from the polymer matrix because of their electrical charge which is identical to that of the fixed

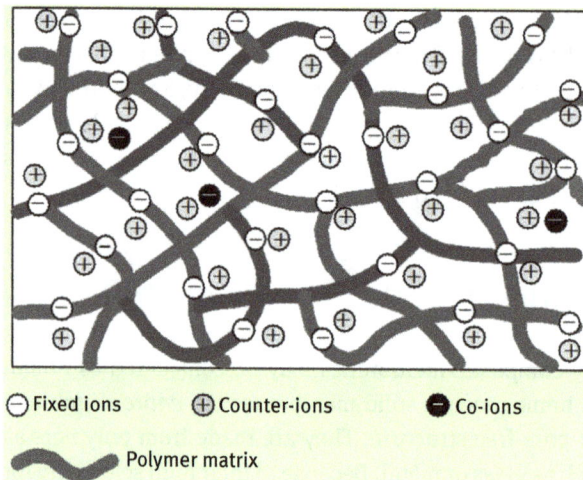

Fixed ions   Counter-ions   Co-ions

Polymer matrix

Figure 7.34: Ion-exchange membrane.

ions. Due to the exclusion of the co-ions, a cation-exchange membrane permits transfer of cations only. Anion-exchange membranes carry positive charges fixed on the polymer matrix. Therefore, they exclude all cations and are permeable to anions only.

The properties of ion-exchange membranes are determined by two parameters:
– The basic polymer matrix (which determines to a large extent the mechanical, chemical, and thermal stability of the membrane)
– The type and concentration of the fixed ionic moiety (which determine the permselectivity and the electrical resistance of the membrane, but they also have a significant effect on the mechanical properties of the membrane).

Most commercial ion-exchange membranes are either homogeneous or heterogeneous. In homogeneous membranes, the ion-exchange groups are homogeneously distributed throughout the polymer, whereas in heterogeneous membranes the ion-exchange particles are dispersed in a neutral polymer matrix.

For the practical preparation of ion-exchange membranes two rather different procedures are used. A quite simple technique is based on mixing an ion-exchange resin and a binder polymer, such as polyvinylchloride and extruding the mixture as a film at a temperature above the melting point of the polymer. The result is a heterogeneous membrane with relatively large domains of ion-exchange material and no conductive regions of the binder polymer. To obtain ion-exchange membranes with satisfactory conductivity, the fraction of ion-exchange resin must be in excess of 50–70 wt%. This often leads to rather high swelling and to poor mechanical stability of the membrane. Furthermore, the size of the ion-exchange particles should be as small as possible, that is <2–20 μm in diameter to be able to make thin membranes with low electrical resistance and high permselectivity.

More recently, homogeneous ion-exchange membranes are produced by either a polymerization of monomers that carry anionic or cationic moieties or by introducing these moieties into a polymer which may be in an appropriate solution or a pre-formed film.

7.19 Characterization of membranes

The performance, stability and durability of a membrane are determined largely by their chemical composition and physical morphology. These properties can be characterized by a wide range of analytical methods. The characterization of membrane properties helps not only to choose the right membrane for a given application but also:
–   to guide the design of membranes with desired properties and
–   to gain a better understanding of their preparation methods and on the selectivity and fouling mechanisms.

The choice of a membrane will depend on (i) its surface physicochemical and chemical properties (e.g., hydrophobic/hydrophilic nature, charge, etc.) as well as on its (ii) structural and transfer characteristics. The first one will allow fouling and interactions among the different types of molecules at the membrane surface to be predicted up to a certain extent and will play a role in the transport mechanism. The second one provides information on how the membrane will perform in the intended membrane process (e.g., the extent of the permeate flow, the type of rejected compounds, etc.).

Since membranes are very different in their properties and applications, a large number of different techniques are required for their characterization. Some characterization methods are specifically used to characterize porous membranes whereas other methods are more typically used for dense membranes. Indeed, some characterization techniques can be applied both on porous and dense membranes.

In general, membrane characterization becomes progressively more difficult as the pore size decreases. The following sections provide general fundamental principles and knowledge of membrane characterization techniques to obtain the essential information correlating membrane properties with their performance.

7.20 Characterization of porous membranes

Microfiltration or ultrafiltration membranes are porous, which induce the separation by discriminating between particle sizes. High selectivity can be obtained

when the solute size or particle size is larger than membrane pore size. Therefore, porous microfiltration or ultrafiltration membranes are generally characterized in terms of their:
- flux or pure water permeability
- pore size
- pore size distribution
- molecular mass/weight cutoff (MWCO)
- membrane thickness.

Other important characteristics are the nature of the membrane material (e.g., hydrophilic or hydrophobic or carrying positive or negative charges), the structure of the membrane (e.g., symmetric or asymmetric) and their mechanical, chemical, and thermal stability.

7.21 Pure water permeability

"Pure water flux" or "pure water permeability" (PWP) is a characteristic property of microfiltration and ultrafiltration membranes. According to Darcy's law, the flux J through a porous medium is proportional to the transmembrane pressure, and inversely proportional to the solvent viscosity:

$$J_v = \frac{L_p \Delta p}{\eta}$$

where J_v is the membrane flux, L_p is the hydrodynamic solvent permeability, η is the viscosity of the solution passing through the pores of the membrane, and Δ_p is the applied pressure.

From this law, one can deduce the hydraulic permeability and the resistance of the membrane (the resistance of the membrane is defined as the reciprocal of the hydraulic permeability).

The method itself is very simple: the water flux through the membrane is measured experimentally as function of the applied pressure (Figure 7.35). The hydrodymamic permeability is obtained from the slope of the plot.

Typical pure water fluxes of microfiltration membranes vary from 500 to 50,000 L h m^{-2}bar^{-1}, while pure water fluxes of ultrafiltration membranes vary from 50 to 800 L h m^{-2}bar^{-1}. In practical application, fluxes are generally lower and a steep flux decline is observed during the first period of operation due to membrane fouling.

Monitoring the permeability allows membrane fouling to be quantified in terms of additional resistance [31]. Conversely, the permeability can be used to evaluate the efficiency of chemical or mechanical cleaning.

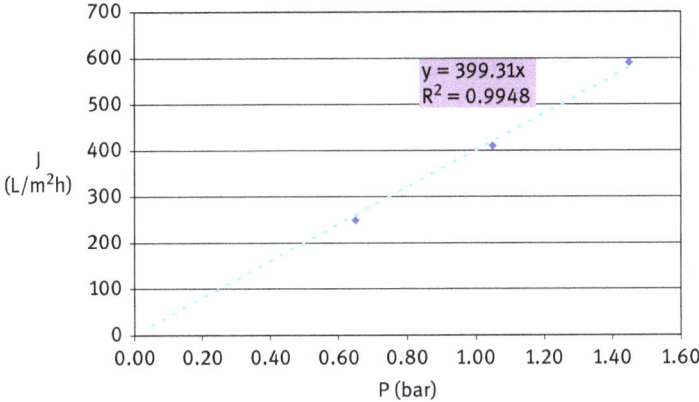

Figure 7.35: Flux versus pressure curve for a membrane possessing a uniform pore size.

7.22 Rejection (R) and molecular weight cut-off

As described earlier, both microfiltration and ultrafiltration are pressure-driven membrane processes using porous membranes where the separation of various components is based on a sieving mechanism, that is this membrane retains particles or molecules that are larger in size than the pores of a membrane.

Membrane rejection R^i to a certain component i can be calculated according to the following expression:

$$R^i = \left(1 - C^i_p/C^i_f\right) \cdot 100$$

where C^i_p and C^i_f are the concentration of the component i in the permeate (p) and in the feed (f), respectively.

Molecular weight cut-off (MWCO) is an indirect measure of the membrane retention performance. More precisely, the membrane MWCO is determined as the solute size that is retained by at least 90%. Due to the fact that the MWCO determination is very sensitive to the experimental conditions, the following standard test conditions are recommended: a trans-membrane pressure of 100 kPa, a feed solution concentration of 0.1% and a test temperature of 25°C. MWCO is determined by performing rejection tests using solutes or globular proteins of known sizes with the membrane of interest.

Figure 7.36 shows the comparison between a membrane with a so-called "sharp cutoff" and a membrane with a "diffuse cutoff". Solutes used in MWCO tests should ideally be soluble in water or in a mildly buffered solution, cover a wide range of sizes and should not adsorb to the membrane surface. Solute rejection measurements provide a very simple technique for determining membrane performance.

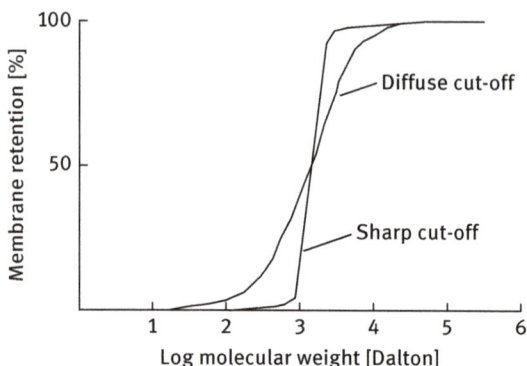

Figure 7.36: Rejection characteristics for a membrane with a "sharp cut-off" compared with that of a membrane with a "diffuse cut off".

Methods based on permeation and rejection performance can also be employed to determine the pore size and pore size distribution of membranes. The pore size can be obtained by measuring the flux through a membrane at a constant pressure using the Hagen–Poiseuille equation:

$$J_V = \frac{\varepsilon r_p^2}{8\mu\tau}\frac{\Delta p}{\Delta x}$$

where J_v is the water flux through the membrane at a driving force of $\Delta p/\Delta x$, with Δp being the pressure difference across the membrane of thickness Δx. The proportionality factor contains the pore radius r_p, liquid dynamic viscosity μ, surface porosity of the membrane ε, and the tortuosity factor τ. The pore size distribution can be obtained by varying the pressure, that is, by a combination of the bubble-point method and permeability methods.

The permeability method has the distinct advantage of experimental simplicity. However, the Hagen–Poiseuille equation assumes that pores are cylindrical, so the geometry is very important and will affect the result. For asymmetric membranes, Δx is the skin layer thickness and must be known to determine pore size.

7.23 Membrane pore size measurement

Bubble point is one of the widely and simple techniques for the characterization of the largest pores in porous membranes. This method measures the pressure needed to blow air through a liquid-filled membrane. The procedure for bubble point measurement is as follows:
1. Wet the membrane with water for hydrophilic or an alcohol/water mixture for hydrophobic membranes.
2. Slowly increase the pressure until the first gas bubble can be detected.

3. Use the correspondent pressure value to calculate the largest pore dimension according to the Laplace equation:

$$r_p = \frac{2\sigma \cos \phi}{p}$$

where r_p is the pore radius, σ is the surface tension of the liquid in contact with air, ϕ is the contact angle between the liquid and the wall of the pore, and p the applied pressure. Since the membrane is supposed to be completely wet by the liquid, the contact angle is assumed as 0, and, therefore, $\cos\phi$ is 1.

The main limitation of this method is that different results will be obtained when different liquids are chosen. Mostly water or isopropanol are used, since the surface tension of water/air is approximately 3.5 times higher than for isopropanol/air. When water is used, then pores down to a few nanometers can be measured [7].

Other factors that influence the measurement are the rate of pressure increase, the pore length, and the affinity between wetting liquid and membrane material. This technique allows estimating pore size distributions when it is performed by a step-wise increase of pressure.

Other techniques to determine the pore size and pores size distribution of a porous membrane that are also based on the Laplace equation, but cover different pores size ranges when the same pressure range is applied, are mercury porosimetry, gas–liquid displacement, and liquid–liquid displacement.

The mercury porosimetry technique is a variation of the bubble point method which uses mercury to gain information on the porous characteristics of solid materials. In this technique, the dry membrane is exposed to a certain volume of liquid mercury. The latter is forced into the membrane by slowly increasing the pressure. Simultaneously, the amount of mercury forced into the porous structure is measured. According to the Laplace equation, the largest pores will be filled first and the required pressure for the mercury to penetrate the porous structure increases with decreasing pore size. Because the interfacial tension of mercury and air is very high, relatively high pressures are required to fill small pores which may damage the membrane's structure.

The disadvantage of this technique is that the apparatus is rather expensive, small pore sizes require high pressure and it does not show a distinction between dead-end pores and inter-connected pores.

Gas–liquid displacement method is identical to the bubble point test:

- The pores of the sample membrane are filled with a lower surface tension liquid;
- The solvent is forced out of the membrane pores by nitrogen gas, which is introduced with increasing pressure;
- As the pressure increases, the liquid will be replaced by nitrogen in the largest pores first and a convective gas flow through these pores will occur;
- The gas flow is measured as a function of the applied pressure.

Again, the Laplace equation describes the relationship between the pore radius and the applied pressure.

Liquid–liquid displacement differs from gas–liquid displacement in the displacing medium: a second liquid instead of a gas displaces the liquid inside the pores. The two liquids applied should be immiscible. A typical liquid pair used is water/isobutanol. Both liquids are first saturated with each other before one of the liquids is used to fill the pores and the other liquid is applied as the replacement liquid. A slow increase in the replacement liquid pressure pushes the liquid out of the largest pores first. With increase in pressure, the liquid in the smaller pores will also be replaced.

The relation between the pore radius and the pressure required to open pores of a certain size is again described by the Laplace equation and the pore size distribution can now be calculated when the flux as a function of the pressure is measured.

Like in the gas–liquid displacement, only active pores contributing to transport are taken into account.

7.24 Microscopy techniques

These techniques provide information on surface topology, roughness and pore size. Several microscopic observation methods are used, which differ by their implementation and their resolution, such as:
-   scanning electron microscopy (SEM)
-   field emission electron microscopy
-   transmission electron microscopy (TEM)
-   atomic force microscopy (AFM)

The advantage of these techniques is that direct visual information of the membrane morphology is obtained.

SEM and TEM are the two most commonly used electron microscopy (EM) methods, with resolutions of the order of 10 nm and 5 nm, respectively (for a membrane sample and in ideal experimental conditions for the instrument) [32].

The operational principle of SEM (figure 7.37) relies on the detection of different scattered electrons by scanning the sample surface with a high energy electron beam. To observe cross sections by SEM, the dried membrane is first fractured at liquid nitrogen temperature, and then fixed perpendicularly to the sample holder. The working principle of SEM is illustrated in Figure 7.38. A beam of electrons is produced at the top of the microscope by an electron gun. The electron beam follows a vertical path through the microscope, which is held within a vacuum. The beam travels through electromagnetic fields and lenses, which focus the beam

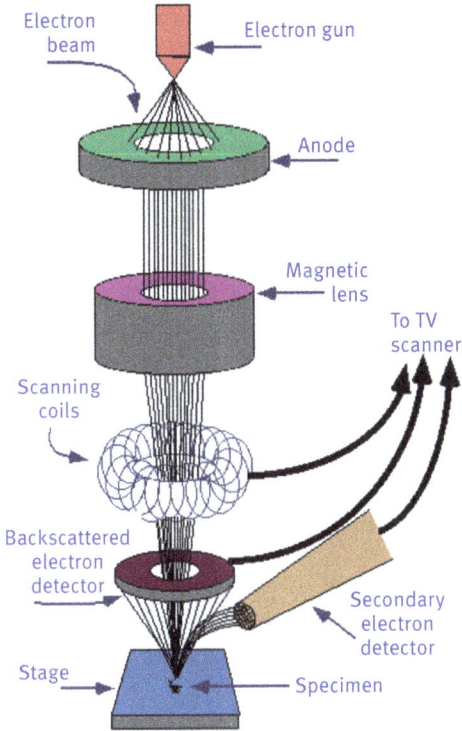

Figure 7.37: Principle of the scanning electron microscope.

down toward the sample. Once the beam hits the sample, electrons and X-rays are ejected from the sample.

In contrast, TEM measurement analyzes the transmitted or forward-scattered electrons through the specimen. The dried sample is first embedded and then sliced by a microtome. An embedding media with no influence on the membrane must be chosen.

Electron microscopy techniques require a high vacuum in the microscope column in order to overcome the slowing down of the electrons due to the presence of gaseous molecules. The vacuum prevents the observation of hydrated objects at room temperature. Therefore, electron microscopy imposes specific techniques of sample preparation, such as [32]:

1. Dehydration of the specimens or immobilization of their water by freezing and preparation of ultrafine sections of the frozen material;
2. Metallization of the membrane surface (after dehydration) in the case of SEM (e.g., through gold, platinum, palladium, or their alloys);
3. Inclusion of resins in the case of TEM so that ultrafine sections, 50–100 nm thick, can be cut.

The surface morphology of membranes can also be examined by AFM. The working principle of AFM is to use a sharp tip to scan the sample surface. When the tip is close to the surface, van der Waals forces will change the vibrating frequency of the tip or cause deflections. By detecting the vibrating frequency or deflections of the tip, a three-dimensional map of membrane surface topography can be obtained. AFM is widely used to characterize membrane surfaces and has the advantage of providing quantitative nanoscale measurements of both lateral and vertical morphology. In addition to morphology mapping, AFM can quantify the interaction force between the membrane surface and the probe used. Thus, much information besides the surface morphology, such as the surface roughness, the fouling propensity and electrical properties, can be revealed. Membrane fouling studies can be performed by analyzing the variation in the roughness of the new, clean membrane with that of the used membrane.

7.25 Mechanical characterization

The mechanical characteristics of a membrane are identified by the following three parameters:
- force, extension and stress at break;
- force, elongation and elastic stress; and
- Young's modulus.

Figure 7.38 shows how these various parameters are determined graphically. A curve of stress versus deformation is obtained in tension (or compression) tests, in which the sample is drawn out at constant speed until it breaks. The sample is stretched unidirectionally. Data are acquired by an extensometer connected to a computer.

Young's modulus (E), being the ratio between the stress σ and the strain ε, is determined from the slope of the tangent to the curve in zero:

$$E = \frac{\sigma}{\varepsilon}$$

Other parameters such as as the stress or the elongation at break can be also determined graphically as indicated in Figure 7.38.

For a given stress, a material having a high E will be deformed less than a material with a low E. Moreover, the faster the deformation takes place, the greater the mechanical resistance of the polymer will be, even at high temperatures.

The stress versus strain diagram gives also information concerning the plastic or elastic deformation of a membrane: at relatively low strain the membranes show elastic deformation, with increasing strain the membranes show plastic deformation, and at a certain point they break. Since the mechanical properties of membranes generally change drastically with the water content of the

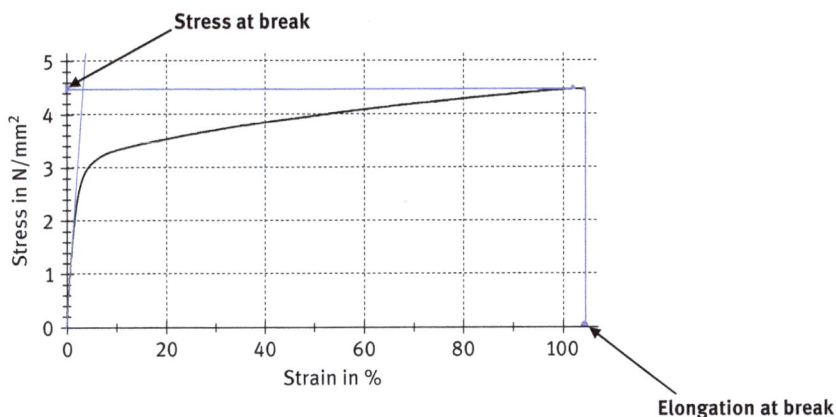

Figure 7.38: Stress versus strain diagram of a membrane sample indicating the elongation and the breakpoint.

membranes, they must be determined with dry membranes and with membranes equilibrated in water or different solutions similar to those used in practical applications.

The swelling of membranes depends on a number of different parameters such as the nature of the basic membrane polymer (e.g., the degree of crystallinity of the polymeric matrix, etc.), the concentration of fixed charges, the nature of the ion-exchange groups and their concentration in the membrane, the counter ions, the cross-linking density, the homogeneity of the membrane and on the composition of the solution with which the membrane is in contact. The structure of membranes based on a highly crystalline polymer is studied using TEM and small angle X-ray diffraction. Infrared spectroscopy measurements can provide some information on the type and degree of cross-linking of ion-exchange membranes.

The total water uptake of the membrane in equilibrium with an electrolyte solution can be determined by measuring the weight difference between a membrane in the wet and dry state. To determine the water content of a membrane, a sample is equilibrated in a test solution. After removing the surface water from the sample, the wet weight of the swollen membrane is determined. The sample is then dried at an elevated temperature over phosphorous pentoxide under reduced pressure until a constant weight is obtained. The water content of a membrane is obtained in weight percent by using the formula:

$$wt\% \ swelling = \frac{W_{wet} - W_{dry}}{W_{wet}} \times 100$$

where W_{wet} and W_{dry} are the weight of a membrane sample in the wet and the dry state.

7.26 Characterization of homogeneous membranes

Homogeneous membranes are generally used for the separation of low molecular mass materials such as gases, salts or solvents. The most important factor of homogeneous membranes are their chemical nature, morphology and the interaction between the membrane material and the permeants. The transport mechanism in these membranes is based on the solution and diffusion of components within the membrane matrix. The separation is achieved either by differences in solubility and/ or diffusivity. Therefore, characterization methods used with microporous structures (devoted mainly to determine pore size and pore size distribution) have to be replaced with other procedures devoted to determine the physical properties related to the chemical structure of the membrane, such as:
- sorption and diffusion measurements
- determination of the glass transition temperature and crystallinity of the basic membrane material
- surface analysis

In the case of ion-exchange membranes, the measurement of electrokinetic properties is required.

A detailed discussion of all methods used to characterize *homogeneous polymeric membranes* is beyond the scope of this book and a more in-depth analysis can be found in literature [7, 33].

One of the principal and simplest methods for characterizing a nonporous membrane is to determine its permeability toward gases and liquids. Permeability measurements can be done using a very simple set-up where a cell containg the homogeneous membrane is pressurized with a known gas. The extent of gas permeation thorugh the membrane is measured by means of a mass flow meter or by a soap bubble meter. The gas permeability P (or permeability coefficient) can be determined using the following equation:

$$J = P/l$$

where J is the gas flow per unit pressure and l is the membrane thickness.

The diffusion coefficient can be also determined from the initial part of the permeation experiment by using the so-called time-lag method [33, 34, 35].

7.27 Differential scanning calorimetry/differential thermal analysis

Various techniques can be used to characterize the parameters that affect the membrane permeability. Such methods mainly determine the membrane morphology.

Crystallinity and glass transition temperature are the parameters that strongly affect membrane permeability.

Differential scanning calorimetry (DSC) and differential thermal analysis (DTA) are identical techniques used to measure transitions or chemical reactions in a polymer sample. DSC determines the energy necessary (dQ/dt) to counteract any temperature difference between the sample and the reference, whereas DTA determines the temperature difference (ΔT) between the sample and the reference upon heating or cooling. A schematic diagram of a DSC curve for a semi-crystalline polymer is shown in Figure 7.39.

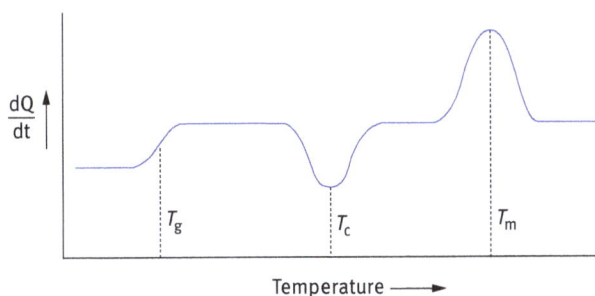

Figure 7.39: Schematic diagram of a DSC curve for a semi-crystalline polymer.

The glass transition temperature can be determined from Figure 7.39. The degree of crystallinity can be obtained from the area under the peak corresponding to melting per unit weight of polymer. This gives the enthalpy of fusion (H_s). To calculate the crystallinity, the enthalpy of fusion for the 100% crystalline material (H_{100}) must be known

$$X_c = 100 \cdot H_s/H_{100}$$

Various other methods can be used for characterization of the membrane structure, such as X-ray diffraction, plasma etching, Fourier-transform infrared (FT-IR) spectroscopy, etc. X-ray diffraction [36, 37] is particularly useful for obtaining information about the size and the shape of crystallites and about the degree of crystallinity in solid polymers. Plasma etching [7] allows measuring the thickness of the top layer in asymmetric and composite membranes; FT-IR [38, 39] allows determining the surface properties by surface analysis.

7.28 Determination of hydrophilic/hydrophobic nature of membranes

The hydrophilic/hydrophilic nature of a material is a very important parameter as it conditions the solute–membrane and solvent–membrane interactions. In some applications, the hydrophobic character of the membrane is the essential requisite for performing the process (e.g., in membrane distillation, membrane crystallization, etc.); in others (e.g., reverse osmosis), hydrophilic membranes are more efficient than hydrophobic ones, these latter being more exposed to fouling when hydrophobic molecules or particles are present in the stream to be filtered.

The hydrophilic/hydrophobic nature of a membrane is determined by measuring the contact angle.

7.28.1 Contact angle measurements

Contact angle measurement is easily performed by establishing the tangent (angle) of a liquid drop with a solid surface at the base [40]. The attractiveness of using contact angles θ to estimate the solid–vapor and solid–liquid interfacial tensions is due to the relative ease with which contact angles can be measured on suitably prepared solid surfaces (Figure 7.40).

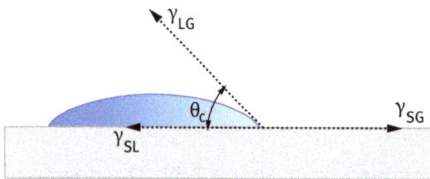

Figure 7.40: Schematic of a sessile-drop contact angle system.

The possibility of estimating solid surface tensions from contact angles relies on a relation that has been recognized by Young [41] in 1805. The contact angle of a liquid drop on a solid surface is defined by the mechanical equilibrium of the drop under the action of three interfacial tensions (Figure 7.41): solid–vapor, γ_{sg}, solid–liquid, γ_{sl}, and liquid–vapor, γ_{lg}. This equilibrium relation is known as Young's equation:

$$\gamma_{LG}\, \cos\theta_c = \gamma_{SG} - \gamma_{SL}$$

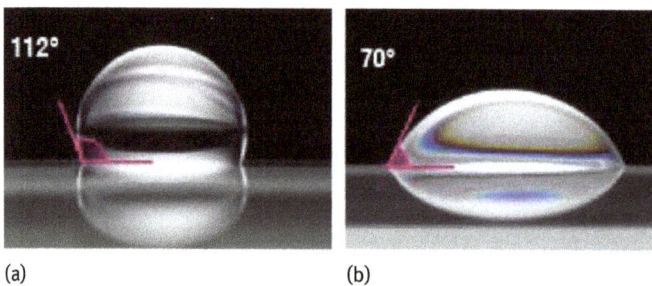

(a)   (b)

Figure 7.41: (a) Hydrophobic membrane and (b) hydrophilic membrane.

where θ_c is the Young contact angle, that is, a contact angle which can be inserted into Young's equation.

Contact angle values go from $0°$ (completely wetted surface) to $180°$ (ideally non-wetted surface). In reality, there are super-hydrophobic membranes that reach contact angle values up to $150°$.

7.29 Characterization of ion-exchange membranes

The methods used to determine the structural properties of ion-exchange membranes are similar to those used for characterizing other polymer membranes. Because ion-exchange membranes are used mainly in separation processes with electrical potential driving forces, their permeability and selectivity are determined under experimental conditions closely related to their practical use, and their properties are expressed in terms commonly used in electrochemistry (e.g., electricalresistance, ion-transfer numbers, charge densities) [42].

Therefore, the most interesting properties of ion-exchange membranes are:

- the electrical charge;
- the electrical resistance in different electrolyte solutions;
- the type and density of fixed charges and their distribution in the membrane matrix;
- the permselectivity of the membrane for different ions;
- the transport rate of neutral components, especially water; and
- the mechanical and chemical stability and the swelling in different electrolyte solutions.

The electrical charge of an ion-exchange membrane can be determined qualitatively by using indicator solutions. A drop of a 0.05% solution of methylene blue and methyl orange on a membrane sample stains a yellow spot on top of an anion-exchange membrane and a deep blue spot on top of a cation-exchange membrane, respectively.

Hydraulic permeability measurements provide information on the diffusive or convective transport of components through a membrane under a hydrostatic pressure driving force. The hydraulic permeability of the membrane is determined at room temperature using deionized water and a hydrostatic pressure driving force in a conventional filtration cell as used in reverse osmosis or ultrafitration experiments [43].

For what concerns the ion-exchange capacity, this is a crucial parameter that affects almost all other membrane properties. It is usually expressed in milli-equivalent per gram dry membrane. Experimentally, the ion-exchange capacity of strong acidic or strong basic ion-exchange membranes is readily determined by titration with

NaOH or HCl, respectively. For these tests, cation- and anion-exchange membranes are equilibrated in 1N HCl or 1N NaOH, respectively, and then rinsed free from chloride ions or sodium ions with deionized water. The ion-exchange capacity of the samples is determined by back titration with 1 N NaOH or 1 N HCl, respectively. Weak base anion-exchange membranes are characterized by equilibration in 1 N NaCl and titrated with standardized 0.1 N AgNO$_3$ solution. The samples are then dried, and the ion-exchange capacity is calculated based on the dry membrane. The accuracy of the measurement depends on the complete exchange of ions in the membrane, which can take some time.

Electrical resistance of ion-exchange membranes is expressed as: Ω cm, Ω m, Ω cm^2, Ω m^2 (more useful from the engeneering point of view). It is determined by the ion-exchange capacity and the mobility of the ion within the membrane matrix. The area resistance of ion-exchange membranes can be determined by direct current (DC) or alternating current (AC) measurements. In DC measurements, the membrane is installed in a cell that consists of two chambers containing the test solution separated by the membrane. Two electrodes are used to provide the electrical potential driving force. The test solution is Na$_2$SO$_4$ 0.5^{-1} M.

The potential drop across the membrane is determined with calomel electrodes attached to Haber-Luggin capillaries placed close to the membrane. The potential drop between the Haber-Luggin capillaries is measured with and without the membrane in the test cell as a function of the current density passing through the cell. The resistance is given by the slope of the current versus the voltage drop curve. To obtain the membrane resistance, the resistance of the cell without the membrane is subtracted from the resistance of the cell with the membrane.

The area resistance is given by:

$$r_{m+s} = R\,A_m = \frac{U}{i} \qquad (7.4)$$

The membrane resistance is:

$$r_m = r_{m+s} - r_s \qquad (7.5)$$

where R is the resistance, A_m is the area of the membrane, U is the voltage drop measured between the Haber-Luggin capillaries and I is the current density, r_{m+s} and r_s are the area resistances of the cell with and without the membrane between the Haber-Luggin capillaries, and r_m is the area resistance of the membrane.

In AC measurements, the membrane resistance is determined from resistance measurements in a cell with and without membrane. The area resistance r_m is related to the specific resistance by:

$$r_m = \rho_{m+s}\,(d_m + d) - \rho_s d \qquad (7.6)$$

where r_m, is the area resistances of the membrane, ρ_{m+s} and ρ_s are the specific resistances of the cell with and without the membrane, and d_m and d are the thickness of the membrane and the distance between the electrodes.

The specific resistance ρ is:

$$\rho = R\frac{A_m}{d} \tag{7.7}$$

where A_m is the cross-sectional area of the cell and R is the resistance measured between the electrodes.

A rather simple method of measuring electrical resistance is based on impedance spectroscopy (IS) or electrochemical impedance spectroscopy (EIS). The difference between the alternating current resistance measurements and the impedance spectroscopy is that, in the first case the frequency of the alternating current is kept constant while in impedance spectroscopy the frequency of the alternating current is changed and the response to the changing frequency is determined by a spectrometer.

The properties of the ion-exchange membrane, the solution and the electrodes can be described electrically by an equivalent circuit (Figure 7.42) where the entire system, that is, the membrane, the electrolyte and the electrodes are treated as a "black box."

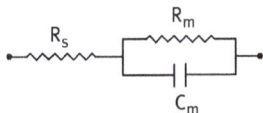

Figure 7.42: Equivalent circuit for an ion-exchange membrane in an electrolyte solution. Here C_m is membrane capacitance, R_m is membrane resistance and R_s is resistance of electrodes and the electrolyte solution.

In analogy to Ohm's law, the impedance is defined [44] as:

$$Z_{(\omega)} = \frac{U_{(\omega)}}{I_{(\omega)}} \tag{7.8}$$

where $Z(\omega)$ is the impedance, $U(\omega)$ is the voltage drop, $I(\omega)$ is the current, and they depend on the circular velocity or circular frequency ω as follows:

$$U_{(\omega)} = U_o \cos \omega t \tag{7.9}$$

$$I_{(\omega)} = I_o \sin \omega t + \varphi \tag{7.10}$$

where $\omega = 2\pi\upsilon$ and υ is the frequency, t is the time, φ is the phase shift between voltage and current, and the subscript o refers to the amplitude of voltage and current.

The impedance be also rearranged as follows:

$$Z_{(\omega)} = \frac{U_{(\omega)}}{I_{(\omega)}} = Z\,cos\varphi - i\,Z\,sin\varphi \qquad (7.11)$$

Equation (7.9) indicates that the impedance is composed of two parts, that is, the real part given by $Z\,cos\,\varphi$ and the imaginary part given by $i\,Z\,sin\,\varphi$.

The real part of the impedance is the resistance; the imaginary part is called reactance.

The impedance related to an electric resistance and to a capacitance is different. For an electric resistance, the imaginary part of the impedance is zero since the current and voltage are in phase and the real part is frequency independent.

$$Z = \frac{U_o}{I_o} = R \qquad (7.12)$$

For a capacitance, the impedance is given by:

$$Z = \frac{1}{i\omega C} \qquad (7.13)$$

The total impedance for a resistance and a capacitance in series is:

$$Z_{(\omega)} = R + \frac{1}{i\omega C} \qquad (7.14)$$

According to eq. (7.14), the imaginary part will disappear at very high frequencies if resistance and capacitance are in series and the impedance is identical to the ohmic resistance. At very low frequencies, the impedance of the capacitance increases with decreasing frequency and becomes infinitely high in direct current.

For resistance and capacitance in parallel, the total impedance is given by:

$$Z_{(\omega)} = \frac{R}{1 + \omega^2 R^2 C^2} - i\,\frac{\omega R^2 C}{1 + \omega^2 R^2 C^2} \qquad (7.15)$$

According to eq. (7.15), the imaginary part of the impedance disappears at very low and at very high frequencies.

In a system composed of an ion-exchange membrane, an electrolyte and two electrodes, there may be both resistances and capacitances in series and in parallel and the impedance can be rather complex and it is not always easy to determine the membrane resistance from the obtained diagram, and mathematical models may be used to obtain reliable data for the membrane resistance.

7.29.1 Permselectivity of ion-exchange membranes

Before introducing perselectivity of ion-exchange membranes, it is advisable to remember that:

- in *CEMs* fixed anions (co-ions) permit transfer of cations (counter-ions) and
- in *AEMs* fixed cations (co-ions), permit transfer of anions (counter-ions).

The permselectivity of a membrane is determined by the ratio of the flux of specific components to the total mass flux through the membrane under a given driving force. In ion-exchange membranes, the permselectivity is related to the transport of electric charges by the counterions. The permselectivity of an ion-exchange membrane is determined by the concentrations of counter-ions and co-ions in the membrane, and the ion concentration in the outside solutions because of the Donnan exclusion.

In ion-exchange membranes, the permselectivity is related to the transport of electric charges by the counter-ions.

The permselectivity of a membrane is given by:

$$\Psi^m = \frac{T_{cou}^m - T_{cou}}{T_{co}}$$

The transport numbers are defined by:

$$T_i = \frac{z_i J_i}{\sum\limits_i z_i J_i}$$

where Ψ is the permselectivity, T is the transport number, z is the valence, and J is the flux; the subscript i refers to cation or anion, the subscripts cou and co refer to counter-ions and co-ions and the superscript m refers to ion-exchange membrane.

An ideal permselective cation-exchange membrane would be permeable for positively charged cations (counterions) only. The permselectivity of a membrane approaches zero when the transport numbers of the ions within the membrane are identical to those in the electrolytic solution.

References

[1]   Guillen, G. R., Pan, Y., Li, M. and Hoek, E. M. Preparation and characterization of membranes formed by nonsolvent induced phase separation: A review. Ind. Eng. Chem. Res. (2011); 50(7), 3798–3817.
[2]   Drioli, E., Giorno, L., (Eds.). Comprehensive Membrane Science and Engineering, Elsevier B.V., Oxford, United Kingdom, Volume 1, (2010).
[3]   Drioli, E., Giorno, L. and Fontananova, E. (Eds.). Comprehensive Membrane Science and Engineering, Elsevier B.V., Oxford, United Kingdom, Volume 1., (2017).
[4]   Bastani, D., Esmaeili, N. and Asadollahi, M. Polymeric mixed matrix membranes containing zeolites as a filler for gas separation applications: A review. J. Ind. Eng. Chem. (2013); 19(2): 375–393.
[5]   Bottino, A., Camera-Roda, G., Capannelli, G. and Munari, S. The formation of microporous polyvinylidene difluoride membranes by phase separation. J. membr. sci. (1991); 57(1): 1–20.

[6]   J. Hildebrand, R. Scott. Solubility of Nonelectrolytes, Reinhold Publishing, New York, 1950.

[7]   Mulder, J. Basic Principles of Membrane Technology, Springer Science & Business Media, (2012).

[8]   Buonomenna, M. G., Figoli, A., Jansen, J. C. and Drioli, E. Preparation of asymmetric PEEKWC flat membranes with different microstructures by wet phase inversion. J. Appl. Polym. Sci. (2004); 92(1): 576–591.

[9]   A. J. Reuvers. Ph.D Thesis, Membrane formation: diffusion induced demixing processes in ternary polymeric systems, University of Twente, 1987

[10]   Jansen, J. C., Buonomenna, M. G., Figoli, A. and Drioli, E. Asymmetric membranes of modified poly (ether ether ketone) with an ultra-thin skin for gas and vapour separations. J. Membr. Sci. (2006); 272(1–2): 188–197.

[11]   Jung, J. T., Kim, J. F., Wang, H. H., di Nicolo, E., Drioli, E. and Lee, Y. M. Understanding the non-solvent induced phase separation (NIPS) effect during the fabrication of microporous PVDF membranes via thermally induced phase separation (TIPS). J. Membr. Sci. (2016); 514, 250–263.

[12]   Lloyd, D. R., Kim, S. S. and Kinzer, K. E. Microporous membrane formation via thermally-induced phase separation. II. Liquid–liquid phase separation. J. Membr. Sci, (1991). 64(1–2), 1–11.

[13]   Jung, J. T., Kim, J. F., Wang, H. H., di Nicolo, E., Drioli, E and Lee, Y. M. Understanding the non-solvent induced phase separation (NIPS) effect during the fabrication of microporous PVDF membranes via thermally induced phase separation (TIPS). J. Membr. Sci. (2016); 514: 250–263.

[14]   Kim, J. F., Jung, J. T., Wang, H. H., Lee, S. Y., Moore, T., Sanguineti, A., and Lee, Y. M. Microporous PVDF membranes via thermally induced phase separation (TIPS) and stretching methods. J. Membr. Sci. (2016); 509: 94–104.

[15]   Castejón, P., Habibi, K., Saffar, A., Ajji, A., Martínez, A. B. and Arencón, D. Polypropylene-Based Porous Membranes: Influence of Polymer Composition, Extrusion Draw Ratio and Uniaxial Strain. Polymers. (2017); 10(1): 33.

[16]   Bastani D., Esmaeili N., Asadollahi M. Polymeric mixed matrix membranes containing zeolites as a filler for gas separation applications: A review. J. Ind. Eng. Chem. 2013; 19: 375–393

[17]   Lau, W. J., Ismail, A. F., Misdan, N. and Kassim, M. A. A recent progress in thin film composite membrane: A review. Desalination. (2012); 287: 190–199.

[18]   Yeo-Jin Kim, Seong-Joong Kim, Jeong-Kim, Yeong-Hoon Jo, Hosik Park, Pyung-Soo Lee, You-In Park, Ho-Bum Park, and Seung-Eun Nam. Characterization of ceramic membranes by gas-liquid displacement porometer and liquid-liquid displacement porometer. Membrane Journal. (2017); Vol.27 No.(3): pp.263–272.

[19]   Wu, P., Xu, Y., Huang, Z. and Zhang, J. A review of preparation techniques of porous ceramic membranes. J. Ceram. Process. Res. (2015); 16(1): 102–106.

[20]   Jeon, M. Y., Kim, D., Kumar, P., Lee, P. S., Rangnekar, N., Bai, P.,and Basahel, S. N. Ultra-selective high-flux membranes from directly synthesized zeolite nanosheets. Nature. (2017); 543(7647): 690.

[21]   J.G. McCarty, H. Wise. Applications of perovskites – Perovskite catalysts for methane combustion. Catal. Today.(1990); 8: p. 231

[22]   Zeng, Y., Lin, Y. S., and Swartz, S. L. Perovskite-type ceramic membrane: Synthesis, oxygen permeation and membrane reactor performance for oxidative coupling of methane. J. Membr. Sci. (1998); 150(1): 87–98.

[23]   Y.S. Lin, Y. Zeng. Catalytic properties of oxygen semipermeable perovskite-type ceramic membrane materials for oxidative coupling of methane. J. Catal. (1996); 164: p. 220

[24]   T. Nitadori, T. Ichiki, M. Misono. Bull. Chem. Soc. Jpn. (1988); 61:, p. 621

[25] Mizusaki, Y. Mima, S. Yamauchi, K. Fukei H,. Tagawa Nonstoichiometry of the perovskite-type oxides $La_{1-x}Sr_xCoO_{3-\delta}$. J. Solid State Chem. (1989); 80: p. 102.

[26] Z. Zhang, X.E. Verykios M,. Baerns Effect of electronic properties of catalysts for the oxidative coupling of methane on their selectivity and activity. Catal. Rev.-Sci. Eng. (1994); 36: p. 507.

[27] E.N. Voskresneskaya, V.G. Roguleva A.G,. Anshits Oxidant activation over structural defects of oxide catalysts in oxidative methane coupling. Catal. Rev.-Sci. Eng. (1995); 37: p. 101.

[28] W.J. Dawson, S.L. Swartz, US Patent 5 112 433 (12 May 1992)

[29] Kim, J. F., Kim, J. H., Lee, Y. M. and Drioli, E. Thermally induced phase separation and electrospinning methods for emerging membrane applications: A review. AIChE Journal. (2016); 62(2): 461–490.

[30] Lee, K. H., Cho, D. H., Kim, Y. M., Moon, S. J., Seong, J. G., Shin, D. W.,and Lee, Y. M. Highly conductive and durable poly (arylene ether sulfone) anion exchange membrane with end-group cross-linking. Energy. Environ. Sci. (2017); 10(1): 275–285.

[31] Yeh, H. M., Wu, H. P. and Dong, J. F. Effects of design and operating parameters on the declination of permeate flux for membrane ultrafiltration along hollow-fiber modules. J. Membr. Sci. (2003); 213(1–2): 33–44.

[32] C Causserand and P Aimarn. Characterization of Filtration Membranes, chapter 1.15 in Comprehensive Membrane Science and Engineering, vol. 1. Editor: E. Drioli, L. Giorno. Elsevier B.V, 2010.

[33] Rutherford, S. W., and D. D. Do. "Review of time lag permeation technique as a method for characterisation of porous media and membranes." Adsorption 3.4. (1997): 283–312.

[34] Morgan, David, Lee Ferguson, and Paul Scovazzo. "Diffusivities of gases in room-temperature ionic liquids: Data and correlations obtained using a lag-time technique." Ind. Eng. Chem. Res. (2005); 44(13): 4815–4823.

[35] Okamoto, K. I., Fuji, M., Okamyo, S., Suzuki, H., Tanaka, K., and Kita, H. Gas permeation properties of poly (ether imide) segmented copolymers. Macromol. (1995); 28(20): 69506956.

[36] Dikin, D. A., Stankovich, S., Zimney, E. J., Piner, R. D., Dommett, G. H., Evmenenko, G., and Ruoff, R. S. Preparation and characterization of graphene oxide paper. Nature. (2007); 448 (7152): 457.

[37] Kim, D. S., Park, H. B., Rhim, J. W. and Lee, Y. M. Preparation and characterization of crosslinked PVA/SiO2 hybrid membranes containing sulfonic acid groups for direct methanol fuel cell applications. J. Membr. Sci, (2004); 240(1–2): 37–48.

[38] Xiong, Y., Fang, J., Zeng, Q. H. and Liu, Q. L. Preparation and characterization of cross-linked quaternized poly (vinyl alcohol) membranes for anion exchange membrane fuel cells. J. Membr. Sci. (2008); 311(1–2): 319–325.

[39] Tang, C. Y., Kwon, Y. N. and Leckie, J. O. Probing the nano-and micro-scales of reverse osmosis membranes–A comprehensive characterization of physiochemical properties of uncoated and coated membranes by XPS, TEM, ATR-FTIR, and streaming potential measurements. J. Membr. Sci. (2007); 287(1):146–156.

[40] Kwok, D. Y. and Neumann, A. W. Contact angle measurement and contact angle interpretation. Adv. Colloid. Interface. Sci. (1999); 81(3): 167–249.

[41] Young, T. III. An essay on the cohesion of fluids. Philos. Trans. R. Soc. Lond. (1805); 95: 65–87.

[42] Cussler, E. L. Membranes which pump. AIChE Journal. (1971); 17(6): 1300–1303.

[43] Strathmann, H., Giorno, L. and Drioli, E. Introduction to Membrane Science and Technology. (Vol. 544). Wiley-VCH, Weinheim, (2011).

[44] Barsoukov E, Macdonald JR. Impedance spectroscopy. Theory, Experiment, and Applications, 2nd edn. Wiley, New Jersey, (2005)

Alfredo Cassano, Carmela Conidi, Enrico Drioli

8 Membrane-based operations and integrated membrane systems in fruit juice processing

8.1 Introduction

The global market for fruit and vegetable juices is forecasted to reach 72.29 billion liters by the year 2017 due to rising preference among customers for healthy drinks [1].

Fruit juices act as nutritious beverages and can play a significant part in a healthy diet because they offer a variety of nutrients that are naturally found in fruits. Evidence suggests an association between a diet rich in fruits and vegetables and improved health as well as reduced risk of chronic diseases (cancer, cardio-vascular and neurological disorders) [2]. The effects of pure fruit juice intake on outcomes linked to cancer, cardiovascular disease, cognition, hypertension, inflammation, oxidation, platelet function, urinary tract infection and vascular reactivity have been recently reviewed by Hyson [3]. The study of molecular processes influenced by fruit juice intake clearly indicate that pure fruit juices increase the antioxidant capacity in plasma in hours after consumption and in some cases on a longer term basis. Although there are many unanswered questions related to fruit juice and health in humans, the demand for organic, superfruit and 100% natural fruit juices is expected to rise remarkably in the next few years.

Fruit juices are commonly marketed in four forms: (1) freshly squeezed juices, (2) frozen concentrated juices, (3) not-from-concentrate juices and (4) juices from concentrate. Fresh squeezed juices obtained from fresh fruits without being pasteurized are characterized by a shelf life of only a few days. Frozen concentrated juices are generally obtained by removing water from the juice in a vapor form through thermally accelerated short time evaporators (TASTEs) and then stored at −6.6 °C or lower until they are sold or packaged for sale.

Not-from-concentrate juices are processed and pasteurized by flash heating immediately after squeezing the fruit without removing the water content from the juice. They can be stored, freezed or chilled for at least a year. Juices from concentrate are concentrated for preservation, handling and storage and reconstituted for consumption and are diluted back to approximately the same solid level (designated as °Brix or percent soluble solids) of the initial juice.

Several unit operations are involved in converting whole fruits to the desired juice. The generalized flow scheme for the production of fruit juices is depicted in Figure 8.1.

Cassano Alfredo, Conidi Carmela, Drioli Enrico, National Research Council of Italy, Institute for Membrane Technology (ITM-CNR), University of Calabria, Rende, Italy

https://doi.org/10.1515/9783110281392-008

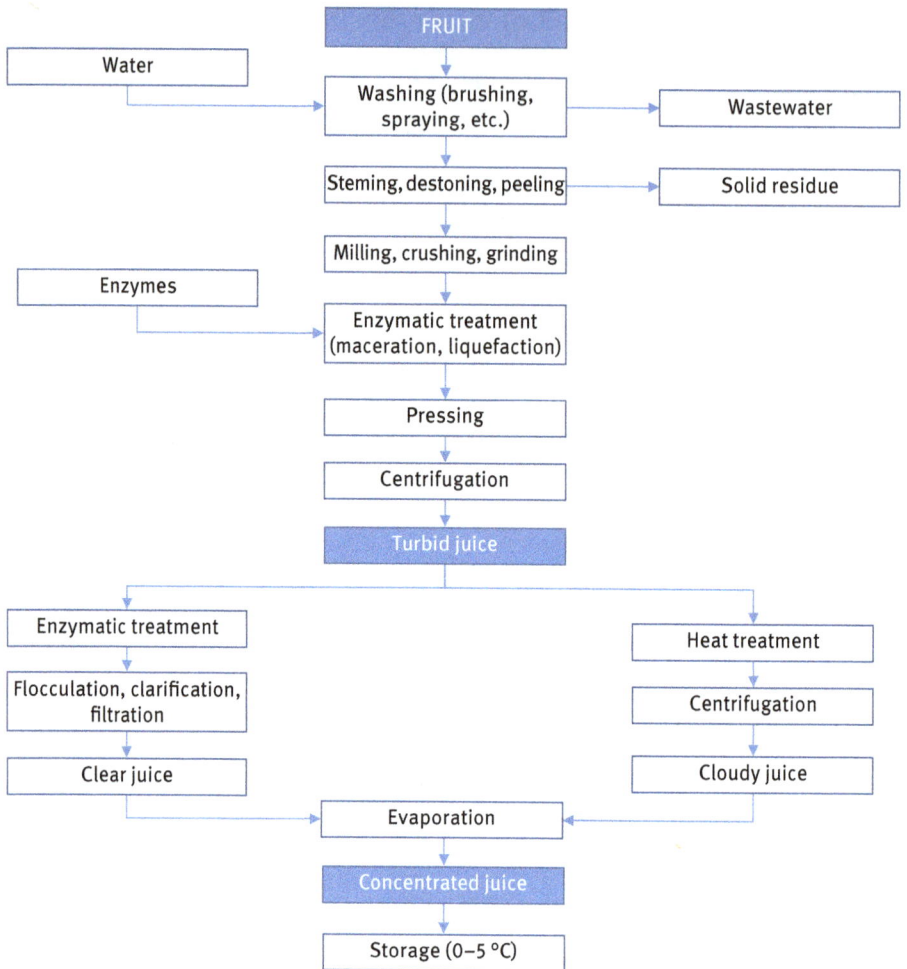

Figure 8.1: General flow sheet of fruit juice processing.

Crushing and juicing are crucial steps in juice manufacture: the goal is to remove as much desirable components from the fruit as possible also without extracting undesirable compounds. Some fruits must be carefully peeled and deseeded or cored prior to juicing.

Macerating enzymes with hydrolytic activity against fibrous plant materials such as pectin, hemicellulose and cellulose can greatly increase the yield and subsequent pressing/clarification steps.

After pressing, fruit juices are naturally cloudy, especially because of the presence of polysaccharides (pectin, cellulose, hemicelluloses, lignin and starch), proteins, tannins and metals. Enzymatic treatment (depectinization), cooling, flocculation (gelatin, silica sol, bentonite and diatomaceous earth), decantation, centrifugation and filtration are typically used to produce clear juices.

The concentration step, usually performed by multistage vacuum evaporation, simplifies juice handling, storage and shipping logistics.

Most of the conventional technologies involved in the production of fruit juices are characterized by some drawbacks related to the quality of the product, the energetic consumption and the environmental impact. For example, the use of fining agents in the clarification step is associated with risks of dust inhalation with consequent health problems because of handling and disposal, environmental problems and significant costs. The thermal treatment by pasteurization inactivates native enzymes, reduces microbial load and affects an increase in juice yield. However, delicate flavors can be destroyed and unacceptable darkening occurs due to enzymatic and nonenzymatic browning. High energy consumption, off-flavor formation, color change and reduction in nutritional values are also typical drawbacks because of thermal effects of evaporation processes for juice concentration.

Membrane processes are interesting alternatives for the use of conventional separation systems for the clarification, fractionation and concentration of liquid foods. They operate at room temperature, exhibit low-energy consumption and high performance and they scale-up easily and reject a wide range of food contaminants [4].

In the field of fruit juice processing membrane, operations cover a wide range of applications. Typically, pressure-driven membrane operations, including microfiltration (MF), ultrafiltration (UF), nanofiltration (NF) and reverse osmosis (RO) represent the state-of-the-art technology for juice clarification, fractionation and concentration [5]. In the past years, other membrane operations such as electrodialysis (ED), osmotic distillation (OD), membrane distillation (MD) and pervaporation (PV) have been used for deacidification, concentration and aroma compounds recovery [6].

This chapter gives an outlook on the most relevant applications of membrane operations in fruit juice processing as an alternative to conventional methodologies. A special attention is paid to the combination of membrane operations in integrated systems that can play a key role in redesigning the traditional flowsheet of fruit processing industry with remarkable benefits in terms of product quality, plant compactness, environmental impact and energetic aspects.

8.2 Fruit juice clarification

Clarifications based on membrane processes, particularly MF and UF, have replaced conventional fining, resulting in the elimination of the use of fining agents and a simplified process for continuous production. In addition, low temperatures used during the process preserve freshness of the fruit juice, aroma and nutritional value. In these processes, the juice is separated into a fibrous concentrated pulp (retentate) and a clarified fraction free of spoilage microorganisms (permeate), improving the microbiological quality of the clarified juice.

Polysulfone (PS), polyvinylidene fluoride (PVDF), polyamide (PA) and polypropylene (PP) membranes are largely used in fruit juice clarification. The major drawback of polymeric membranes is their low stability in drastic conditions of pH and, consequently, limited shelf life for juice processing applications. Ceramic membranes have greater resistance to chemical degradation and much longer shelf life; however, their cost is higher if compared to that of polymeric membranes.

Tubular, hollow-fiber and plate-and-frame membrane modules are the most used configurations for the clarification of fruit juices at industrial level. For pulpy juice with high solid content and viscosity, large bore-tubular modules or plate and frame modules with large spacers are preferred.

The productivity of MF and UF membranes in fruit juice clarification is strongly affected by concentration polarization and fouling phenomena. Concentration polarization is a reversible phenomenon because of accumulation of rejected solutes such as hydrocolloids, macromolecules and other relatively larger solutes at the membrane surface leading to an additional hydraulic resistance to the permeate flux with time in the filtration process.

Membrane fouling is because of the deposition and accumulation of feed components on the surface and/or within the membrane pores leading to an irreversible pore blockage. It is a key factor affecting a membrane system's economic and commercial viability, as it reduces productivity and potentially shortens membrane life. Fouling is a function of time and depends strongly on the operating and fluid dynamic conditions, volumetric reduction factor, nature of the membrane and type of feed solution [7].

Fruit juices contain a high concentration of pectin, cellulose, hemicelluloses and proteins that make the juice highly viscous thereby impeding filtration and subsequent concentration processes. The use of pectinases added to the bulk juice before membrane filtration allows hydrolyzation of pectins that reduce the juice viscosity and the formation of fouling layers on the membrane surface [8]. Pectic enzymes include the following: de-esterifying enzymes (pectinesterases), which catalyze de-esterification of the methoxyl group of pectin forming pectic acid; depolymerizing enzymes (hydrolases and lyases) and protopectinases [9].

The enzymatic treatment of mosambi juice followed by adsorption with bentonite produced a significant improvement in the UF process in comparison with other pretreatment methods, including centrifugation and fining by gelatin or bentonite [10]. An improvement in productivity of MF and UF membranes was also observed in the treatment of depectinized pineapple and cherry juices [11]. An enzymatic concentration of 20 mg/L was considered to be economically advantageous in comparison to higher concentrations (100 and 300 mg/L). Maximum permeate flux in the UF of apple juice was achieved with an enzyme concentration of 300 Ferment Depectinization Unit (FDU)/g pectin. Higher amounts of enzyme did not result in any improvement in the permeate flux [8].

Echavarría et al. [12] found that depectinization increased the permeate flow of apple juice by 53.1% when the pectinolytic enzyme preparation recirculated across tubular UF membranes of 100 and 300 kDa.

The optimal enzymatic concentration for achieving maximum permeate flux and steady-state conditions during filtration of black currant juice using a 100 kDa UF was estimated to be 200 mL/100 L juice [13]. The enzymatic treatment also had a positive effect on the release of anthocyanins and flavonols in the juice.

Immobilization of enzymes on MF and UF membranes has been also investigated. The use of pectinase immobilized on UF membranes is expected to hydrolyze pectin to lower molecular weight compounds at the membrane–permeate interface, thus improving the membrane productivity or extending the membrane operation without cleaning [14].

The immobilization of commercial pectinase on hollow fiber PS membranes was found to retard the rate of formation of gel layer during UF of apple juice. The permeate flux was kept at sufficiently high values to reduce the membrane cleaning steps and extending the filtration time [15].

Permeate fluxes and quality of clarified juices in both MF and UF processes are strongly affected by operating conditions, such as transmembrane pressure (TMP), cross-flow velocity, temperature and volume reduction factor (VRF), nature of the membrane and nature of the feed solution.

At low pressures, the permeate flux is proportional to the applied pressure; at higher pressures, the flux becomes independent of the pressure and it is controlled by the mass transfer where the concentration polarization layer reaches a limiting concentration, according to the gel polarization model [16]. A concentration profile from bulk solution to the membrane surface is generated by the rejected material accumulated on the membrane. The formation of a viscous and gelatinous-type layer is responsible for an additional resistance to the permeate flux in addition to that of the membrane. The TMP limiting value (TMP_{lim}) depends on physical properties of the suspension and feed flow rate. The gel-polarized layer is assumed to be dynamic: higher flow rates tend to remove the deposited material that reduces the hydraulic resistance through the membrane and increases the permeate flux [17].

The effect of TMP on the steady-state permeate flux in the clarification of depectinized kiwifruit juice with a cellulose acetate flat-sheet membrane is illustrated in Figure 8.2. Higher flow rates tend to remove the deposited material that reduces the hydraulic resistance through the membrane and consequently, higher permeate fluxes are obtained [18].

According to the film model [19], the permeate flux increases linearly with temperature because of an increase in mass transfer coefficient and reduction in juice viscosity.

A decrease in the permeate flux is observed when the juice concentration is increased. Besides, the method used for extracting the juice affects the suspended solids in the raw juice and, consequently, the productivity of the process.

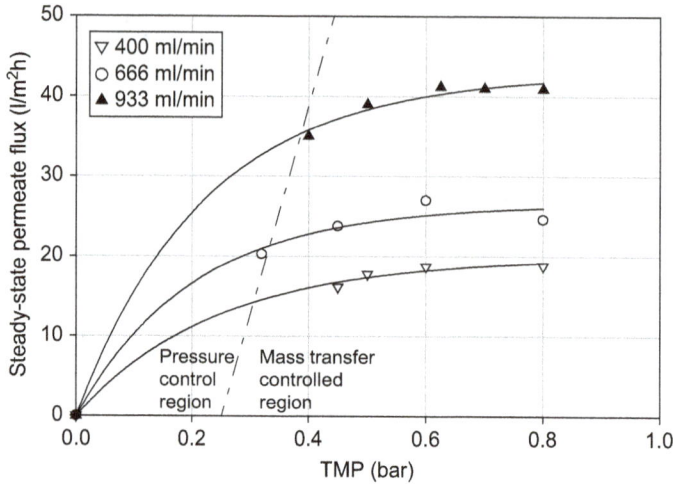

Figure 8.2: UF of kiwifruit juice. Effect of TMP on the steady-state permeate flux at different feed flow-rate (operating temperature, 30 °C) [18].

The permeate flux in the clarification of fruit juices by MF and UF membranes is also affected by the VRF. It is defined as the ratio of initial feed to the final retentate volume. High VRF (>10) can be easily reached with juices with low pulp content such as grape and apple juices. For very pulpy juices, such as tropical fruits, optimum VRF values are less than 5. The yield (1-(1/VRF)) of recovered juice is a crucial item in the process economics.

The effect of VRF on the permeate flux of PS and polyacrylonitrile (PAN) membranes with different molecular weight cut-off (MWCO) in the clarification of blood orange juice is depicted in Figure 8.3. The permeate flux decreases gradually by increasing the VRF because of concentration polarization and gel formation. In particular, the permeate flux versus VRF curve is characterized by an initial period in which a rapid decrease in the permeate flux occurs; a second period, up to VRF 1.5, corresponding to a smaller decrease in permeate flux, and a third period characterized by a steady-state value [20]. As expected, PS membranes with a MWCO of 100 kDa exhibited higher permeate fluxes compared to 50 kDa membranes of the same material.

Various models have been proposed to analyze and predict the flux decline behavior during filtration of fruit juices. Filtration theories based on the formation of a cake layer on the membrane surface that offers a hydrodynamic resistance to permeate flow are commonly used [19]. A practical approach is based on the 'resistance-in-series' model in which the membrane is considered a thickness barrier with an intrinsic hydraulic resistance (R_m) that depends on its manufacturing procedure and respective morphology. An additional resistance (R_c) is related to the concentration polarization and to the cake layer because of high concentration of solutes at the membrane surface. The fouling resistance (R_f) is an extra resistance because of

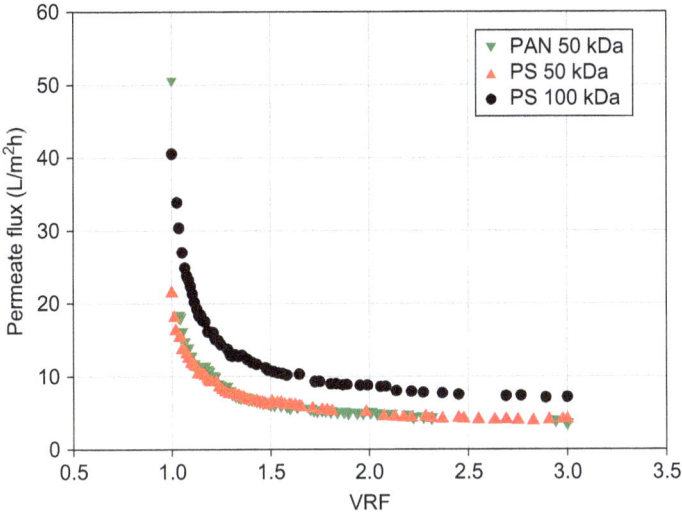

Figure 8.3: UF of blood orange juice. Effect of VRF on permeate flux (batch concentration configuration; TMP, 0.5 bar; Qf, 130 L/h; T, 20 °C) [20].

the irreversible accumulation of juice components on the surface and/or within the membrane pores. Assuming that these resistances – membrane, cake layer and fouling – act in series, the permeate flux can be expressed according to Darcy's law [17]:

$$J = \frac{\text{TMP}}{\mu R_t} \tag{8.1}$$

where J is the permeate flux ($m^3\, s^{-1}\, m^{-2}$), TMP is transmembrane pressure (kPa), μ is cinematic fluid viscosity (Pa s) and R_t is the total resistance (m^{-1}) expressed as follows:

$$R_t = R_m + R_c + R_f \tag{8.2}$$

The intrinsic hydraulic resistance of the membrane can be calculated from the initial water flux (J_w) of the membranes. The resistance related to fouling and concentration polarization can be estimated using water flux measurements taken after rinsing the membrane subsequent to filtration runs.

Juice depectinization allows reduction of resistance becasue of enzyme-induced aggregation with a positive effect on the permeate flux [21, 22]. Similarly, the estimated total resistance after MF of an enzyme-treated acai pulp with 0.8 μm ceramic membranes resulted in 21% lower than the untreated pulp [23]. The resistance rate related to fouling dropped to 27% after enzymatic treatment.

Figure 8.4 shows the influence of TMP on the total, fouling and cake layer resistance analyzed in the pomegranate juice filtration by using modified poly(ether ether ketone) (PEEKWC) and polysulfone (PS) hollow fiber (HF) membranes [24].

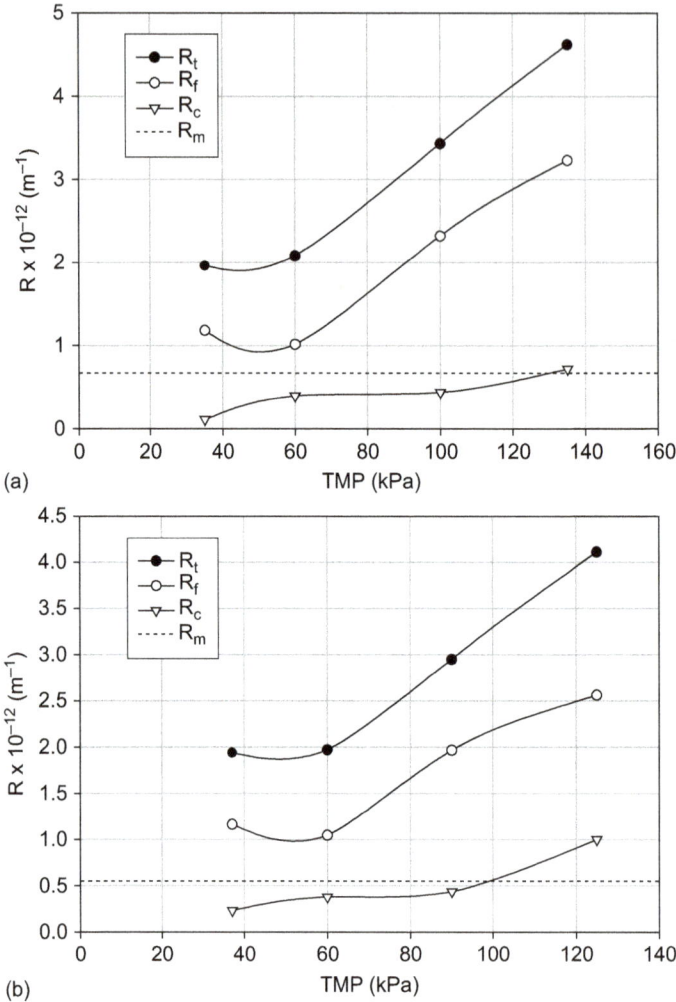

Figure 8.4: Effect of TMP on resistances for (a) PEEK-WC hollow fiber membranes ($T = 25\ °C$; $vf = 3.80\ m\ s^{-1}$) and (b) PS hollow fiber membranes ($T = 25\ °C$; $vf = 3.80\ m\ s^{-1}$) [24].

For PEEKWC membranes, R_t is increased in the range of investigated pressures and the permeate flux was controlled by R_f. Starting from 60 kPa, the contribution of R_f to R_t increased by increasing the operating pressure: this phenomenon was attributed to the enhanced flux and convective flow of the solute toward the membrane. Consequently, more solutes passed through the membrane pores determining an increase in the fouling resistance. In the investigated range of TMP values, the contribution of R_f to the total resistance was always higher than R_c (Figure 8.4a). An increase in TMP produced an increase in the total resistance also for PS membranes (Figure 8.4b). R_f controlled the permeation flux in the range 35–100 kPa.

Above this value the permeation flux was also controlled by R_c (for pressures higher than 100 kPa, R_c was higher than R_m).

Vladisavljević et al. [25] evaluated the effect of operating parameters such as TMP, feed flow rate and temperature on the fouling resistance in the clarification of apple juice with ceramic UF membranes. Both R_t and R_f decreased with feed flow rate because an increase in flow rate enhanced mass transfer coefficient and reduced concentration polarization and accumulation of retained solutes on the membrane surface.

For small TMPs, R_f significantly decreased with increasing feed flow rate, which was because of a higher rate of solute back transfer. At relatively high operating pressures (above 300 kPa), the steady-state R_f was virtually independent on the feed flow rate.

A resistance-in-series model was also proposed to quantify the flux decline during UF of depectinized mosambi juice [26]. Authors found that R_f increased with pressure and decreased with Reynolds number. R_f varied between 6 and 10 times R_m in the selected operating conditions. The contribution of R_f was between 60% and 74% of the overall total resistance.

Recently, Dahdouh et al. [27] evaluated the possibility of predicting the filterability of fruit juices, according to their intrinsic characteristics such as physical and chemical, elemental analysis, particle size distribution, rheological behavior variables and capillary suction time. Partial least squares (PLS) method was particularly used for the analysis of relationship between the filterability of fruit juices and characteristics of the juice studied.

Autoregressive integrated moving average (ARIMA) models have been also recently investigated to predict the permeate flux in the UF of fruit juices during 6 h of continuous operation. Models were constructed with the filtration data of bergamot, kiwifruit and pomegranate juices clarified with different membranes [28].

The treatment of fruit juices by MF and UF membranes produces a great reduction in haze and viscosity without modifying the acidity and the soluble solid content of the juice.

Color and clarity of the juice are improved after filtration because of the removal of suspended colloidal particles and higher molecular weight soluble solids [29].

Suspended solids were completely removed from blood orange juice by using tubular PVDF membranes with a MWCO of 15 kDa [30]. The reduction of ascorbic acid and total antioxidant activity (TAA) in clarified juice with respect to fresh juice was 8.41% and 1.5%, respectively. The rejection of the UF membrane toward total anthocyanins was 9.4%. Total soluble solids (TSS) appeared to be higher in the retentate than in the recovered permeate: this phenomenon was attributed to the presence of suspended solid content in the pulpy products that can interfere with the measurement of the refractive index. These observations were in agreement with results obtained in the clarification of melon juice [31], cactus pear juice [32], orange juice [33] and acerola juice [34] by MF and UF membranes.

Bioactive compounds of kiwifruit juice, such as glutamic, folic, ascorbic and citric acids were recovered in the ultrafiltered juice obtained by using a 30 kDa cellulose acetate membrane. A linear relationship between the VRF of the process and the recovered amount of bioactive compounds in the clarified juice was observed [35].

Recently, Galiano et al. [36] evaluated the functional properties of pomegranate juice (*Punica granatum* L.) clarified by using PVDF and polysulfone (PS) hollow fiber (HF) membranes. PVDF membranes presented a lower retention toward healthy phytochemicals (total phenols, total flavonoids and total anthocyanins) in comparison to PS membranes. Accordingly, the juice clarified with PVDF membranes showed the best antioxidant activity. These results were explained by assuming the key role of membrane material on membrane–solute interactions, and consequently, on membrane selectivity. Indeed, PS is made of a carbon chain alternating aromatic and aliphatic units that are responsible for the hydrophobic character of the polymer, while oxygen and sulfur dioxide subunits provide hydrophilic character to the polymer. On the other hand, PVDF is made of alternating units of CH_2 and CF_2, conferring a hydrophobic nature to the material. The hydrophilic subunits of PS polymer could be, therefore, more prone to create hydrogen bonds and Van der Waals interactions with the hydroxyl groups exhibited by polyphenols, flavonoids and anthocyanins with consequent adsorption of these components at membrane surface. PVDF membranes, on the contrary, were less susceptible to fouling phenomena and highly permeable to the considered compounds.

PVDF UF membranes with a MWCO of 15 kDa showed a higher rejection (80%–90%) toward esters, such as methyl butanoate, ethyl butanoate, ethyl benzoate and methyl benzoate, and a lower rejection toward more polar compounds such as alcohols (3-hexen-1-ol, (E)-2-hexen-1-ol and 1-hexanol) of kiwifruit juice [37]. 25 kDa UF membranes are retained over 80% of the flavored compounds in the passion fruit juice [38]. Aroma compounds of orange juice, mainly terpenic hydrocarbons, were largely retained by MF ceramic membranes with a pore size of 0.2 μm because of their apolar properties and association with insoluble solids of the juice. Oxygenated compounds such as alcohols, esters, aldehydes and terpenols were mainly recovered in the clarified juice [33].

8.3 Fruit juice concentration

Most of fruit juices that are sold commercially worldwide are made from juice concentrates. They can be frozen and sold as frozen concentrate, or they can be reconstituted with water and sold as "from concentrate" juices.

The concentration of fruit juices allows reducing the juice volume with consequent reduction in transport, storage, handling and packaging costs. In addition, concentrated juices are more stable, presenting higher resistance to microbial and chemical deterioration than the original juice as a result of reduction in water activity. Finally, concentrated juices can be easily handled for a final drying treatment.

Traditionally, fruit juice concentration is carried out by multistage vacuum evaporation. Unfortunately, the thermal processing promotes significant changes in the sensory and nutritional quality of the product because these characteristics are conferred by thermosensitive compounds such as aromas and vitamins. Besides, color changes and "cooked" notes recognized as off-flavors are recurring issues in thermally concentrated juices [39].

Membrane technologies show great potential in mild low-temperature treatment of fruit juices with low energy consumption. Nanofiltration (NF), reverse osmosis (RO), direct osmosis (DO), MD and OD are typical membrane operations investigated for juice concentration on both laboratory and industrial scale [40].

8.3.1 Nanofiltration

NF is an intermediate membrane process between UF and RO. It is typically used for the separation of multivalent ions and uncharged organic solutes with molecular weight in the range of 100 and 1,000 Da. NF membranes are characterized by pore diameters in the range of 1–3 nm. They operate at lower pressures (generally in the range of 3–30 bar) than RO membranes. NF membranes are characterized by higher fluxes and lower energy consumption than RO membranes, and better retention than UF membranes for lower molar mass molecules such as sugars, natural organic matters and ions [41].

The concentration of apple and pear juice by using different NF membranes in tubular and flat-sheet configuration at low pressures (8–12 bar) was investigated by Warczok et al. [42]. Results indicated that both retention and permeation values play a key role in the selection of a specific membrane and irreversible fouling of fruit juices is relatively low.

The performance of different NF membranes in terms of permeate flux and retention of anthocyanins from clarified açai juice was investigated by Couto et al. [43]. Among the investigated membranes, the NF 270 membrane is a composite membrane composed of a polyamide top layer and a polysulfone microporous support that presented the highest productivity as well as the highest retention toward anthocyanins (above 99%).

Watermelon juice was concentrated by using a PVDF NF membrane in a spiral-wound configuration with a MWCO of 150–300 Da in selected operating conditions (HL2521TF, GE Osmonics) [44]. The content of lycopene, flavonoids and total phenolic in concentrated samples increased by increasing the VRF. Average permeation fluxes were 2.3 L/m^{2}h with continuous extraction of the concentrate at a VRF of 3. Lycopene showed the highest rejection coefficient (0.99), followed by flavonoids

(0.96) and total phenolic content (0.65). The same membrane was used to concentrate bioactive compounds of strawberry juice at an operating temperature of 20 °C and a TMP of 6 bar [45]. Average permeation fluxes of 4.0 L/m²h and 3 L/m²h were obtained for natural juice and microfiltered juice, respectively. For both juices, the total phenolic content, the anthocyanins content and the antioxidant activity increased by increasing the VRF. The major anthocyanin detected by HPLC, pelargonidin-3-O-glycoside, presented retention up to 95%.

Recently, NF membranes with MWCO around 200 Da were evaluated for the concentration of phenolic compounds in blackberry juice [46]. Experimental results indicated an increased retention of total anthocyanins (>94%) and TSS (44%–97%) by increasing the TMP. Among the investigated membranes, the NF270 membrane showed the highest potential for concentration of phenolic compounds and juice deacidification (at high pressures sugars were completely retained and retention of acids was under 90%).

Similarly, NF membranes were used to produce concentrated phenolic fractions from clarified bergamot juice [47]. Commercial NF membranes (NFPES 10, N30F and NF270) with different MWCO and polymeric material were evaluated in terms of productivity and selectivity toward compounds of interest (flavonoids and total phenolic compounds), TAA and sugars. For all investigated membranes, the rejection of polyphenols increased by increasing the TMP (Figure 8.5). According to rejection data toward sugars and flavonoids (naringin, hesperidin and neohesperidin), the NF PES10 membrane exhibited the best performance in terms of purification of flavonoids from sugars (the largest gap between the rejection coefficients toward sugar

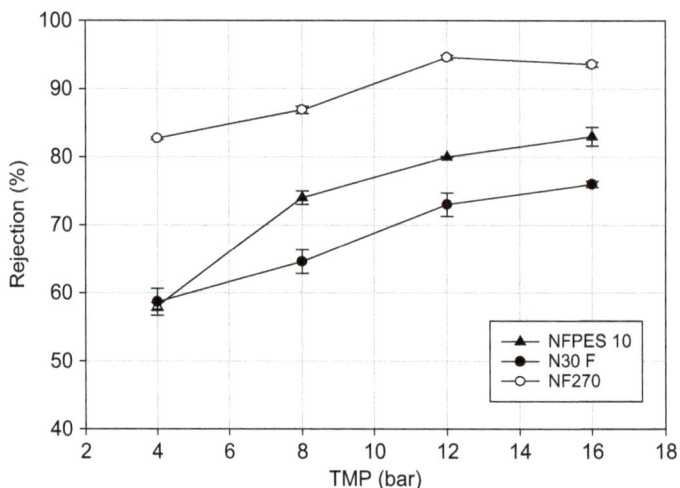

Figure 8.5: NF of clarified bergamot juice with different polymeric membranes. Polyphenols rejection at different TMP values [47].

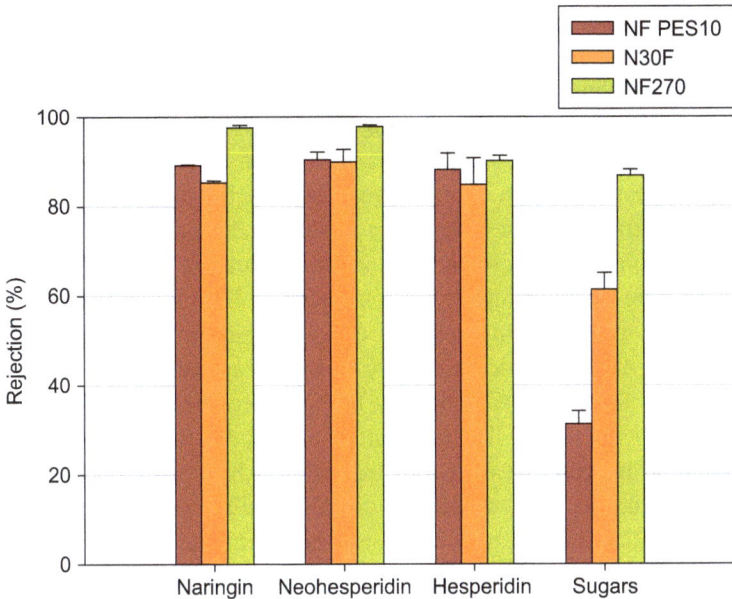

Figure 8.6: NF of clarified bergamot juice with different polymeric membranes. Rejection of NF membranes toward flavonoids and sugars [47].

and phenolic compounds) (Figure 8.6). The produced retentate fractions, characterized by high antioxidant activity, were considered of interest for nutraceutical applications. The hydrophobic character of NF PES10 and N30F membranes accounted for their higher fouling index when compared to the NF270 membrane.

8.3.2 Reverse osmosis

RO membranes mainly separate water from the juice rejecting nutritional, aroma and flavored compounds. Typical advantages of RO over traditional evaporation are in terms of minimal heat damage of the quality properties, maintenance of sensorial and nutritional properties of the product, absence of caramelization reactions, use of compact installations, reduced energy consumption and lower capital investments [48]. However, the use of high operating pressures is needed to overcome the osmotic pressure of the juice (ranging from 10 to 200 bar).

The retention of juice constituents and the permeate flux are strictly correlated with the type of membranes and the operating conditions used during the process.

Grape juice was concentrated by using a plate and frame RO module composed of HR98PP thin film composite membranes to 30 ° Brix [49]. The quality parameters of the concentrated juice were not modified in comparison to those of the single strength juice independentely by the process conditions. The selected conditions for the concentration of grape juice were 30 ºC and 60 bar due to adequate value of the permeate flux, the maintenance of the product quality and membrane integrity.

Jesus et al. [50] evaluated the quality of orange juice concentrated by plate and frame RO membranes, taking into account the concentration factor, soluble solids content and vitamin C concentration. The batch process carried out at 60 bar presented a greater concentration factor (5.8) and resulted in a 30 °Brix final product. The concentrated juice presented higher acidity, vitamin C content and viscosity in comparison to the single strength juice as a consequence of increased pulp and soluble solids contents. It best maintained the characteristic aroma of the fresh juice when compared to the juice concentrated by thermal evaporation.

The permeate flux in RO is affected mainly by TMP, followed by temperature and feed flow rate. The permeate flux in the treatment of apple juice through a tubular polyamide RO membrane increased by increasing the TMP. The high permeate flux was obtained at high feed velocities, but it considerably diminished for values higher than 1.5–2 m/s. The permeate flux decreased by increasing the soluble solid content of the juice due to the increased osmotic pressure and viscosity. An increase in membrane productivity is also obtained when temperature is raised due to increased diffusivity coefficient of the feed solution and the corresponding decrease in the juice viscosity [51].

The drawbacks of RO are its inability to reach the concentration of standard products produced by evaporation because of the limitation of high osmotic pressure. For cellulosic and noncellulosic membranes, the most efficient flux and solute recovery were obtained at a concentration lower than 30 °Brix [52]. For these limitations, RO can be considered an advantageous technique as a preconcentration step before a final concentration of the juice with other technologies (freeze concentration, thermal evaporation, MD and OD).

8.3.3 Membrane distillation and osmotic distillation

MD and OD have attracted considerable interest in the concentration of thermosensitive solutions, such as fruit juices, because they work under atmospheric pressure and room temperature, thus avoiding thermal and mechanical damage of the solutes [53].

These processes are based on water vapor transfer promoted by a vapor pressure difference generated between the two sides of a macroporous hydrophobic membrane. In the OD process, the membrane separates two liquid phases at different solute concentration: a dilute solution on one side and a hypertonic salt solution on the opposite side. As stripping solutions, organic solvents (e.g., polyglycerol and glycerol) or inorganic salts (e.g., $CaCl_2$, NaCl, $MgCl_2$, and $MgSO_4$) can be applied.

The hydrophobic nature of the membrane prevents penetration of the pores by aqueous solutions, creating air gaps within the membrane. The difference in solute concentration, and consequently in water activity of both solutions generates a vapor pressure gradient across the membrane causing a transfer of water vapor across the pores from the high-vapor pressure phase to the low one at the vapor–liquid interface [54].

In MD, the physical origin of the vapor pressure difference is a temperature gradient rather than a concentration gradient: the feed is maintained at high temperature while cold water is used as a stripping permeate [55]. In these conditions, a net pure water vapor flux from the warm side to the cold one occurs.

Typical hydrophobic membranes for MD and OD applications are those realized in polypropylene (PP), polyethylene (PE), PVDF and polytetrafluoroethylene (PTFE) with high porosity (70%–80%) and a thickness of 10–300 µm.

OD has been investigated on both laboratory and pilot scale to concentrate several fruit juices, including kiwifruit [56], pineapple [57–59], grape [60, 61], passion fruit [62], camu-camu [63, 64], noni [65], orange [33, 66] and apple [61, 66] juices. These studies demonstrate that OD is capable of the concentration of fruit juices up to a high TSS content. This high concentration prevents the juice from deterioration.

Hollow fiber membrane contactors are preferred when compared to flat-sheet membranes for industrial-scale OD processes because of their high active surface area, low manufacturing cost and easy scale up.

Feed and brine flow rates as well as brine concentration are the most important parameters that affect the water flux in OD. Flow rates directly affect the thickness of the boundary layer at the membrane surface that presents a resistance to the mass transfer, whereas the concentration of brine affects the vapor pressure gradient through the membrane, which is directly related to the magnitude of the driving force. The contribution of concentration polarization on transmembrane flux is more prominent when compared to that of temperature polarization [57].

Higher evaporation fluxes are obtained for clarified juices in comparison with single strength juices indicating a clear effect of pulp on the performance of the OD process [58].

The concentration of clarified pomegranate juice by OD was investigated by Cassano et al. [67]. An OD membrane module containing PP hollow fiber membranes and $CaCl_2$ as stripping solution were used to concentrate the juice from 162 g/kg^{-1} to 520 g TSS/kg. The time evolution of evaporation flux, TSS and brine concentration is illustrated in Figure 8.7. In the first part of the process (range 0–80 min), the decline in evaporation flux (Figure 8.7a) can be attributed to the decrease in brine concentration (Figure 8.7b). The dilution of the brine solution leads to a decrease in vapor pressure of the osmotic agent reducing the driving force of water transport through the membrane [68]. In the second part of the process (range 80–140 min), the evaporation flux is declined by 30.2% corresponding to a decrease in brine concentration of 7.4%. This result can be attributed to the strong influence of the concentration level on the evaporation flux. Indeed, the increase in TSS content of the juice results in an exponential increase in juice viscosity. Consequently, at low TSS of the feed juice, the flux decay is more attributable to the dilution of the stripping solution; at higher TSS

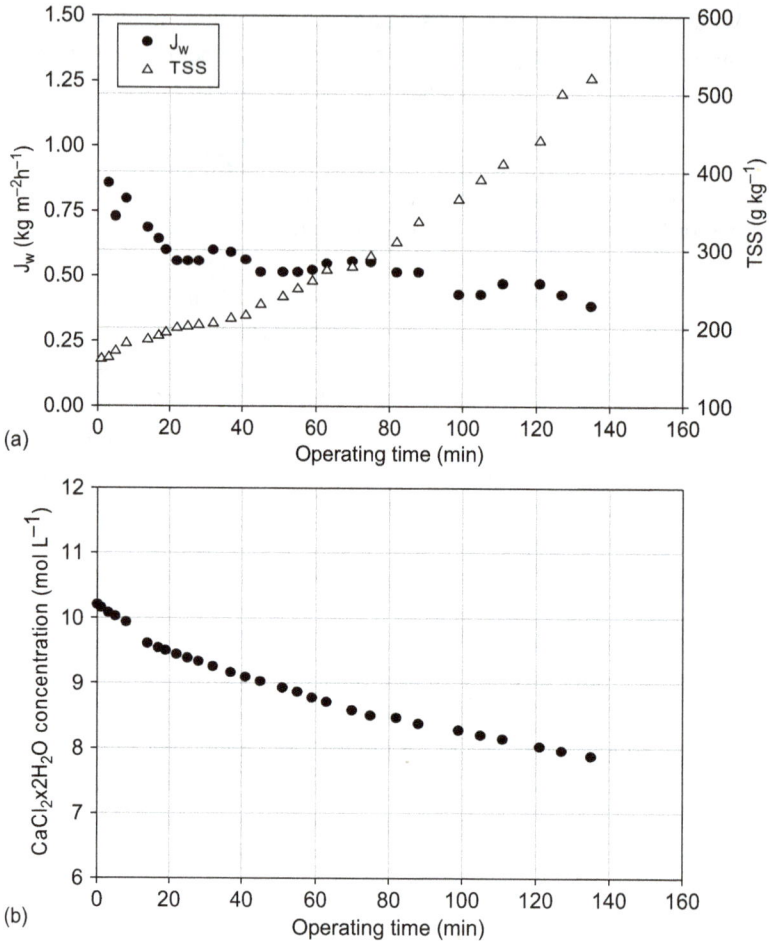

Figure 8.7: Concentration of clarified pomegranate juice by OD. Time course of (a) evaporation flux and TSS; (b) concentration of brine solution. (Operating conditions: $T = 25 \pm 2\ °C$; $Q_j = 500$ mL/ min; $Q_b = 333$ mL/min; TMP = 0.4 bar) [67].

values, it mainly depends on juice viscosity (viscous polarization) and, consequently, on juice concentration and temperature. This behavior was also observed in the concentration of different fruit juices [56, 62] and sucrose solutions [69] by OD.

As reported in most investigated applications, the OD process is very efficient in maintaining the nutritional quality of the original fresh juice through the preservation of thermosensitive compounds (i.e., vitamins, phenolic compounds) and antioxidant activity [33,56,61,63,65].

In terms of sensorial quality, some losses of aroma compounds were observed during the initial concentration of clarified orange juice by OD [33]. These losses can be drastically limited by preconditioning the OD membrane with the clarified juice from the clarification unit and by avoiding thermal regeneration of brine during concentration.

The OD concentration of pasteurized pineapple juice produced a 51 °Brix concentrate, retaining an average of 62% of the volatile compounds present in the initial juice [59]. OD membranes with relatively large pore sizes at the surface exhibited higher organic volatile retention per unit water removal than those with smaller surface openings [70]. This phenomenon was attributed to a greater intrusion of the liquid feed and brine streams with a resulting increase in the thickness and resistance of the boundary layer entrance of larger pores.

The main technical and economic limitation of OD is the dilution of the extraction brine, because the reconcentration and reutilization of this solution involve important costs. However, these limitations could be solved by the high commercial price of the juice and the interest to offer a high quality product [65].

As with OD, the MD process can take place at atmospheric pressure and at a temperature lower than the normal boiling point of the feed solutions preserving thermosensitive compounds during juice concentration. In direct contact membrane distillation (DCMD), both sides of the membrane are charged with liquid-hot feed water on the evaporator side and cooled permeate on the permeate side. This configuration offers some key advantages because of its suitability for applications in which the volatile component is water [71].

Commercial plate PVDF membranes with a nominal pore size of 0.1 µm were used in the concentration of orange juice by DCMD [72]. These membranes exhibited a very good retention toward soluble solids, sugars and organic acids, with a 100% rejection in sugars and organic acids. An increase in permeate flux was observed by pretreating the juice by UF.

Black currant juice was concentrated by MD up to 58.2 °Brix at different temperature differences by using a laboratory-size hollow fiber hydrophobic PP membrane module [73]. Analytical measurements showed that density, total anthocyanins and acidity of the juice increased proportionally to the increase of the TSS in the measured range.

A few degree centigrade increase in the driving force (ΔT from 15 °C to 19 °C) influenced the distillate flux significantly and the operation time of the MD process: steady state fluxes at ΔT of 15 °C and 19 °C were of 0.45 and 0.8 kg/m²h, respectively.

Higher evaporation fluxes (about 9 L/m²h) were measured by Gunko et al. [74] in the concentration of ultrafiltered and depectinized apple juice by using PVDF membranes with a mean pore size of 0.45 µm at a TSS content of 50 °Brix. Further concentration of the juice up to 60–65 °Brix resulted in a reduced productivity (3.8–3.0 L/m²h).

Recently, a two-step DCDM process has been investigated for the concentration of orange juice. The clarified juice at 9.5 °Brix was preconcentrated up to 24 °Brix and then concentrated up to 65 °Brix by using a PP hollow fiber membrane module (Enka

Microdyn MD-020-2N-CP) having a nominal pore size of 0.2 mm and a membrane surface area of 0.1 m² [75].

The performance of the MD process in terms of evaporation flux (Figure 8.8a) revealed that in the preconcentration step the flux decay was strictly correlated with the reduction in the average temperature difference between the feed and permeate side (Figure 8.8b). On the other hand, the decrease in evaporation flux observed during the subsequent concentration step was mainly because of the increase in the juice viscosity (Figure 8.8d).

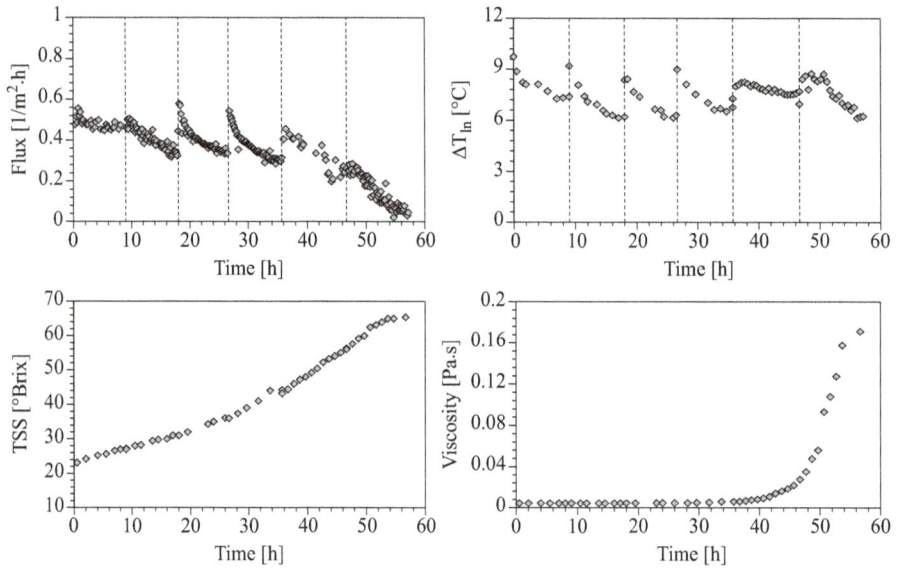

Figure 8.8: Concentration of blood orange juice by DCMD. Time evolution of (a) evaporation flux, (b) thermal gradient, (c) total soluble solids content and (d) viscosity [75].

Typical advantages of MD compared to other traditional technologies for the concentration of fruit juice are as follows: high quality of concentrates, low operating temperatures (24 °C–48 °C), possibility to achieve high contents of dry substances (60%–70 %) and low energy costs.

Membrane pore wetting, low evaporation fluxes and flux decay represent typical disadvantages that still limit the development of MD on large scale [76].

Alves and Coelhoso [77] compared the performance of OD and MD processes in terms of water flux and aroma retention in the concentration of orange juice by using a hollow fiber membrane contactor. By applying a similar overall driving force, the MD flux resulted less than half of the value observed in OD. This was attributed to the temperature gradient created between the bulk and the membrane interface, which reduces the driving force for water transport in MD. Regarding the transport of aroma compounds (ethyl butyrate and citral) through the membrane, a higher retention per amount of water removal was observed in OD.

Owing to evaporation at the feed side and the condensation at the brine side, a temperature difference occurs in OD and this thermal effect reduces the driving force of the water transport through the membrane. In order to enhance water flux, OD can be combined with MD in a way where brine and feed solutions are thermostated separately at different temperatures. This coupled operation is known as *membrane osmotic distillation* (MOD). In the approach investigated by Bélafi-Bakó and Koroknai [78], clarified apple juice was concentrated by using an osmotic agent (CaCl$_2$) recirculated in the lumen side of tubular PP membranes at lower temperature than the shell side (temperature difference was maintained at 10, 15 and 20 °C). The combination of driving forces resulted more effective than MD or OD alone producing higher evaporation fluxes and faster juice concentration. Similarly, red fruit juices (chokeberry, red currant and cherry juices), previously clarified by UF, were concentrated by MOD maintaining different bulk temperatures on each side of PP hollow fiber membranes and using an osmotic salt solution as the receiving phase [79]. Produced concentrated juices showed an excellent preservation of the TAA (more than 97%) of the feedstock, confirming mildness and effectiveness of coupled operation.

Onsekizoglu et al. [80] evaluated the impact of different concentration processes, including MD, OD, MOD and conventional thermal evaporation on the quality of apple juice up to 65 °Brix. The fresh apple juice, with an initial TSS content of 12 °Brix was previously clarified by a combined application of fining agents (gelatine and bentonite) and UF and then concentrated up to 65 °Brix.

The concentrated juice presented nutritional and sensorial quality similar to that of the original juice, especially regarding the retention of the bright color and pleasant aroma, which are lost during thermal evaporation. In addition, the formation of 5-hydroximethilfurfural, an indicator of the Maillard reaction, resulted only in thermally evaporated juices. The content of phenolic and organic acids and sugars was preserved during the different concentration processes, whereas the aroma profile, studied in terms of trans-2-exenal, was remarkably lost in apple juice concentrates produced by thermal evaporation. A higher retention of trans-2-exenal was observed in the OD process in comparison to MD. The MOD process allowed to retain the aroma profile of the initial juice was proposed as the most promising alternative to the conventional thermal evaporation technique.

A comparison of key factors of conventional evaporation and membrane concentration processes is reported in Table 8.1. The low relatively permeation fluxes attainable with OD and MD make these processes economically uncompetitive when compared with thermal evaporation and pressure-driven membrane operations. When, however, the product to be concentrated is a solution containing macrosolutes or colloids that is sensitive to shear or thermal degradation MD and OD have important advantages.

Table 8.1: Key factors of conventional evaporation and membrane concentration techniques.

Process	Maximum achievable concentration (°Brix)	Product quality	Evaporation rate or flux	Possibility of treating different products with the same installation	Operating cost	Capital investment	Energy consumption
Evaporation	60–70	Poor	200–300 L/h	No	Moderate	Moderate	Very high
Reverse osmosis	25–30	Very good	5–10 L/m^{2}h	No	High	High	High
Direct osmosis	50	Good	1–5 L/m^{2}h	Yes	High	Moderate	Low
Membrane distillation	60–70	Good	1–10 L/m^{2}h	Yes	High	Moderate	Low
Osmotic distillation	60–70	Very good	1–3 L/m^{2}h	Yes	High	High	Low

8.4 Aroma recovery

The aroma in fruit juices is formed by a mixture of hundreds of different organic compounds present at very low concentrations (typically ppm or ppb levels) characterized by extreme volatility. Consequently, these compounds are changed or lost during thermal processing of fruit juices such as concentration or pasteurization. A possible way of minimizing the flavor quality of processed juices is to use separation techniques for the recovery of aroma compounds and their addition back to the concentrated juice before packing.

Currently, volatile aroma compounds are removed in the vapor phase in the first step of the evaporation process and the aroma fraction is then trapped by condensation in a separate aroma recovery unit. This approach leads to undesirable reduction of the aroma content and modification of the aroma profile of the juice (for conventional processes, the yield of aroma recovery is limited to 40%–65%). In addition, the evaporation step is the most energy requiring step in industrial fruit juice processing.

Membrane processes, due to their high selectivity and possibility to operate at moderate temperatures, offer an interesting alternative to conventional procedures for aroma recovery.

PV is a membrane process for separating dilute species in liquid solutions extensively studied for aroma recovery. In this process, components of a liquid mixture permeate selectively through a dense membrane under a chemical potential gradient obtained by partial pressure reduction on the permeate side and can be collected by a cold trap [81, 82].

Compared to traditional processes, PV has many advantages including minimum loss of aroma compounds, no heat damage to heat-sensitive aromas, low energy consumption and no additional separation treatment for added solvents or absorbents.

In most PV studies, aroma compounds are concentrated in the permeate by means of preferential permeation through organophilic membranes, which reduce both the membrane area and the required condensation duty for the permeate. Polydimethylsiloxane (PDMS) is by far the most used material for aroma recovery by PV. Other polymers include polyoctylmethyl siloxane (POMS), polyether-block-amide (PEBA), polyurethane, acrylate, ethylene propylene diene terpolymer (EPDM), ethylene-vinyl acetate copolymer (EVA) and styrene-block-styrene (SBS) [83].

Pereira et al. [84] evaluated the performance of various membrane materials with binary and quaternary synthetic aqueous solutions of typical tropical fruit aroma compounds, such as ethyl acetate, ethyl butanoate, ethyl hexanoate and 1-octen-3-ol, as well as with single strength and clarified pineapple juices. Composite flat membranes having selective layers EPDM or EVA and composite hollow fiber membrane having selective layer EPDM were used for comparison. The authors concluded that more permeable membranes are preferred if the feed concentration is high enough to induce phase separation in the permeate after its condensation. Composite EPDM hollow fiber membrane showed the best performance for PV of synthetic and single-strength pineapple juices.

Feed flow rates had no significant effect on the performance of the process. On the other hand, a decreasing permeate pressure increases both permeation flux and enrichment factor, whereas an increase in temperature increases the water flux more significantly than the aroma compounds flux, resulting in lower enrichment factor [85, 86].

An increase in both flux and selectivity was also observed with an increase in feed temperature and decrease in downstream pressure in binary aqueous solutions of typical aroma compounds of strawberry juice [87].

Concentration polarization represents one of the problems to be overcome for the industrial application of PV in aroma recovery. In addition, fluxes of aroma compounds and water (of the order of 0.1 kg/m²h) are much lower than fluxes obtainable with other membrane technologies such as vacuum membrane distillation (VMD) and sweep gas membrane distillation (SGMD).

In VMD, the feed solution is separated into a retentate and a permeate by means of a partial pressure gradient induced through a microporous hydrophobic membrane by the vacuum on the permeate side. The liquid stream vaporizes at the membrane surface and the vapor diffuses through the gas phase inside the membrane pores [88].

High partial pressure gradients and consequently high fluxes are created because the permeate pressure is much lower than the vapor pressure of the diffusing species [89]. The most suitable material for VMD membranes include polymers such as PTFE, PVDF and PP.

Diban et al. [90] investigated the performance of VMD in the separation of ethyl 2,4-decadienoate (the main pear aroma compound) from a model tricomponent containing water/ethanol as solvent by using PP hollow fiber microporous membranes. According to experimental results, temperature and downstream pressure showed a stronger influence on the aroma enrichment factor: the highest values of the enrichment factor, up to 15, were obtained by working at lower temperatures and higher downstream pressures. A strong and reversible sorption of the aroma compound onto the PP material was also observed.

The recovery of black currant aroma compounds in a lab-scale VMD setup at feed flow rate from 100 to 500 L/h at 30 °C gave concentration factors from 4 to 15. Concentration factors increased with decreased juice temperature and increased feed flow rate. At 5 vol.% feed volume reduction, the amount of recovered aroma was between 68 vol.% and 83 vol.% of the highly volatile compounds and between 32 vol.% and 38 vol.% of the poorly volatile compounds [91].

In SGMD, a cold inert gas sweeps the permeate side of the membrane carrying the volatile molecules. In this case, condensation occurs outside the membrane module [92].

Bagger-Jørgensen et al. [93] compared the performance of SGMD and VMD in the recovery of aroma compounds from berry juice. The SGMD flux increased by increasing the temperature, feed flow rate and sweeping gas flow rate. Concentration factors increased by increasing the temperature and feed flow rate. A 73–84 vol.% aroma recovery of the most volatile compounds of black currant juice was obtained at a volume reduction of 13.7% (vol.%), a temperature of 45 °C and a feed flow rate of 400 L/h. The aroma recovery in SGMD was less influenced by the feed flow rate but more influenced by the temperature when compared to VMD. Higher fluxes were achieved during concentration by VMD that allows reduction in the operation time and degradation of polyphenolic compounds in the juice.

8.5 Integrated membrane operations

Integration of membrane operations have been suggested for replacement of conventional juice processing unit operations such as clarification, concentration and aroma recovery.

A first combination of UF and RO membranes for juice clarification and concentration was patented by Lawhon and Lusas [94] in 1987. Authors described the use of UF to separate juice into pulp and serum prior to heat treatment and helps to minimize flavor loss and deterioration. The clarified juice containing almost all the flavor and aroma components is concentrated by RO to levels above 42 °Brix. The UF retentate is recombined with the RO concentrate after heating to inactivate the spoilage of microorganisms. Based on a similar approach, the Fresh Note system was then proposed to concentrate orange juice up to 60 °Brix, whereas almost retaining the fresh juice flavors. The process involves a two-stage RO system employing high- and low-rejection membranes after juice clarification by MF or UF [95].

A process design for an annual production of 3,500 ton apple juice concentrate based on the use of membrane operations and conventional separation systems was proposed by Alvarez et al. [96]. The process involves the enzymatic treatment of the raw juice combined with ultrafiltration, the preconcentration of the clarified juice up to 25 °Brix by RO, the aroma compounds recovery by PV and a final concentration up to 72 °Brix using conventional evaporation (Figure 8.9). A reduction in the total capital investment of 14% and an increase in process yield of 5% was planned in relation to the conventional process. The return of investment was calculated as 34% higher for the membrane-based process.

Kozák et al. [97] investigated a multistep membrane process on both laboratory and large scale for the concentration of black currant juice. Operating parameters were optimized on laboratory scale. In large-scale experiments, the depectinized juice was prefiltered through a 100 µm bag filter and then preconcentrated through a RO flat-sheet membrane module (MFT-Köln) at an operating pressure of 51 bar and a temperature of 24 °C. The concentration was carried out by using a PP hollow fiber membrane module (MD 150 CS 2N, Microdyn) with an average pore size of 0.2 µm. The sensory analysis showed a little loss of aroma compounds in the reconstituted juice if

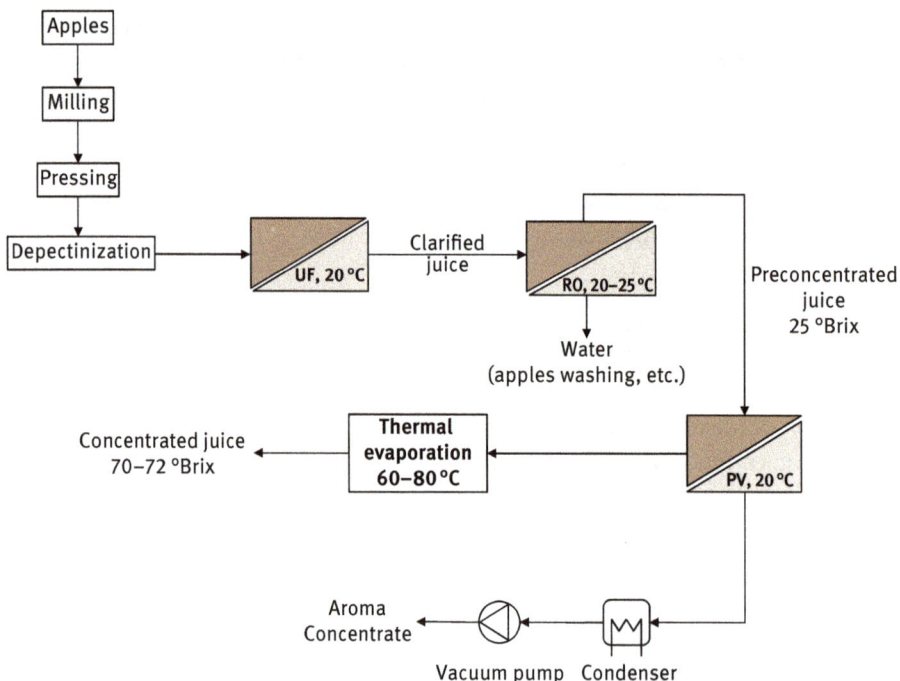

Figure 8.9: Production of apple juice concentrate and apple juice aroma by integration of membrane-based operations and thermal evaporation (adapted from [96]).

compared to the raw juice, whereas the color intensity and the acidic flavor intensity remained unchanged.

Integrated membrane systems based on the combination of UF and OD have been studied on laboratory scale for the clarification and concentration of different fruit juices, including kiwifruit [56, 98], cactus pear [99], orange [100], blood orange [101], pomegranate [67] and bergamot [102] juice. In particular, clarified juices were concentrated up to 60–65 °Brix by using a Liqui-Cell Extra-Flow 2.5×8-in. membrane contactor (Membrana, Charlotte, USA) containing microporous PP hollow fibers (having external and internal diameters of 300 μm and 220 μm, respectively) with an average pore diameter of 0.2 μm and a total membrane surface area of 1.4 m². The clarified juice was recirculated in the shell side of the OD membrane module, whereas calcium chloride dehydrate solution used as stripping solution was pumped through the fiber lumens (tube side) in a countercurrent mode. In all these applications, the nutritional and sensorial characteristics of the fresh juice were well preserved in the final OD retentate. Similarly, the antioxidant activity of the juice was well preserved independently by the final content of soluble solids in the concentrated juice. The composition of anthocyanins in the clarified and concentrated pomegranate juice by UF and OD, respectively, is reported in Table 8.2. In comparison with the clarified

Table 8.2: Anthocyanin compounds composition of pomegranate juice clarified and concentrated by integrated UF/OD process. From Cassano et al. [67].

Operation	Sample	TSS (°Brix)	Cyanidin 3,5-diglucoside (mg L^{-1})	Delphinidin 3-glucoside (mg L^{-1})	Cyanidin 3-glucoside (mg L^{-1})	Pelargolidin 3-glucoside (mg L^{-1})	Total anthocyanins (mg L^{-1})
UF	Feed	16.2	46.95 ± 0.96	20.6 ± 1.64	30.4 ± 1.11	5 ± 0.44	102.8 ± 3.18
	Permeate	16.2	44.26 ± 0.95	17.7 ± 2.00	25.81 ± 2.44	4.2 ± 0.38	90.7 ± 0.71
	Retentate	–	45.82 ± 1.55	20.28 ± 1.52	28.74 ± 1.02	5 ± 0.29	100.6 ± 3.53
OD	Feed	16.2	43.2 ± 1.60	12 ± 0.48	23.4 ± 1.70	4.3 ± 0.43	87.7 ± 5.56
	Retentate t=55 min	25.0	39.4 ± 0.84[a]	12.25 ± 0.40[a]	18.7 ± 1.17[a]	4.5 ± 0.11[a]	74.6 ± 2.34[a]
	Retentate t=88 min	33.6	39 ± 0.43[a]	12 ± 0.66[a]	20 ± 0.95[a]	4.5 ± 0.15[a]	75.55 ± 1.87[a]
	Retentate t=135 min	52.0	39 ± 0.70[a]	11.5 ± 0.60[a]	20.9 ± 0.78[a]	4.4 ± 0.05[a]	75.85 ± 1.74[a]

[a] value referred to a TSS content of 16.2 °Brix

juice, the reduction of total anthocyanins in OD retentates at different content of TSS is about 14%, much lower than typical values measured in thermal evaporation.

The conceptual design proposed for the production of high quality fruit juices, as alternative to conventional processes based on the use of thermal evaporation, is illustrated in Figure 8.10. The raw juice, after an optional depectinization step (depending on the type of treated juice), is previously clarified by UF and then concentrated by OD. Alternatively, the clarified juice can be preconcentrated by RO. The UF retentate can be processed for microbiological stabilization (pasteurization). Being the retentate composition less sensitive to heat than small aroma molecules, vitamins and sugars, it can be pasteurized to inactivate enzymes and microorganisms and then added in adequate proportions to the concentrated juice together with water.

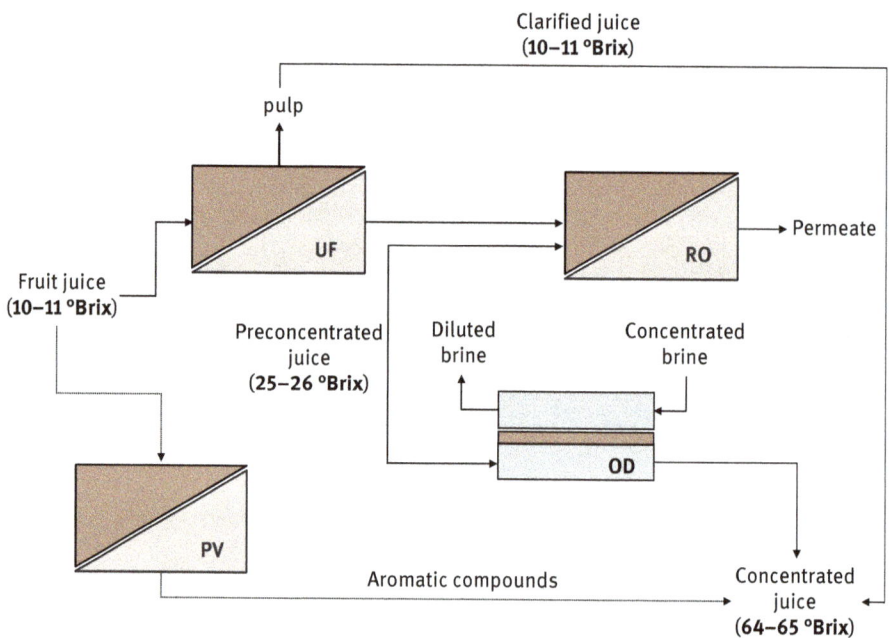

Figure 8.10: Conceptual process design for the production of concentrated fruit juices by integrated membrane operations.

A combination of MF and OD for the clarification and concentration of melon juice was investigated by Vaillant et al. [31]. The juice was clarified by a ceramic multichannel membrane (Membralox® 1P10-40, Pall-Exekia) with an average pore diameter of 0.2 μm and then concentrated by using PP hollow fiber OD membranes. Calcium chloride, used as a brine solution, was circulated in the shell side of the OD module. Physicochemical analyses indicated that suspended solids and carotenoids were totally concentrated in the MF retentate. The retention of β-carotene by the MF membrane was attributed to its strong association with membrane and wall structures of the cell fragments. Microbiological analyses showed that MF can ensure microbiological stability of the juice in a single step. The main physicochemical and nutritional properties of the fresh juice were well preserved in the concentrated juice at 55 °Brix.

The concentration of fruit juices combined with the recovery of aroma compounds through the implementation of integrated membrane systems have been also investigated. Cassano et al. [37] evaluated the potential of an integrated membrane system for the production of concentrated kiwifruit juice based on the use of UF and OD membranes combined with the recovery of aroma compounds by PV. For the majority of the aroma compounds detected, the enrichment factor in the permeate of the fresh juice was higher than the clarified and concentrated juice, respectively (Figure 8.11). Therefore, the use of PV was suggested for the removal of aroma compounds directly from the fresh juice, before clarification and concentration steps.

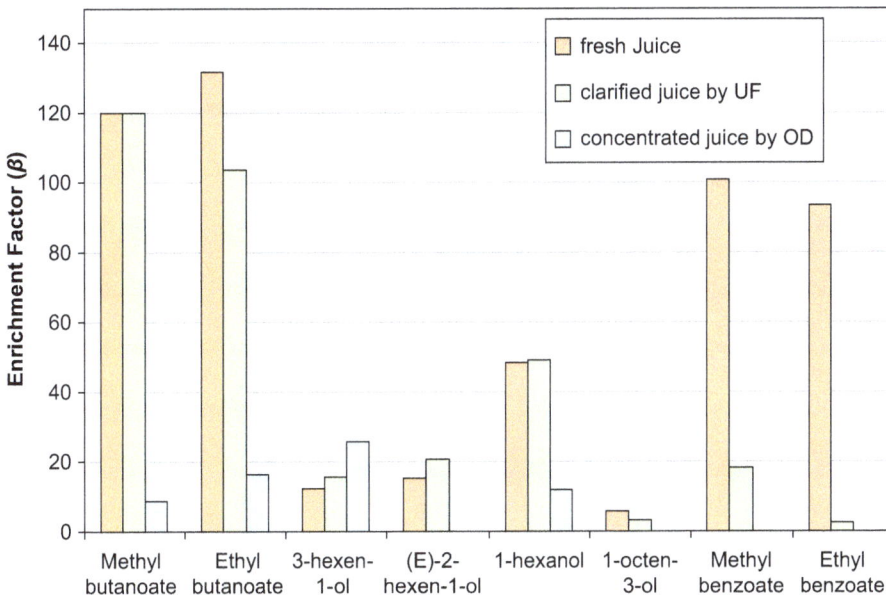

Figure 8.11: Enrichment factor of aroma compounds recovered by PV at 20 °C in the permeate of fresh, clarified and concentrated kiwifruit juice [37].

A conceptual process design for the production of black currant juice concentrate through the integration of membrane-based operations was proposed by Sotoft et al. [103]. The proposed process is based on the recovery of aroma compounds by VMD followed by a first concentration of the juice up to 45 °Brix by using a combination of NF and RO membranes: the raw juice was first treated by RO with a dense membrane having a high degree of sugar retention (99.7%) and then processed by NF. The high rejection of RO membranes and the high concentration factor of NF membranes allowed to overcome the high osmotic pressure limitations typically encountered in RO. The NF retentate was finally concentrated by DCMD producing a concentrated juice with a TSS content of 65–70° Brix. A flowsheet of the integrated process is illustrated in Figure 8.12.

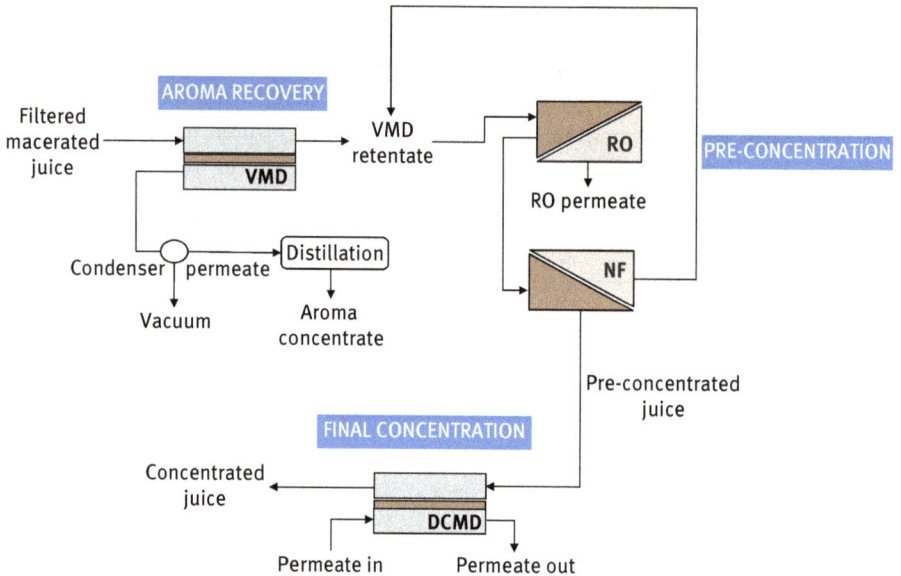

Figure 8.12: Process layout for aroma recovery and juice concentration based on membrane processes (adapted from [103]).

The annual production scale was fixed at 17,283 ton of 66 °Brix concentrated juice/ year with an estimated production price of 0.40 €/kg of concentrated juice based on a membrane lifetime of 1 year. The calculated costs related to membrane lifetimes of 2 and 3 years were 0.30 and 0.27 €/kg of concentrated juice, respectively.

An integrated process based on a combination of VMD and OD process was also investigated in order to overcome the loss of aroma compounds by recovering them from OD brine [104]. In particular, a model solution containing four common aroma compounds was concentrated by using an OD membrane module containing PP hollow fiber membranes (Celgard Liquicel G542) and $CaCl_2$ as stripping solution. This solution was recycled through the shell side of the VMD membrane module, whereas vacuum was applied on the lumen side. Aroma compounds absorbed in the extraction brine were extracted by using the VMD system under vacuum and collected into a cold trap. Experimental results showed a clear effect of feed flow rate on the aroma extraction indicating that the feed side boundary layer resistance mainly dominates the transfer of aroma compounds. On the other hand, brine flow rates did not show a clear correlation with aromatic extraction.

The increase in brine concentration accelerated the OD process but did not influence the extraction percentage of aroma compounds. The total recovery of aroma compounds resulted in an average of 75% (higher than that generally observed in traditional thermal evaporation or in OD). In addition, the process of aroma removal and recovery resulted faster than the concentration process of fruit juices by OD. Therefore the combination of OD and VMD can be exploited to decrease the energy requirements for a given production capacity.

8.6 Conclusions

Membrane-based operations offer several advantages over conventional clarification, concentration and aroma recovery methodologies employed in fruit juice processing industries, including improved product quality, easy scale-up and low energy consumption.

Well-developed process engineering, including pretreatment, selection of appropriate membrane modules and membrane materials, optimization of operating and fluid-dynamic conditions are key factors to overcome intrinsic limitations related to fouling and short lifespan of the membranes. Further developments in this area are expected through the implementation of integrated membrane systems as clearly demonstrated by several applications on laboratory and semi-industrial scale. In this view, the set up and design of innovative integrated membrane systems will lead to redesigning of traditional flowsheets of fruit juice processing industries providing an essential contribution to the concept of process intensification together with a sustainable industrial growth in this field.

References

[1]   Memon NA. Fruit and vegetable juices global market. Pakistan Food J. 2012; September-October: 20–23.
[2]   Slavin JL, Lloyd B. Health benefits of fruits and vegetables. Adv Nutr. 2012; 3: 506–516.
[3]   Hyson DA. A review and critical analysis of the scientific literature related to 100% fruit juice and human health. Adv Nutr. 2015; 6: 37–51.
[4]   Jing L, Howard AC. Applications of membrane techniques for purification of natural products. Biotechnol Lett. 2010; 32: 601–608.
[5]   Echavarría AP, Torras C, Pagan J, Ibarz A. Fruit juice processing and membrane technology application. Food Eng Rev. 2011; 3: 136–158.
[6]   Strathmann H, Giorno L, Drioli E. An Introduction to Membrane Science and Technology. Consiglio Nazionale delle Ricerche; Roma; 2006.

[7] Cassano A, Marchio M, Drioli E. Clarification of blood orange juice by ultrafiltration: analyses of operating parameters, membrane fouling and juice quality. Desalination. 2007; 212: 15–27.

[8] Alvarez S, Alvarez R, Riera FA, Coca J. Influence of depectinization on apple juice ultrafiltration. Colloid Surface A. 1998; 138: 377–382.

[9] Hoondal G, Tiwari R, Tewari R, Dahiya N, Beg Q. Microbial alkaline pectinases and their industrial applications: A review. Appl Microbiol Biotechnol. 2002; 59: 409–418.

[10] Rai P, Majumdar GC, Das Gupta S, De S. Effect of various pretreatment methods on permeate flux and quality during ultrafiltration of mosambi juice. J Food Eng. 2007; 78: 561–568.

[11] Barros S, Mendes E, Peres L. Influence of depectinization in the ultrafiltration of West Indian cherry (*Malpighia glabra* L.) and pineapple (*Ananas comosus* juices). Ciênc Tecnol Aliment 2004. 2004; 24: 194–201.

[12] Echavarría A, Pagán J, Ibarz A. Effect of previous enzymatic recirculation treatment through a tubular ceramic membrane on ultrafiltration of model solution and apple juice. J Food Eng. 2011; 102: 334–339.

[13] Pap N, Mahosenaho M, Pongrácz E, Mikkonen H, Jaakkola M, Virtanen V, Myllykoski L, Horváth-Hovorka Z, Hodúr C, Vatai G, Keiski RL. Effect of ultrafiltration on anthocyanin and flavonol content of black currant juice (*Ribes nigrum* L.). Food Bioprocess Tech. 2001; 5: 921–928.

[14] Alkorta I, Garbisu C, Llama MJ, Serra JL. Viscosity decrease of pectin and fruit juices catalyzed by pectin lyase from *Penicillium Italicum* in batch and continuous-flow membrane reactors. Biotechnol Techniques. 1995; 9: 95–100.

[15] Carrin ME, Ceci L, Lozano JE. Effects of pectinase immobilization during hollow fiber ultrafiltration of apple juice. J Food Process Eng. 2000; 23: 281–298.

[16] Mulder M. Basic Principles of Membrane Technology. Kluwer Academic Publishers, Dordrecht, 1991.

[17] Cheryan M. Ultrafiltration and Microfiltration Handbook. Technomic Publishing Company, Lancaster, 1998.

[18] Cassano A, Donato L, Conidi C, Drioli E. Recovery of bioactive compounds in kiwifruit juice by ultrafiltration. Innov Food Sci Emerg Technol. 2008; 9: 556–562.

[19] Fane AG, Fell CJD. A review of fouling and fouling control in ultrafiltration. Desalination. 1987; 62: 117–136.

[20] Conidi C, Destani F, Cassano A. Performance of hollow fiber ultrafiltration membranes in the clarification of blood orange juice. Beverages. 2015; 1: 341–353.

[21] Araújo MCP, Gouvêa ACMS, Couto DS, Cabral LMC, Godoy RLO, Freitas SP. Effect of enzymatic treatment on the viscosity of raw juice and anthocyanins content in the microfiltrated blackberry juice. Desalination Water Treat. 2011; 27: 37–41.

[22] Yu J, Lencki RW. Effect of enzyme treatments on the fouling behavior of apple juice during microfiltration. J Food Eng. 2004; 63: 413–423.

[23] Machado RMD, Haneda RN, Trevisan BP, Fontes SR. Effect of enzymatic treatment on the cross-flow microfiltration of acai pulp: Analysis of the fouling and recovery of phytochemicals. J Food Eng. 2012; 113: 442–452.

[24] Cassano A, Conidi C, Tasselli F. Clarification of pomegranate juice (*Punica Granatum* L.) by hollow fibre membranes: analyses of membrane fouling and performance. J Chem Technol Biotechnol. 2015; 90: 859–866.

[25] Vladisavljević GT, Vukosavljević P, Bukvíc B. Permeate flux and fouling resistance in ultrafiltration of depectinized apple juice using ceramic membranes. J Food Eng. 2003; 60: 241–247.

[26] Rai P, Rai C, Majumdar GC, DasGupta S, De S. Resistance in series model for ultrafiltration of mosambi (*Citrus sinensis* (L.) *Osbeck*) juice in a stirred continuous mode. J Membr Sci. 2006; 283: 116–122.

[27] Dahdouh L, Wisniewski C, Kapitan-Gnimdu A, Servent A, Dornier M, Delalonde M. Identification of relevant physicochemical characteristics for predicting fruit juices filterability. Sep Purif Technol. 2015; 141: 59–67.

[28] Ruby-Figueroa R, Saavedra J, Bahamonde N, Cassano A. Permeate flux prediction in the ultrafiltration of fruit juices by ARIMA models. J Memb Sci. 2017; 524:108–116.

[29] Cassano A, Tasselli F, Conidi C, Drioli E. Ultrafiltration of Clementine mandarine juice by hollow fibre membranes. Desalination. 2009; 241: 302–308.

[30] Cassano A, Drioli E, Galaverna G, Marchelli R, Di Silvestro G, Cagnasso P. Clarification and concentration of citrus and carrot juices by integrated membrane processes. J Food Eng. 2003; 57: 153–163.

[31] Vaillant F, Cisse M, Chaverri M, Perez A, Dornier M, Viquez F, Dhuique-Mayer C. Clarification and concentration of melon juice using membrane processes. Innov Food Sci Emerg. 2005; 6: 213–220.

[32] Cassano A, Conidi C, Drioli E. Physico-chemical parameters of cactus pear (*Opuntia ficus-indica*) juice clarified by microfiltration and ultrafiltration processes. Desalination. 2010; 250: 1101–1104.

[33] Cissé M, Vaillant F, Perez A, Dornier M, Reynes M, The quality of orange juice processed by coupling crossflow microfiltration and osmotic evaporation. Int J Food Sci Technol. 2005; 40: 105–116.

[34] Matta VM, Moretti RH, Cabral LMC. Microfiltration and reverse osmosis for clarification and concentration of acerola juice. J Food Eng. 2004; 61: 477–482.

[35] Cassano A, Donato L, Conidi C, Drioli E. Recovery of bioactive compounds in kiwifruit juice by ultrafiltration. Innov Food Sci Emerg Technol. 2008; 9: 556–562.

[36] Galiano F, Figoli A, Conidi C, Menichini F, Bonesi M, Loizzo MR, Cassano A, Tundis R. Functional properties of *Punica granatum* L. juice clarified by hollow fiber membranes. Processes. 2016; 4: 1–16.

[37] Cassano A, Figoli A, Tagarelli A, Sindona G, Drioli E. Integrated membrane process for the production of highly nutritional kiwifruit juice. Desalination. 2006; 189: 21–30.

[38] Yu ZR, Chiang BH, Hwang LS. Retention of passion fruit juice compounds by ultrafiltration. J Membr Sci. 1986; 51: 841–844.

[39] Saénz C, Sepúlveda E, Araya E, Calvo C. Colour changes in concentrated juices of prickly pear Opuntia ficus indica during storage at different temperatures. LWT-Food Sci Technol. 1993; 26:417–421.

[40] Jiao B, Cassano A, Drioli E. Recent advances on membrane processes for the concentration of fruit juices: A review. J Food Eng. 2004; 63: 303–324.

[41] Conidi C, Cassano A, Drioli E. Recovery of phenolic compounds from orange press liquor by nanofiltration. Food Bioprod Process. 2012; 90: 867–874.

[42] Warczok J, Ferrando M, Lopez F, Guell C. Concentration of apple and pear juices by nanofiltration at low pressure. J Food Eng. 2004; 63: 63–70.

[43] Couto DS, Dornier M, Pallet D, Reynes M, Dijoux D, Freitas SP, Cabral LMC. Evaluation of nanofiltration membranes for the retention of anthocyanins of acai (*Euterpe oleracea* Mart.). juice. Desalin Water Treat. 2011; 27: 108–113.

[44] Arriola NA, dos Santos GD, Prudencio ES, Vitali L, Petrus JCC, Castanho Amboni RDM. Potential of nanofiltration for the concentration of bioactive compounds from watermelon juice. Int J Food Sci Technol. 2014; 49: 2052–2060.

[45] Arend GD, Adorno WT, Rezzadori K, Di Luccio M, Chaves VC, Reginatto FH, Cunha Petrus JC. Concentration of phenolic compounds from strawberry (*Fragaria X ananassa Duch*) juice by nanofiltration membrane. J Food Eng. 2017; 201: 36–41.

[46] Acosta O, Vaillant F, Perez AM, Dornier M. Concentration of polyphenolic compounds in blackberry (*Rubus Adenotrichos Schltdl*.) juice by nanofiltration. J Food Process Eng. 2017; 40: 1–7.

[47] Conidi C, Cassano A. Recovery of phenolic compounds from bergamot juice by nanofiltration membranes. Desalin Water Treat. 2015; 56: 3510–3518.

[48] Girard B, Fukumoto LR. Membrane processing of fruit juice and beverages: A review. Crit Rev Food Sci Nutr. 2000; 40: 91–157.

[49] Santana I, Guraka PD, da Matta VM, Pereira Freitas S, Cabral LMC. Concentration of grape juice (*Vitis labrusca*) by reverse osmosis process. Desalin Water Treat. 2011; 27: 103–107.

[50] Jesus DF, Leite MF, Silva LFM, Modesta RD, Matta VM, Cabral LMC. Orange (*Citrus sinensis*) juice concentration by reverse osmosis. J Food Eng. 2007; 81: 287–291.

[51] Álvarez S, Riera FA, Álvarez R, Coca J. Permeation of apple aroma compounds in reverse osmosis. Sep Purif Technol. 1998; 14: 209–220.

[52] Medina BG, Garcia A. Concentration of orange juice by reverse osmosis. J Food Process Eng. 1998; 10: 217–230.

[53] Kostantinos BP, Harris NL. Osmotic concentration of liquid foods. J Food Eng 2001; 49: 201–206.

[54] Hogan PA, Canning RP, Peterson PA, Johnson RA, Michaels AS. A new option: Osmotic distillation. Chem Eng Prog 1998; 94:49–61.

[55] Wang P, Chung TS. Recent advances in membrane distillation processes: Membrane development, configuration design and application exploring. J Membr Sci. 2015; 474: 39–56.

[56] Cassano A, Drioli E. Concentration of clarified kiwifruit juice by osmotic distillation. J Food Eng. 2007; 79: 1397–1404.

[57] Ravindra Babu B, Rastogi NK, Raghavarao KSMS. Concentration and temperature polarization effects during osmotic membrane distillation. J Memb Sci. 2008; 322: 146–153.

[58] Hongvaleerat C, Cabral LMC, Dornier M, Reynes M, Ningsanond S. Concentration of pineapple juice by osmotic evaporation. J Food Eng. 2008; 88: 548–552.

[59] Shaw PE, Lebrun M, Ducamp MN, Jordan MJ, Goodner KL. Pineapple juice concentrated by osmotic evaporation. J Food Quality. 2002; 25: 39–49.

[60] Versari A, Ferrarini R, Tornielli GB, Parpinello GP, Gostoli C, Celotti E. Treatment of grape juice by osmotic evaporation. J Food Sci. 2004; 69: 422–427.

[61] Cissé M, Vaillant F, Bouquet S, Pallet D, Lutin F, Reynes M, Dornier M. Athermal concentration by osmotic evaporation of roselle extract, apple and grape juices and impact on quality. Innov Food Sci Emerg Technol. 2011; 12: 352–360.

[62] Vaillant F, Jeanton E, Dornier M, O'Brien GM, Reynes M, Decloux M. Concentration of passion fruit juice on industrial pilot scale using osmotic evaporation. J Food Eng 2001; 47: 195–202.

[63] Rodrigues RB, Menez HC, Cabral LMC, Dornier M, Rios GM, Reynes M. Evaluation of reverse osmosis and osmotic evaporation to concentrate camu-camu juice (*Myciaria dubia*). J Food Eng. 2004; 63: 97–102.

[64] Souza ALR, Pagani MM, Dornier M, Gomes FS, Tonon RV, Cabral LMC. Concentration of camu–camu juice by the coupling of reverse osmosis and osmotic evaporation processes. J Food Eng. 2013; 119: 7–12.

[65] Valdés H, Romero J, Saavedra A, Plaza A, Bubnovich V. Concentration of noni juice by means of osmotic distillation. J Membr Sci. 2009; 330: 205–213.

[66] Rehman WU, Zeb W, Muhammad A, Ali W, Younas M. Osmotic distillation and quality evaluation of sucrose, apple and orange juices in hollow fiber membrane contactor. Chem Ind Chem Eng Q. 2017; 23: 217–227.

[67] Cassano A, Conidi C, Drioli E. Clarification and concentration of pomegranate juice (*Punica granatum* L.) using membrane processes. J Food Eng. 2011; 107: 366–373.

[68] Alves VD, Koroknai B, Bélafi-Bakó K, Coelhoso IM. Using membrane contactors for fruit juice concentration. Desalination. 2004; 162: 263–270.

[69] Courel M, Dornier M, Henry JM, Rios GM, Reynes M. Effect of operating conditions on water transport during the concentration of sucrose solutions by osmotic distillation. J Membr Sci. 2000; 170: 281–289.

[70] Barbe AM, Bartley JP, Jacobs AL, Johnson RA. Retention of volatile organic flavour/fragrance components in the concentration of liquid foods by osmotic distillation. J Membr Sci. 1998; 145: 67–75.

[71] Lawson KW, Loyd DR. Membrane distillation II. Direct contact MD. J Membr Sci. 1996; 120: 123–133.

[72] Calabrò V, Jiao B, Drioli E. Theoretical and experimental study on membrane distillation in the concentration of orange juice. Ind Eng Chem Res. 1994; 33: 1803–1808.

[73] Kozák A, Békássy-Molnár, E, Vatai G. Production of black-currant juice concentrate by using membrane distillation. Desalination. 2009; 241: 309–314.

[74] Gunko S, Verbych S, Bryk M, Hilal N. Concentration of apple juice using direct contact membrane distillation. Desalination. 2006; 190: 117–124.

[75] Quist-Jensen CA, Macedonio F, Conidi C, Cassano A, Aljlil S, Alharbi OA, Drioli E. Direct contact membrane distillation for the concentration of clarified orange juice. J Food Eng. 2016; 187:37–43.

[76] El-Bourawi MS, Ding Z, Ma R, Khayet M. A framework for better understanding membrane distillation separation process. J Membr Sci. 2006; 285: 4–29.

[77] Alves VD, Coelhoso IM. Orange juice concentration by osmotic evaporation and membrane distillation: a comparative study. J Food Eng. 2006; 74: 125–133.

[78] Bélafi-Bakó K, Koroknai B. Enhanced water flux in fruit juice concentration: Coupled operation of osmotic evaporation and membrane distillation. J Membrane Sci. 2006; 269: 187–193.

[79] Koroknai B, Csanádi Z, Gubicza L, Bèlafi-Bakó K. Preservation of antioxidant capacity and flux enhancement in concentration of red fruit juices by membrane processes. Desalination. 2008; 228: 295–301.

[80] Onsekizoglu P, Savas Bahceci K, Jale Acar M. Clarification and concentration of apple juice using membrane processes: A comparative quality assessment. J Membrane Sci. 2010; 352: 160–165.

[81] Karlsson HOE, Trägårdh G. Applications of pervaporation in food processing. Trends Food Sci Technol. 1996; 7: 78–83.

[82] Feng X, Huang RYM. Liquid separation by membrane pervaporation: a review. Ind Eng Chem Res. 1997; 36: 1048–1066.

[83] Pereira CC, Ribeiro CP, Nobrega R, Borges CP. Pervaporative recovery of volatile aroma compounds from fruit juices. J Membr Sci. 2006; 274: 1–23.

[84] Pereira CC, Rufino JRM, Habert AC, Nobrega R, Cabral LMC, Borges CP. Aroma compounds recovery of tropical fruit juice by pervaporation: membrane material selection and process evaluation. J Food Eng. 2005; 66: 77–87.

[85] Aroujalian A, Raisi A. Recovery of volatile aroma components from orange juice by pervaporation. J Membr Sci. 2007; 303: 154–161.

[86] Rafia N, Aroujalian A, Raisi A. Pervaporative aroma compounds recovery from lemon juice using poly(octyl methyl siloxane) membrane. J Chem Technol Biotechnol. 2011; 86: 534–540.

[87] Isci A, Sahin S, Sumnu G. Recovery of strawberry aroma compounds by pervaporation. J Food Eng. 2006; 75: 36–42.

[88] Bandini S, Gostoli C, Sarti C. Separation efficiency in vacuum membrane distillation. J. Membrane Sci. 1992; 73: 217–229.

[89] Izquierdo-Gil MA, Jonsson G. Factors affecting flux and ethanol separation performance in vacuum membrane distillation (VMD). J Membr Sci. 2003; 214: 113–130.

[90]   Diban N, Voinea OC, Urtiaga A, Ortiz I. Vacuum membrane distillation of the main pear aroma compound: experimental study and mass transfer modelling. J Membrane Sci. 2009; 326: 64–75.

[91]   Bagger-Jørgensen R, Meyer AS, Varming C, Jonsson G. Recovery of volatile aroma compounds from black currant juice by vacuum membrane distillation. J Food Eng. 2004; 64: 23–31.

[92]   Khayet M, Godino MP, Mengual JI. Theory and experiments on sweeping gas membrane distillation. J Membr Sci. 2000; 165: 261–272.

[93]   Bagger-Jørgensen R, Meyer AS, Pinelo M, Varming C, Jonsson G. Recovery of volatile fruit juice aroma compounds by membrane technology: sweeping gas versus vacuum membrane distillation. Innov Food Sci Emerg. 2011; 12: 388–397.

[94]   Lawhon JT, Lusas EW. Method of producing sterile and concentrated juices with improved flavor and reduced acid. US Patent 4,643,902, 1987.

[95]   Walker JB. Membrane process for the production of superior quality fruit juice concentrate. In Proceedings of the 1990 International Congress on Membrane and Membrane Processes, Chicago, USA, 20–24 August 1990, p. 283.

[96]   Alvarez S, Riera FA, Alvarez R, Coca J, Cuperus FP, Bouwer S, Boswinkel G, van Gemert RW, Veldsink JW, Giorno L, Donato L, Todisco S, Drioli E, Olsson J, Trägårdh G, Gaeta SN, Panyor L. A new integrated membrane process for producing clarified apple juice and apple juice aroma concentrate. J Food Eng. 2000; 46: 109–125.

[97]   Kozák Á, Bánvölgyi Sz., Vincze I, Kiss I, Békássy Molnár E, Vatai G. Comparison of integrated large-scale and laboratory scale membrane processes for the production of black currant juice concentrate. Chem Eng Proc. 2008; 47: 1171–1177.

[98]   Cassano A, Jiao B, Drioli E. Production of concentrated kiwifruit juice by integrated membrane processes. Food Res Int. 2004; 37: 139–148.

[99]   Cassano A, Conidi C, Timpone R, D'Avella M, Drioli E. A membrane-based process for the clarification and the concentration of the cactus pear juice. J Food Eng. 2007; 80: 914–921.

[100]   Galaverna G, Di Silvestro G, Cassano A, Sforza S, Dossena A, Drioli E, Marchelli R. A new integrated membrane process for the production of concentrated blood orange juice: Effect on bioactive compounds and antioxidant activity. Food Chem. 2008; 106: 1021–1030.

[101]   Destani F, Cassano A, Fazio A, Vincken JP, Gabriele B. Recovery and concentration of phenolic compounds in blood orange juice by membrane operations. J Food Eng. 2013; 117: 263–271.

[102]   Cassano A, Conidi C, Drioli E. A membrane-based process for the valorization of the bergamot juice. Sep Sci Technol. 2013; 48: 537–546.

[103]   Sotoft LF, Christensen KV, Andrénsen R, Norddahl B. Full scale plant with membrane based concentration of blackcurrant juice on the basis of laboratory and pilot scale tests. Chem Eng Proc. 2012; 54: 12–21.

[104]   Hasanoğlu A, Rebolledo F, Plaza A, Torres A, Romero J. Effect of the operating variables on the extraction and recovery of aroma compounds in an osmotic distillation process coupled to a vacuum membrane distillation system. J Food Eng. 2012; 111: 632–641.

Enrico Drioli, Juntae Jung, Aamer Ali

9 Blue energy

9.1 Introduction

Electrochemical potential difference between solutions of different ion concentrations can be used to obtain energy from a salinity gradient [1]. In fact, the difference in ionic gradient between the seawater and fresh water is higher than that for biological solutions, and the energy density for such systems is much higher than the natural ones. The power potentially obtainable when the 37,300 km^3 annual global river discharge meets the sea, for example, is estimated to be about 2 TW, enough to supply a significant percentage of the global energy demand. By exploiting the salinity gradient between seawater and freshwater, theoretically, 0.8 kW/m^3 can be extracted, which is equivalent to a waterfall falling from a height of 225 m [2, 3]. In addition to the sea, natural or industrial salt brine can be utilized as saline sources [4, 5].

Until now, several approaches have been proposed to harvest the energy from salinity, including technologies such as pressure-retarded osmosis (PRO) [6, 7, 8], reverse electrodialysis (RED) [9, 10], capacitive mixing [11, 12], vapor pressure difference utilization [13] and hydrogels [14]. Among those technologies, PRO and RED have been well recognized as the most advanced and promising ones [1]. Both technologies are based on the membrane separation process; the energy acquisition process is changed depending on the properties of the membranes. In the following sections, the concept of blue energy as well as the principle, prospects and limitations of two major energy-harvesting technologies from salinity, PRO and RED, will be discussed.

9.2 Blue energy

The current energy production mainly relies upon the fossil fuels and, therefore, suffers from vulnerability, imminent scarcity and production of greenhouse gases. The current climate changes are believed to be largely due to the burning of fossil fuels. These facts provide the strong motivation to cut down or eliminate the use of fossil fuel and find the alternative green resources of energy. Blue energy or the salinity-gradient energy represents an entirely untapped source of clean and renewable energy. It refers to the energy extracted by controlled mixing of two streams with different salinity levels.

Enrico Drioli, Aamer Ali, National Research Council of Italy, Institute for Membrane Technology (ITM-CNR), University of Calabria, Rende, Italy
Juntae Jung, Department of Energy Engineering, Hanyang University, Seoul, South Korea

https://doi.org/10.1515/9783110281392-009

The concept of blue energy can be better explained in form of Gibb's free energy of mixing. Gibb's free energy of mixing for an ideal dilute solution can be expressed as follows:

$$\Delta G_{mix} = \Delta G_b - (\Delta G_c + \Delta G_d) \tag{9.1}$$

where the subscript b represents the bulk solution emerging from the mixing of concentrated and dilute solutions represented by the subscripts c and d, respectively.

Equation (9.1) can be expressed in form of entropy of the mixing as follows:

$$\Delta G_{mix} = (n_c + n_d)T\Delta S_b - (n_c T\Delta S_c + n_d T\Delta S_d) \tag{9.2}$$

where n is the amount of particles expressed in moles, T is the temperature in K and S is the molar entropy that can be expressed as follows:

$$\Delta S_{mix} = -R\sum_i x_i ln x_i \tag{9.3}$$

where R is the universal gas constant and x_i is the mole fraction of component i.

Equations (9.1)–(9.3) allow calculating the free energy of mixing when two solutions with different salinities are mixed. The theoretical energy produced can be varied depending on the pair of concentrated and diluted solution pairing. For instance, mixing fresh river water from the Mississippi with seawater from the Gulf of Mexico would gain 1.4 MJ; mixing fresh sea water from the Red Sea with hypersaline water from the Dead Sea would gain 10 MJ. In addition, brine from an industrial area such as seawater reverse osmosis(RO) and salt mining can be useful candidates to increase the salinity gradient. Mixing fresh injection water from a vacuum salt mining industry with hypersaline brine from a salt cavern would gain 15 MJ. In total, about 2 TW of blue energy might be available globally from the river water falling into the sea. Wastewater discharged into the sea can generate additional 18 GW of energy. The actual quantities are anyhow lower due to the inherent irreversible energy dissipation; however, this can change the outlook of contribution of current various sources of energy in overall energy production across the globe as shown in Figure 9.1. According to some optimistic estimation, 80% of the world's power consumption can be generated from the salinity gradient, thus reducing emission of greenhouse gases [15].

9.2.1 The principles of harvesting blue energy using PRO and RED

The transport across the membrane is determined by the chemical potential gradient across the membrane. In the case of blue energy, the salinity gradient across the membrane acts as a main driving force for the transport. However, when two solutions with different salinity were bounded by the membranes, the component transported is varied depending on the characteristics of the membrane. For example, if the membrane selectively allows passage of water but rejects the salt molecules or ions, only the

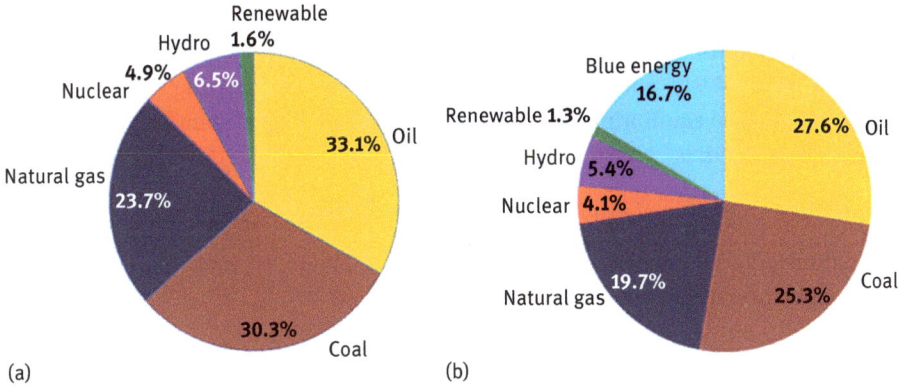

Figure 9.1: (a) Current and (b) perspective (after including blue energy) contribution of different energy sources in net global energy production. From Post [2]. (Figure freely accessible online at http://edepot.wur.nl/12605). Last access on 7 May 2018.)

water is selectively transported through the membrane. This kind of phenomenon is called osmosis that is the transport of water across a selectively permeable membrane from a region of higher water chemical potential to a region of lower water chemical potential. On the other hand, if the membrane exclusively allows transport of ions with certain charge, the ion can be selectively moved. This kind of membrane is called an ion-exchange membrane, which is traditionally used in electrodialysis or diffusion in the power of an electrochemical potential gradient.

Approaches for harvesting blue energy generation are made by the PRO and RED, which utilize the osmosis phenomenon and selective ion transport through the ion-exchange membranes, respectively. Both technologies have the same driving force in a way that energy is released when the two solutions with different salinity are mixed. However, when viewed from a thermodynamic approach, the way of extracting energy and parameters affecting the process are different.

The following equation is the molar free energy for an ideal solution:

$$\mu_i = \mu_i^0 + v_i \Delta p + RT\ln x_i + z_i F \Delta\varphi \tag{9.4}$$

where μ_i^0 refers to molar free energy under standard conditions, v_i refers to partial molar volume, Δp refers to pressure difference, x_i refers to molar fraction of the component, z_i refers to valence of an ion, F refers to Faraday's constant, $\Delta\varphi$ refers to electrical potential difference and R and T refer to gas constant and temperature, respectively.

There is no transport of ions and pressure change in the PRO process; hence, the free energy difference can be theoretically presented by the chemical potential change of the system before and after the mixing:

$$E = \sum_i (E_{i,c} + E_{i,d} - E_{i,b}) = \sum_i (C_{i,c} V_c RT\ln X_{i,c} + C_{i,d} V_d RT\ln X_{i,d} - C_{i,b} V_b RT\ln X_{i,b}) \tag{9.5}$$

where E refers to free energy under standard conditions, c refers to molar concentration, V_i refers to the volume, the subscripts c and d refer to concentrated and diluted solution, respectively, and b refers to brackish solution after mixing.

Assuming the equilibrium state of NaCl solutions in PRO system and no pressure difference applied, $\mu_{H_2O,c} = \mu_{H_2O,d}$ and $\Delta p = 0$; eq. (9.4) is now converted to

$$v_{H_2O}\Delta\pi + RT\ln x_{H_2O,c} = RT\ln x_{H_2O} \tag{9.6}$$

For sodium chloride solutions, $\ln x_{H_2O} = \ln(1-2x) \approx 2\ln(1-x)$, $and = v_{H_2O} = v$; hence, the osmotic pressure can be reduced as follows:

$$\Delta\pi = \frac{2RT}{v}\ln\frac{1-x_{H_2O,d}}{1-x_{H_2O,c}} \tag{9.7}$$

On the other hand, RED is based on the transport of ions, and electrical current is generated directly from the flow of ions. Accordingly, the driving force for RED can be presented differently from that for PRO. The ion concentration difference across the ion-exchange membranes generates a Nernst potential; thereby the direct permeation of ions occurs from concentrated to diluted solution. Assuming equilibrium state of NaCl solutions in RED system and no pressure difference is applied, $\mu_{NaCl,c} = \mu_{NaCl,d}$ and $\Delta p = 0$, eq. (9.1) is now converted to

$$\frac{RT}{Z_{Na}F}\ln x\text{Na, c} + \frac{RT}{Z_{Cl}F}\ln x\text{Cl, c} = \frac{RT}{Z_{Na}F}\ln x\text{Na, c} + \frac{RT}{Z_{Cl}F}\ln x\text{Cl, c} + \Delta\varnothing \tag{9.8}$$

(where $\Delta\varnothing$ is electro chemical potential difference)

For sodium chloride solutions, $Z_{Na} = Z_{Cl} = 1$ and $x_{Na} = x_{Cl} = x$; hence, the Nernst potential of NaCl solutions across the ion-exchange membrane can be reduced to

$$\Delta\varnothing = \frac{2RT}{F}\ln\frac{x_c}{x_d} \tag{9.9}$$

By comparing eqs. (9.7) and (9.9), one can recognize that the driving force of each energy harvesting technology is similar; however, due to the dissimilar working principles and disparate membranes use, PRO and RED require different membrane performance. Water permeability, salt permeability and structure parameter are critical factors in determining power density (PD) and specific energy obtained in PRO. Internal area resistance and permselectivity, described as the ability of the membrane to prevent the passage of coions, are crucial parameters in determining PD and specific energy of RED.

9.2.2 PRO for blue energy

In order to understand PRO, it is crucial to understand osmotic processes. Figure 9.2 illustrates different osmotic processes between diluted and concentrated saline water across the semipermeable membrane. The osmotic pressure is driven by a difference in solute concentrations across the membrane that allows passage of water, but rejects most solute molecules or ions. Osmotic pressure (π) is the pressure that, if applied to the more concentrated solution, would prevent transport of water across the membrane. Forward osmosis (FO) uses π as the driving force, resulting in concentration of a feed stream and dilution of a highly concentrated stream (referred to as the draw solution). On the contrary, in RO, pressure exceeding π has to be applied on saline solution to make the freshwater permeate through the semipermeable membrane. PRO can be viewed as an intermediate process between FO and RO, where hydraulic pressure is applied in the opposite direction of the osmotic pressure gradient (similar to RO). However, the net water flux is still in the direction of the concentrated draw solution (similar to FO).

Water transport through a semipermeable membrane in FO, RO and PRO can be described with the following general equations:

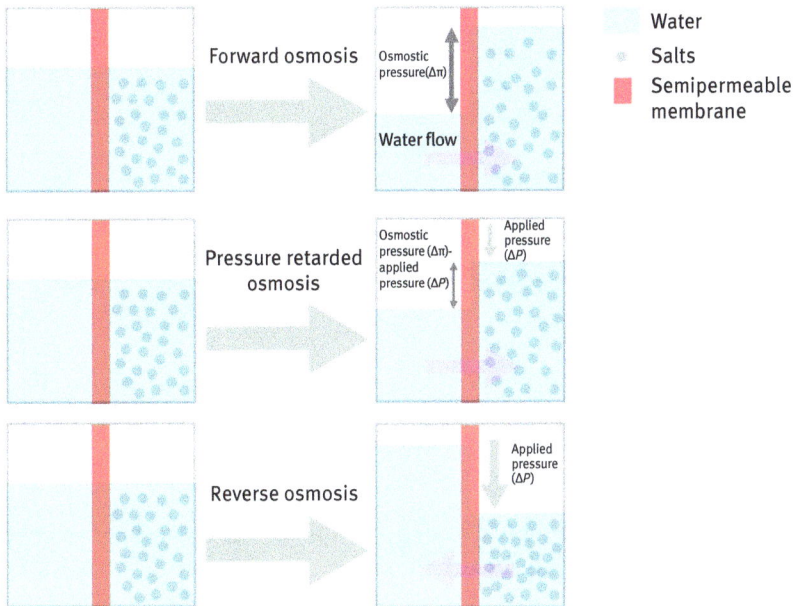

Figure 9.2: Schematic representation of osmotic pressure. From Han et al. [8].

$$J_w = A(\sigma\,\Delta\pi - \Delta P) \tag{9.10}$$

where J_w is the water permeation flux (m/s), A the water permeability constant of the membrane (m³/m²Pas), σ the reflection coefficient and ΔP (Pa) is the applied pressure.

PRO is one of the way of generating electricity using the osmosis phenomenon. It comprises of the flow of water from low salinity water to high salinity water through a semipermeable membrane to produce pressurized water that generates electricity through the mechanical turbine. In other words, when a low salinity solution (feed or draw solution) is separated from a high salinity solution (draw solution) confined in a fixed volume container with a semipermeable membrane, the permeation of fresh-water from the former causes an increase in pressure on the latter side. The semiperme-able membrane ideally retains all the ions and allows the flow of water from low concentration to high concentration, increasing the pressure of this stream. A part of high pressure stream is used to run a turbine that generates electricity and the rest is used to pressurize saltwater by using a pressure exchanger. A schematic illustration of the process is shown in Figure 9.3. The high-and low-concentrated solutions can be seawater and fresh water or seawater and brine from desalination, respectively.

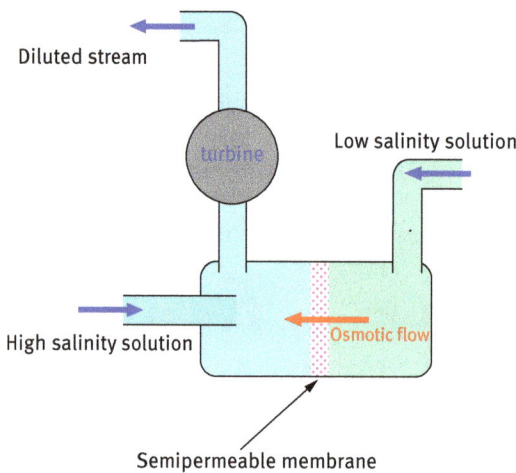

Figure 9.3: Schematic diagram of the PRO process.

The productivity of the PRO process can be evaluated using two performance metrics. First is the PD that indicates the amount of power produced per unit membrane area, A_m (W/m^2) [16]. The system power output can be calculated by multiplying flow rate, Q (m^3/s) and pressure, ΔP (Pa), across the membrane. Accordingly, by combining eq. (9.10), the PD can be calculated as follows:

$$W = \Delta P \times \frac{Q}{Am} = \Delta P \times J_w = \Delta P \times A(\sigma \Delta \pi - \Delta P) \qquad (9.11)$$

A PRO process with high PD means a higher power output can be achieved under the identical membrane area, which gives driving force for cut down capital cost and operating cost of the PRO process. The achievable theoretical maximum PD of a PRO process can be calculated by differentiating eq. (9.11) with respect to ΔP. PD reaches its maximum when the ΔP is equivalent to half of Δπ. The maximum PD can be given as follows:

$$W_{max} = A \times \frac{(\Delta\pi)^2}{4} \qquad (9.12)$$

Equations (9.11) and (9.12) indicate that the prerequisite for an efficient PRO process is the membrane permeability; a low permeability membrane would render the conversion of even a high salinity gradient hopelessly inefficient.

The projected PD as a function of concentration difference between the draw and feed solutions is shown in the form of a contour plot in Figure 9.4. The concentration of the high-salinity draw solution ranges from 1 to 2 M NaCl, while that of the low-salinity feed solution varies from 0 to 0.5 M NaCl. As it can be expected, the maximum PD is obtained for the minimum feed concentration and the highest draw solution concentration (right lower corners of the graphs). For the lower hydraulic pressure difference (ΔP =10 bar), the contour lines vary from 0.84 to 8.6 W/m² for the concentrations differences considered in the analysis. In this case, the maximum PD cannot be obtained over the entire range of salinity gradient because the optimal hydraulic pressure difference for peak PD varies depending on the salinity gradient. Only the left top point indicating 1 M draw and 0.5 M feed represents the peak PD at ΔP = 10 bar. On the other hand, the contour lines of projected PD vary from 0 to 19.1 W/m² at a hydraulic pressure difference of 32.5 bar. The red triangle in (b) indicates a meaninglessregion as the calculated results are based on the water fluxes at hydraulic pressure differences above the flux reversal point (ΔP > Δπ).

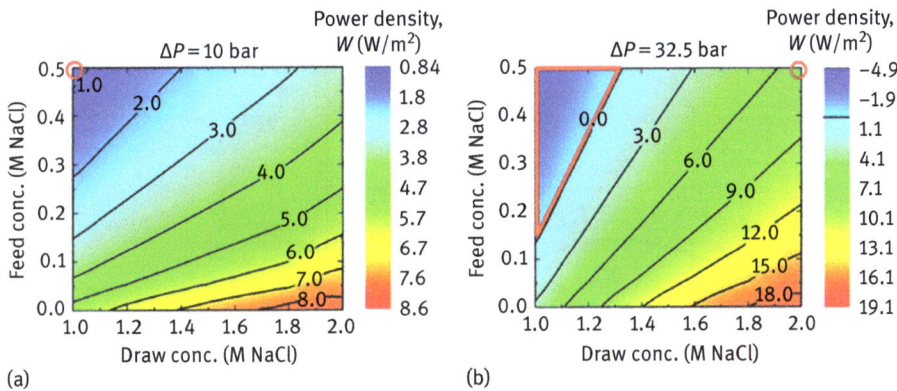

Figure 9.4: A contour plot of power densities as a function of feed and draw solution concentration under applied hydraulic pressure differences of 10 bar (left) and 32.5 bar (right). From Kim and Elimelech [17]. Reprinted with permission.

In addition to increasing salinity gradient across the membrane, advanced membrane performance and optimized module design can also highly affect the PRO performance. Contrary to theoretical estimation, in the practical applications, the maximum PD occurs at the hydraulic pressures difference less than half of the osmotic pressure difference due to the detrimental effect of concentration polarization (external concentration polarization [ECP] and internal concentration polarization [ICP]) and reverse salt diffusion. The effect of ECP, ICP and reverse salt diffusion cause detrimental effect on J_w and PD as reported by Yip and Elimelech [18]. The modeled flux and power densities as a function of applied hydraulic pressure for various combinations of concentration differences between the feed and draw solutions are shown in Figure 9.5.

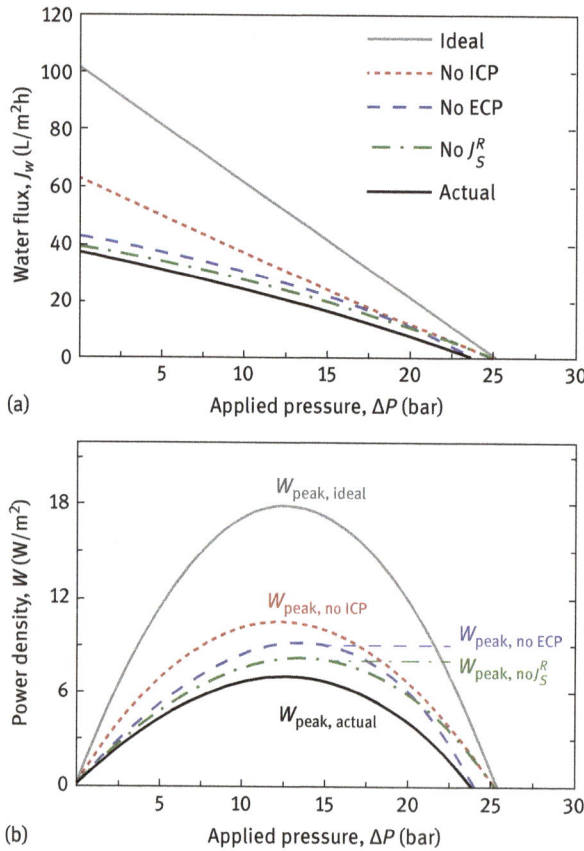

(a)

(b)

Figure 9.5: Representative plots of (a) water flux J_w and (b) PD W as a function of applied hydraulic pressure difference ΔP. The ideal J_w and W without any detrimental effect are indicated by the solid gray line and calculated using eqs. (9.10) and (9.11), respectively. From Yip and Elimelech [18]. (Reprinted with permission from [Yip, N. Y., & Elimelech, M. (2011). Performance limiting effects in power generation from salinity gradients by pressure retarded osmosis. Environmental Science & Technology, 45(23), 10273–10282. Copyright 2011 American Chemical Society.)

In conclusion, the current power densities obtained for 0.5 M NaCl draw solutions are significantly lower than the minimum value (5 W/m^2) considered economically viable for power generation through PRO at commercials scale due to effect of ECP, ICP and salt flux across the membrane. However, as shown in Figures 9.4 and 9.5, the maximum achievable PD can be developed by increasing the salinity gradient across the membrane or mitigating the effect of ECP, ICP and salt flux. In this regard, several factors including the concentration of high and low salinity solutions, permeability and selectivity of the membrane, process design and the ICP inside the support layer of the membranes are currently being discussed.

Concentration polarization (ICP) in PRO

In general, concentration polarization in membrane processes refers to the relative increase in solute concentration at solution–membrane interface due to the selective permeation of solvent through the membrane. Concentration polarization in osmotically driven membrane processes is more severe than that in the pressure-driven membrane processes and can be divided into ICP and ECP. The former and latter types of concentration polarization refer to the concentration polarization developed inside the membrane support layer and at the membrane–solution interface, respectively. ICP significantly impacts the effective osmotic pressure gradient across the membrane, which can considerably decrease the permeability of the membrane. ECP occurs at both sides of the membrane. Relatively higher concentration of solute at the feed–membrane interface due to retention is referred as the concentrative ECP. The solution at membrane–draw solution interface gets diluted due to permeation of solvent from the other side, giving rise to dilutive ECP. Different types of concentration polarization are shown schematically in Figure 9.6. Both types of concentration polarization deteriorate the process performance. ECP can be controlled by the selection of appropriate process conditions. To alleviate the ICP, an appropriate designing of a support layer is required.

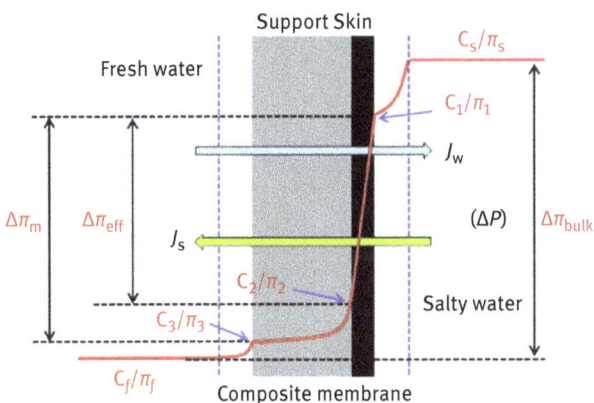

Figure 9.6: Concentration polarization in PRO. From Han et al. [8]. Reprinted with permission.

Membranes for PRO

The inherent characteristics for membranes for PRO are described by the water-permeability coefficient (A), solute permeability coefficient (B) and the membrane characteristic parameter (S).

The solute permeability coefficient can be defined as follows:

$$B = J_w \left(\frac{1-R}{R} \right) \exp \left(-\frac{J_w}{k} \right) \tag{9.13}$$

where k is the mass transfer coefficient for a given membrane cell. The structure parameter (S) describes the support layer:

$$S = \frac{x.\tau}{\varphi} \tag{9.14}$$

where x is the thickness of the support layer, and τ and φ are the tortuosity and porosity, respectively.

A membrane for PRO should exhibit high water permeability and minimum reverse salt diffusion. The ICP should be reduced by minimizing the structural parameter. The effect of these parameters on PD is described in Figure 9.7. For a

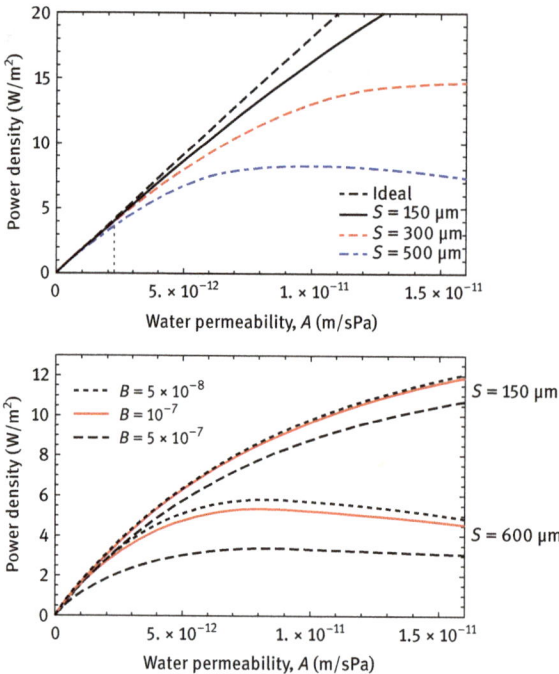

Figure 9.7: The effect of water permeability on PD for membranes with various values of structure parameter and the solute permeability coefficient (from Ramon et al. [19]).

given water permeability, the PD increases by decreasing the membrane structural parameter. Similarly, for any given water permeability and structure parameter, the PD increased by decreasing the solute permeability coefficient.

Additionally, the membranes for PRO must exhibit good mechanical strength to withstand against the applied hydraulic pressure and minimum fouling tendency. Several membrane materials have been tested for the fabrication of the membranes for PRO. A major milestone achieved in the preparation of PRO membranes is the formation of thin-film composite hollow fiber membranes [20], which enable to achieve PD as high as $10.6\,W/m^2$ by using the seawater brine and wastewater brine as the draw and feed solution, respectively. An example of the morphology of these membranes is shown in Figure 9.8. The support layer offers the required mechanical strength at the expense of minor resistance. The separation function is performed by the dense layer.

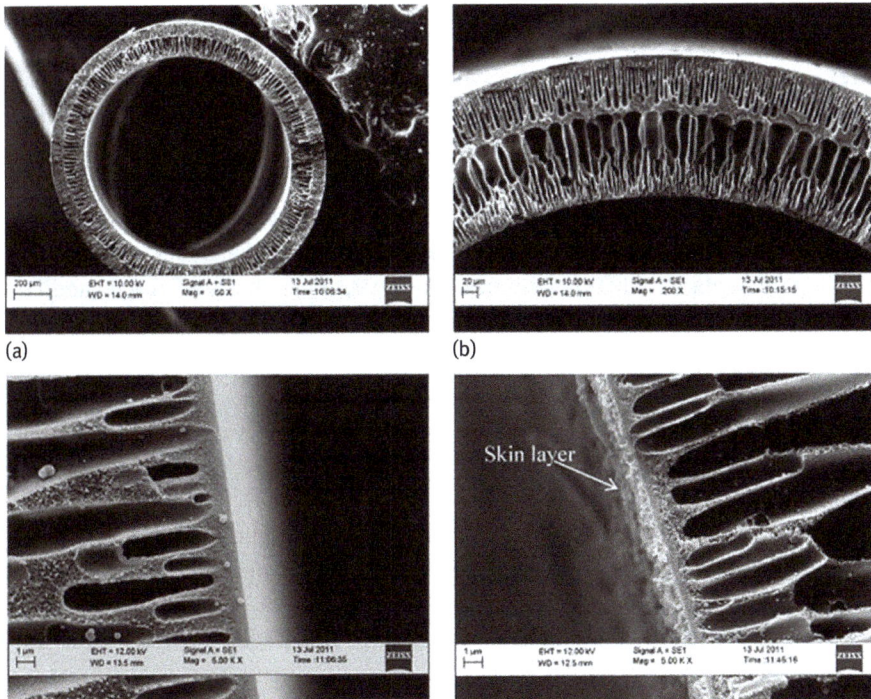

Figure 9.8: Morphology of hollow fibers: (a) cross-section of substrate at 50×, (b) substrate enlarged at 200×, (c) enlarged lumen side of substrate at 5,000× and (d) enlarged lumen side of TFC hollow fibers at 5,000×. From Chou et al. [20]. Reprinted with permission.

Membranes for power generation by PRO can be found in [21]. Cellulose acetate is the world's most abundantly available polymeric material with well-known toughness and smoothness, which makes it an attractive candidate for membrane synthesis. Its hydrophilic nature makes it further attractive for membranes for osmotically driven membrane processes including PRO. The wetting of the membrane reduces concentration polarization and increases the water flux. Similarly, polysulfone and polyethersulfone are good candidates for the fabrication of a support layer due to their good chemical resistance and mechanical properties. The hydrophilic nature of these materials, however, is undesirable for osmotically driven membrane processes. Similarly, polyamide is a good candidate for membrane synthesis for PRO.

Innovative materials and fabrication techniques with exciting performance are being explored for applications in PRO. These materials have the potential to extend the performance of PRO beyond the traditional performance limits. Membranes made of one-and two-dimensional (1D and 2D) materials are emerging as attractive candidates in separation processes. These materials are characterized by their 1D or 2D atomic layer that offers extremely low resistance to mass transfer across the membrane. These materials include carbon nanotubes, metal oxide nanotubes, layered zeolites, layered aluminophosphates and porous graphene. Their unique shape, size and structure render them the highly desirable properties for separation applications. However, their fabrication at large scale at economically feasible cost is still a challenge. Understanding and development of reliable structure–properties relationship in these membranes is another challenge.

Biomimetic membranes represent another potentially interesting class for PRO applications [22]. These membranes have Aquaporin water channels in their structure, which give very high permeability and selectivity to these membranes – the basic requirements for PRO membranes. However, the stability of these modified membranes under the real hydraulic pressure is a challenge.

Commercialization of PRO

The successful commercial implementation of PRO for power generation has been hampered due to the following facts:
- unavailability of low cost and robust membranes with minimum ICP;
- potential environmental impact caused due to the disruption of natural flow of water;
- extensive pretreatment required for the streams involved.

A commercial plant from Norwegian company Statkraft based on PRO [23] produced $1 W/m^2$ of electricity using cellulose acetate membranes, which is quite low than the feasible value of $5 W/m^2$, although some lab-made membranes have approached this figure. Considerable amount of energy is consumed in pretreating the streams; it has been estimated that in a PRO plant with a 50% overall efficiency, the actual extractable energy from the mixing of river water and sea water is $0.3–0.4 kWh/m^3$, leaving about $0.1 kWh$ of drivable useful energy from per cubic meter of fresh water.

Increasing the water permeability of the membrane skin layer and optimizing the thickness of the support layer can further enhance the efficiency of the process.

9.2.3 Reverse Electrodialysis for production of blue energy

In electrodialysis, the salt ions are transported from one solution to another solution through an ion-selective membrane under the influence of an applied electric potential difference [24]. In RED, the potential difference is generated due to the movement of ions [25]. Similar to the electrodialysis, concentrated and dilute solutions are separated through ion-exchange membranes (Figure 9.9). Cation-and anion-exchange membranes are used to transfer the respective ions from high-to low-concentration solution flowing through alternative channels. The salinity gradient results in the potential difference over each membrane, known as the membrane potential. The transport of the ions is governed by the difference in chemical potential between the two solutions. In RED, a number of cells are staked between the anode and cathode. The electric potential difference between the outer compartments of the membrane stack is the sum of electric potential difference over each membrane. The positive ions permeate through the cation-exchange membrane in the direction of the cathode, while the negative ions permeate through the anion-exchange membranes in the direction of the anode. Redox reactions occurring at the electrodes ensure the electroneutrality of the solution in the anode and cathode compartments.

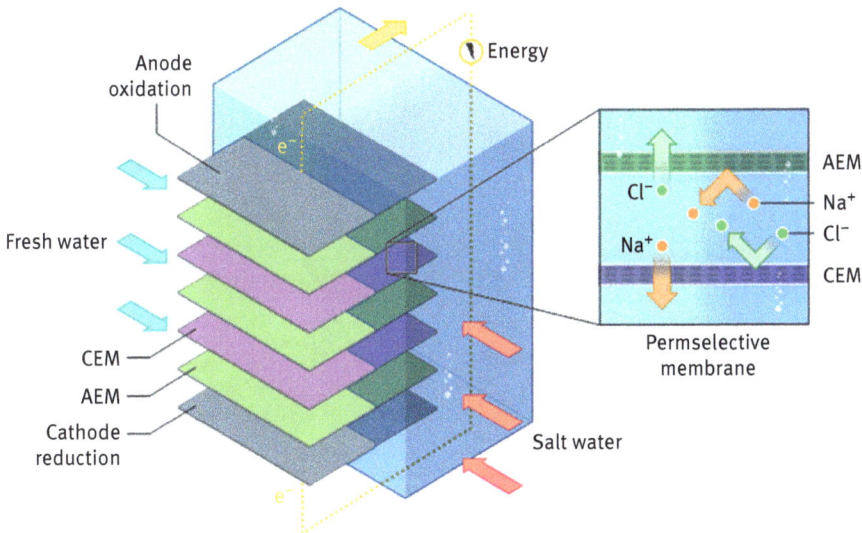

Figure 9.9: Schematic diagram of RED stack and a cell. CEM and AEM represent cation-exchange membrane and anion-exchange membrane, respectively.

RED technology was mainly matured in the laboratory of a Dutch company Wetsus, which eventually resulted into the creation of spin-off company REDSTACK [26]. This company now operates the world's first RED pilot plant located in the Netherland since 2014.

9.3 Parameters affecting RED performance

9.3.1 Solutions properties

The performance of a RED system drops significantly when more realistic solutions are used instead of the simple model solutions. An example of such scenario is shown in Figure 9.10(a) where replacing NaCl solution with brackish water and exhaust brine from solar pond in low-and high-salinity compartment, respectively, decreases the maximum gross PD from 3.04 to 1.13 W/m^2. The nature of ions present in the solution may also severely affect the performance of RED system. Sata [27] has reported a 2- to 3-fold increase in electric resistance of various commercial cation-exchange membranes when NaCl solution was replaced with MgCl$_2$ solution.

Tufa et al. [28] have also studied the effect of various ions on the performance of RED system in the presence of NaCl as shown in Figure 9.10(b). The negligible effect of HCO$_3^{-1}$ on RED performance was associated with its very low concentration in low-(LCC, 8.5×10^{-6} M) and high-concentration compartments (HCC, 8.5×10^{-4} M). Na^{+1} and K^{+1} have similar electrochemical properties (ionic radii of unhydrated Na$^+$ and K$^+$

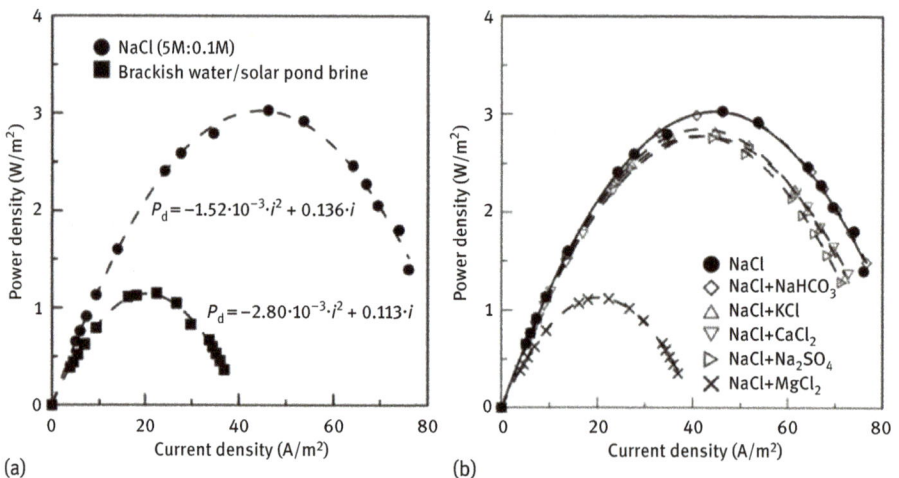

Figure 9.10: (a) Voltage vs current and (b) gross PD as a function of current density for two different types of solution. From Tufa et al. [28]. (Figure freely accessible online at http://pubs.rsc.org/en/content/articlehtml/2014/ra/c4ra05968a . Open Access Article. Last access on 7 May 2018.)

are 1.17 °A and 1.64 °A, respectively; ionic radii of hydrated Na^+ and K^+ are 3.58 °A and 3.32 °A, respectively; 18 ion diffusion coefficients in water for Na^+ and K^+ are 1.334×10^{-9} m^2/s and 1.957×10^{-9} m^2/s, respectively); therefore, the effect of K^+ ions on gross PD was limited despite of its significant concentration (0.32 M) in the solution. Similarly, the presence of Ca^{+2} did not significantly affect the performance of the process due to its low concentration in both LCC and HCC (0.004 M/0.096 M $CaCl_2$, respectively). The presence of Mg^{+2}, which significantly increases the membrane resistance, dropped the gross PD drastically.

The variations of voltage with respect to current and gross PD with respect to current density for various solution concentrations are shown in Figures 9.11(a) and 9.11(b), respectively. In accordance with Ohm's law, variation of voltage as a function of current at various solution compositions in HCC is linear. A decrease in HCC concentration from 5.4 to 4 M resulted into 38% reduction in the highest voltage observed. The corresponding reduction in the maximum gross PD was recorded as −62.5%.

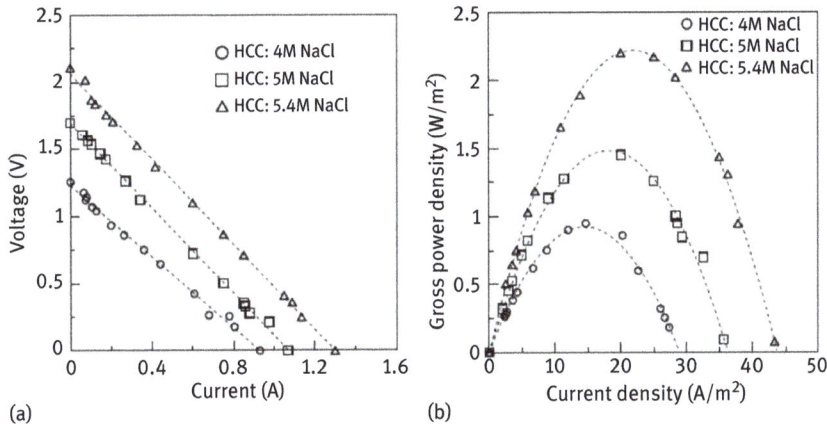

(a)   (b)

Figure 9.11: (a) Voltage with respect to current (b) Gross PD as function of current density at various HCC concentrations. From Tufa et al. [29]. Reprinted with permission.

Although increasing the HCC concentration increases the power generation, the risk of membrane fouling and scaling also increases at the same time. At such high solution concentration, some salts may precipitate when operating the system with real solutions. A proper pretreatment of feed solution and use of proper antifouling and antiscaling strategies may be necessary to avoid these complications.

9.3.2 Effect of solution temperature on performance

The temperature of high-and low-concentration solutions used in RED can vary significantly due to seasonal changes. In addition, if the high concentration is achieved by using some thermal desalination technique, the temperature of stream entering into HCC can be different than the ambient temperature. Tufa et al. [29] have examined the effect of feed temperature on performance of RED system. As shown in Figure 9.12(a), the open circuit voltage increases from 1.67 to 1.74 V when the temperature of brine is increased from 10 °C to 50 °C. As shown in Figure 9.12(b), the current density, on the other hand, almost doubles by changing the brine temperature from 10 °C to 50 °C. By warming the brine from room temperature to 50 °C, the gross PD increases by 44%. These observations can be explained on the basis of resistance to ion transport through the membrane as well as through the interface decreases with temperature due to increased ionic mobility.

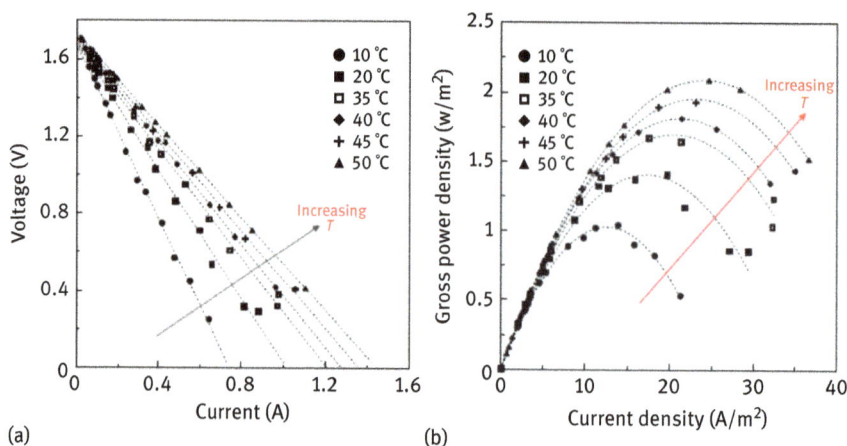

Figure 9.12: Variation of (a) voltage vs current and (b) gross PD vs current density, at different temperatures (HCC: 5 M NaCl, LCC: 0.5 M NaCl). From Tufa et al. [29]. Reprinted with permission.

9.3.3 Membrane properties

Performance of RED is a direct function of membrane properties. Excellent physico-chemical properties, thermal and chemical stability, mechanical strength and long-term performance stability are the essential membrane characteristics for successful real applications of RED. Membrane swelling, defined by the water uptake of the membrane under given conditions, is generally considered an adverse effect in RED applications. This is particularly true if it decreases the ion-exchange capacity of the membrane and does not decrease the membrane resistance at the same time. This effect is generally quantified in terms of fixed charged density defined as the ratio of ion-exchange capacity to swelling degree. The ionic resistance determines the energy loss in an operating RED stack and, therefore, affects the power output of the system. Permselectivity, described as the ability of the membrane to prevent the passage of coions, is controlled by the fixed charged number of the membrane and concentration of the external solution. Ideally, the permselectivity of an ion-exchange membrane should be one; however, according to Donnan's theory, a certain number of co-ions can contribute to the transport current. Thus, the perms-electivity becomes lower than 1 for concentrated solutions. A detail about properties of various commercially available and custom-made ion-exchange membranes and their performance can be found in the review article by Hong et al. [30].

9.4 Process design

A RED stack at effect scale may consist of 20–50 cells. Thus at this scale, it is not only the membrane properties that govern the overall process but also the proper process designing contributes significantly. A better design for electrodes and spacers combined with improved pretreatment can significantly improve the overall performance. The effect of concentration polarization must be minimized. Fouling has been mentioned as a major problem while using real freshwater and seawater in RED [31]. The fouling severely affects the process performance in this case, reducing the power output by 40% during the first day of the operation. Anion-exchange membranes are particularly prone to colloidal and organic fouling. Thus, the better fouling control strategies are crucial in ensuring the stable and long-term performance of RED. Therefore, an effective pretreatment to alleviate the fouling issues should be an integral part of the process design for RED.

9.5 Future challenges for RED

For feasible upscaling of RED, several challenges have to be addressed. The most fundamental challenge is the availability of the appropriate membranes at a reasonable cost. At present, the membranes for RED are 2–3 times more expensive than those for RO desalination. However, the cost reduction can be foreseen with an increase in global demand of these membranes as witnessed in the case of RO membranes where significant cost reduction has been achieved during last two decades due to the development in membrane materials and fabrication techniques. A clear vision on stack design with respect to the feed pretreatment and friction losses needs to be established. The challenges related to pretreatment of river water and seawater and hydrodynamic aspects of the process have not been explored in detail yet. The nature and intensity of membrane fouling for various solutions requires further investigations. The data from real scale RED units are not available. In this context, more pilot studies, similar to Dutch pilot facility located at the Afsluitdijk, are required [32]. While considering RED for large-scale power production, the environmental legislations and regulations must be considered. The effect of changing nutrient flows, sediment transport, changing local salinity and building of the power plant on local environment should be thoroughly analyzed.

References

[1]   Logan, Bruce E., and Menachem Elimelech. "Membrane-based processes for sustainable power generation using water." Nature 488.7411 (2012); 313.

[2]   J.W. Post. Blue Energy: electricity production from salinity gradients by reverse electrodialysis. 2009.

[3]   R.S. Norman. Water salination: a source of energy. Science. (1974); 186: 350–352.

[4]   Tedesco, M., Cipollina, A., Tamburini, A., van Baak, W., & Micale, G. Modelling the Reverse ElectroDialysis process with seawater and concentrated brines. Desalination and Water Treatment. (2012); 49(1–3): 404–424.

[5]   F. Macedonio, E. Drioli, A. A. Gusev, A. Bardow, R. Semiat, M. Kurihara. Efficient technologies for worldwide clean water supply. Chem. Eng. Process: Process Intensification. 2012; 51: 2–17.

[6]   J.H. Kim, S.J. Moon, S.H. Park, M. Cook, A.G. Livingston, Y.M. Lee. A Robust Thin. Film Composite membrane incorporating thermally rearranged polymer support for organic solvent nanofiltration and pressure retarded osmosis. J. Membr. Sci. 2018.

[7]   A.P. Straub, A. Deshmukh, M. Elimelech. Pressure-retarded osmosis for power generation from salinity gradients: is it viable?. Energy & Environmental Science. (2016); 9: 31–48.

[8]   G. Han, S. Zhang, X. Li, T.-S. Chung. Progress in pressure retarded osmosis (PRO) membranes for osmotic power generation. Progress in Polymer science. (2015); 51: 1–27.

[9]   N.Y. Yip, M. Elimelech. Comparison of energy efficiency and power density in pressure retarded osmosis and reverse electrodialysis. Environ. Sci. Technol. 2014; 48: 11002–11012.

[10] N.Y. Yip, D.A. Vermaas, K. Nijmeijer, M. Elimelech. Thermodynamic, energy efficiency, and power density analysis of reverse electrodialysis power generation with natural salinity gradients. Environ. Sci. Technol. 2014; 48: 4925–4936.

[11] D. Brogioli. Extracting renewable energy from a salinity difference using a capacitor. Phys. Rev. Lett. 2009; 103: 058501.

[12] F. La Mantia, M. Pasta, H.D. Deshazer, B.E. Logan, Y. Cui. Batteries for efficient energy extraction from a water salinity difference. Nano letters. 2011; 11: 1810–1813.

[13] A.P. Straub, N.Y. Yip, S. Lin, J. Lee, M. Elimelech. Harvesting low-grade heat energy using thermo-osmotic vapour transport through nanoporous membranes. Nature Energy. 2016; 1: 16090.

[14] X. Zhu, W. Yang, M.C. Hatzell, B.E. Logan. Energy recovery from solutions with different salinities based on swelling and shrinking of hydrogels. Environ. Sci. Technol. 2014; 48: 7157–7163.

[15] Kuleszo, J., Kroeze, C., Post, J., & Fekete, B. M. The potential of blue energy for reducing emissions of CO2 and non-CO2 greenhouse gases. J. Integr. Environ. Sci. 2010; 7 (S1): 89–96.

[16] Ramon, G. Z., Feinberg, B. J., & Hoek, E. M. Membrane-based production of salinity-gradient power. Energy. Environ. Sci. 2011; 4 (11): 4423–4434.

[17] Y.C. Kim, M. Elimelech. Potential of osmotic power generation by pressure retarded osmosis using seawater as feed solution: Analysis and experiments. J. Membr. Sci. 2013; 429: 330–337.

[18] N.Y. Yip, M. Elimelech, Performance limiting effects in power generation from salinity gradients by pressure retarded osmosis. Environ. Sci. Technol. 2011; 45: 10273–10282.

[19] G.Z. Ramon, B.J. Feinberg, E.M. Hoek, Membrane-based production of salinity-gradient power. Energy. Environ. Sci. (2011); 4: 4423–4434.

[20] S. Chou, R. Wang, L. Shi, Q. She, C. Tang, A.G. Fane. Thin-film composite hollow fiber membranes for pressure retarded osmosis (PRO) process with high power density. J. Membr. Sci. 2012; 389: 25–33.

[21] Peinemann, K. V., Gerstandt, K., Skilhagen, S. E., Thorsen, T., & Holt, T. Membranes for power generation by pressure retarded osmosis. Membranes for Energy Conversion, Volume 2. 2008; 263–273.

[22] Wang, H. L., Chung, T. S., Tong, Y. W., Jeyaseelan, K., Armugam, A., Duong, H. H. P.,… & Hong, M. Mechanically robust and highly permeable AquaporinZ biomimetic membranes. J. Membr. Sci. 2013; 434: 130–136.

[23] Skilhagen, S. E., Dugstad, J. E., & Aaberg, R. J. Osmotic power—power production based on the osmotic pressure difference between waters with varying salt gradients. Desalination. 2008; 220(1–3): 476–482.

[24] Strathmann H Electrodialysis, a mature technology with a multitude of new applications. Desalination. 2010; 264: 268–288.

[25] Post, J. W., Veerman, J., Hamelers, H. V., Euverink, G. J., Metz, S. J., Nymeijer, K., & Buisman, C. J. Salinity-gradient power: Evaluation of pressure-retarded osmosis and reverse electrodialysis. J. Membr. Sci. 2007; 288 (1–2): 218–230.

[26] http://www.redstack.nl/en/the-company (Last access: 11 May 2018).

[27] T. Sata. I.-e. Membranes, Royal Society of Chemistry, Cambridge. United Kingdom, 2004.

[28] R.A. Tufa, E. Curcio, W. van Baak, J. Veerman, S. Grasman, E. Fontananova, G. Di Profio. Potential of brackish water and brine for energy generation by salinity gradient power-reverse electrodialysis (SGP-RE). RSC Adv. 2014; 4: 42617–42623.

[29] R.A. Tufa, E. Curcio, E. Brauns, W. van Baak, E. Fontananova, G. Di Profio. Membrane distillation and reverse electrodialysis for near-zero liquid discharge and low energy seawater desalination. J. Membr. Sci. 2015; 496: 325–333.

[30] J.G. Hong, B. Zhang, S. Glabman, N. Uzal, X. Dou, H. Zhang, X. Wei, Y. Chen. Potential ion exchange membranes and system performance in reverse electrodialysis for power generation: A review. J. Membr. Sci. 2015; 486: 71–88.

[31] D.A. Vermaas, D. Kunteng, M. Saakes, K. Nijmeijer. Fouling in reverse electrodialysis under natural conditions. Water Res. 2013; 47: 1289–1298.

[32] http://www.redstack.nl/en/projects/36/afsluitdijk-project (Last access 11 May 2018).

Elena Tocci, Enrico Drioli

10 Molecular modeling of polymer-based membrane processes

10.1 Introduction

The positive aspects of the increase in the standards of the quality of life have drawback aspects like the occurrence of related problems such as water stress, the environmental pollution and the increase of CO_2 emissions into the atmosphere. Membrane technologies have been gradually used for a wide range of applications, including the production of potable water, energy generation, tissue repair, pharmaceutical production, food packaging and the separations needed for the manufacture of chemicals, and have provided feasible alternatives for more traditional purification and separation processes.

The core of every membrane process relies on a nanostructured/functionalized thin interface that controls the exchange between two phases due to external forces, under the effect of fluid properties and based on the intrinsic characteristics of the membrane material [1].

In order to develop advanced membrane technologies, a good understanding of the materials properties and their transport mechanisms as well as the realization of innovative functional materials with improved properties are key issues. In the effort to develop next-generation membranes with enhanced performances, much attention has been put in the new membrane's material and architecture.

Amorphous polymers or nanostructured composites with inorganic components are an important class of materials to solve many of the aforementioned problems. However, the design of these multifunctional materials, based on experimentation and correlative thinking alone, is unreliable, time consuming, expensive and often not successful. Systematic multiscale computer-aided molecular design offers a very attractive alternative. These techniques allow for the very elaborate investigation of complex material behavior with regard to the links between structure, dynamics and relevant properties, which are most important for the penetrant transport and other relevant processes (e.g., selective transport, separation, catalysis, biodegradation, sensor applications) of interest.

During the last decades, computational chemistry had a favorable impact in almost all branches of materials research, ranging from phase determination to structural characterization and property prediction [2–7], as it allows dealing with different types of polymers as well as, for example, with thermal conductivity of composites [8], advanced batteries [9, 10] and so on.

Elena Tocci, Enrico Drioli, National Research Council of Italy, Institute for Membrane Technology (ITM-CNR), University of Calabria, Rende, Italy

https://doi.org/10.1515/9783110281392-010

Moreover, molecular simulations have reached an incredible development in the field of membrane science in the recent years, addressing both the membrane material itself and the transport and sorption phenomena [11–16]. Today, molecular dynamics (MD) simulations can be considered as a chemical engineering tool being part of the "molecular processes-product-process (3PE)" integrated multiscale approach [17].

In this chapter, atomistic modeling tools will be the focus of interest, more specifically, MD simulations.

10.2 Basics of molecular modeling of polymer-based membrane materials

Different levels of atomistic simulations exist, ranging from quantum-mechanical (QM) models to statistical methods. This means numerically solving the classical or QM microscopic equations for the motion of interacting atoms, or even deeper – electrons and nuclei.

QM, or ab initio, methods describe matter at the electronic level, considering the fundamental particles, electrons and protons. The equation from which molecular properties can be derived is the Schrodinger equation and various approximations must be introduced in order to extend the utility of the method to polyatomic systems [18].

Atomistic methods are used to compute molecular properties, which do not depend on electronic effects; the whole atom is modeled just as a soft sphere and obeys the laws of statistical mechanics. Atomistic simulation utilizes analytic potential energy expressions (sometimes referred to as the empirical or classical potentials) to describe the systems.

Assuming that the Born–Oppenheimer approximation holds and the nuclei can be treated as classical particles, the dynamics of the system can be described by Newton's second law. MD is a simulation method that tracks the temporal evolution of a microscopic model system for generating the trajectories by direct numerical integration of Newton's equations of motion, with appropriate specification of an interatomic potential and suitable initial and boundary conditions [19–22]. In classical simulations, the objects are most often described by pointlike centers (with atomic radius R_i and atomic weight m_i) which interact through pair- or multibody interaction potentials. The internal degrees of freedom of a molecule are modeled by a set of parameters and analytical functions that depend on the mutual position of particles in the configuration. The combination of parameters and functions, which describes the potential energy of the system, is called a force field.

Despite classical nature, force fields can mimic the behavior of atomistic systems with an accuracy that approaches the highest level of QM calculations in a fraction of the time. The idea is to replace the true potential function with a simplified model valid in the region being simulated. The underlying assumption is that one can treat the ions and electrons as a single, classical entity. Force fields are constructed by parameterizing the potential function using either experimental data (X-ray and electron diffraction, nuclear magnetic resonance and infrared spectroscopy) or ab initio and semiempirical QM calculations. Different types of force fields were developed during the last years. Among them are large-scale atomic/molecular massively parallel simulator (LAMMPS) [23], nanoscale molecular dynamics (NAMD) [24], groningen machine for chemical simulations (Gromacs) [25] and daresbury laboratory program for molecular dynamics simulations (DLPOLY) [26], which enabled large-scale simulations on complex systems. There are major differences to be noticed for the potential forms. The first distinction is to be made between pair- and multibody potentials. The energy is described in terms of a simple function that accounts for distortion from "ideal" bond distances and angles (in terms of *internal degrees of freedom*), as well as for nonbonded van der Waals and Coulombic interactions between molecules/atoms.

10.2.1 Force fields

The general form of a force field describing the potential energy of the N-atom system can be written as follows:

$$V(\vec{r}_1, \vec{r}_2, ..., \vec{r}_N) = \sum V_1(\vec{r}_i) + \sum_{i,\,j \triangleright i} V_2(\vec{r}_i, \vec{r}_j) + \sum_{i,\,j \triangleright i,\,k \triangleright j} V_3(\vec{r}_i, \vec{r}_j, \vec{r}_k) + ... \qquad (10.1)$$

It includes all types of interactions, that is, bonded and nonbonded. The part of the potential energy V representing bonding interactions will include terms of the following kind:

$$V_{\text{intramolecular}} = \frac{1}{2} \sum_{\text{bonds}} K_{ij}^r (r_{ij} - r_{eq})^2 + \frac{1}{2} \sum_{\substack{\text{bend} \\ \text{angles}}} K_{ijk}^\theta (\theta_{ijk} - \theta_{eq})^2$$

$$+ \frac{1}{2} \sum_{\substack{\text{torsion} \\ \text{angles}}} \sum_m K_{ijkl}^{\varphi,m} (1 + \cos(m\varphi_{ijkl} - \gamma_m)) \qquad (10.2)$$

The "bonds" typically involve the separation $r_{ij} = |\vec{r}_i - \vec{r}_j|$ between adjacent pairs of atoms in a molecular framework, and a harmonic form with specified equilibrium separation has been used, although this is not the only possible type. The "bend angles" θ_{ijk} are between successive bond vectors such as $\vec{r}_i - \vec{r}_j$ and $\vec{r}_j - \vec{r}_k$, and involve three-atom coordinates. Usually this bending term is quadratic in the angular displacement from the equilibrium value, although periodic functions are also used. The "torsion angles" φ_{ijkl} are defined in terms of three connected bonds; hence, four atomic coordinates are used. The part of the potential energy V representing nonbonded interactions between atoms is traditionally split into one-body, two-body and three-body terms:

$$V_{nonbonded}(\vec{r}_1, \vec{r}_2, ..., \vec{r}_N) = \sum_i v(\vec{r}_i) + \sum_i \sum_{j>i} w(\vec{r}_i, \vec{r}_j) + ... \qquad (10.3)$$

The $v(\vec{r}_i)$ term represents an externally applied potential field and describes external force fields (e.g., gravitational field) and external constraining fields (e.g., the "wall function" for particles in a chamber). The pair potential $w(\vec{r}_i, \vec{r}_j) = w(r_{ij})$ neglects three-body (and higher order) interactions. The Lennard–Jones potential is the most commonly used form:

$$w^{LJ}(r) = 4\varepsilon \left[\left(\frac{\sigma}{r}\right)^{12} - \left(\frac{\sigma}{r}\right)^{6} \right] \qquad (10.4)$$

where σ is the diameter and ε is the depth of the potential energy well. If electrostatic charges are present, the appropriate Coulomb potentials are added:

$$w^{Coulomb}(r) = \frac{Q_1 Q_2}{4\pi\varepsilon_0 r} \qquad (10.5)$$

where Q_1, Q_2 are the charges and ε_0 the permittivity of the free space. Typical examples for force field functions are summarized in Table 10.1.

Table 10.1: Examples of potential terms used in force fields.

Potential terms	Graphical description	Mathematical description
Bond stretching		$E_{stretch} = \sum_{1, 2\,pairs} K_b(b - b_0)^2$
Bond bending		$E_{bend} = \sum_{angles} K_\theta(\theta - \theta_0)^2$
Bond torsion/dihedral angle		$E_{torsions} = \sum_{1, 4\,pairs} K_\varphi(1 - \cos(n\varphi))$
6–12 van der Waals Electrostatic		$E_{van\,der\,Waals} = \sum_{nonbonded\,pairs} \left(\frac{A_{ik}}{r_{ik}^{12}} - \frac{C_{ik}}{r_{ik}^{6}} \right) \cdot$ $E_{electrostatic} = \sum_{nonbonded\,pairs} \frac{q_j q_k}{D r_{ik}}$

The force constants K_b, K_θ, K_φ and the equilibrium values b_0, θ_0 and so on are atomic parameters that are experimentally derived from X-ray, nuclear magnetic resonance, infrared, microwave, Raman spectroscopy and ab initio calculations on a given class of molecules (alkanes, alcohols, etc.). The number of different atom types (for one and the same element) then determines the accuracy of a force field concerning different chemical environments for involved atoms.

10.2.2 Equations of motion

The energy, or Hamiltonian of a system composed of N-particle with coordinates $r_N = (r_1; r_2; ... r_N)$ and momenta $p_N = (p_1; p_2; ... p_N)$, may be written as a sum of kinetic and potential terms:

$$H(r_N, m\dot{r}_N) = V(r_N) + K(m\dot{r}_N) \tag{10.6}$$

where it is assumed that the kinetic energy, K, depends only on the momenta (pi) and it is separable from the potential energy, $V(r_N)$, that depends only on atomic positions.

The solution of Newton's equations represents the evolution of a physical system. Particles in MD move naturally under their own intermolecular forces and follow Newton's second law:

$$F_i = m_i \ddot{r}_i = -\frac{\partial V}{\partial r_i} \tag{10.7}$$

where m_i, $\ddot{r}_i$ and r_i are the mass, acceleration and position of particle i, respectively.

From the knowledge of the force on each atom, it is possible to determine the acceleration of each atom in the system.

The necessary starting positions $\vec{r}_i(0)$ of the atoms are in the given case usually obtained from methods of chain packing procedures (see Section 2.3). The starting velocities $\vec{v}_i(0)$ of all atoms are assigned via a suited application of the well-known relation between the average kinetic $K(m\dot{r}_N)$ energy of a polyatomic system and its temperature T:

$$K(m\dot{r}_N) = \sum_{i=1}^{N} \frac{1}{2} m_i \dot{r}^2_i = \frac{3N-6}{2} k_B T \tag{10.8}$$

k_B is the Boltzmann constant. $(3N - 6)$ is the number of degrees of freedom of an N-atom model considering the fact that in the given case the center of mass of the whole model with its six translation and rotation degrees of freedom does not move during the MD simulation.

In MD simulations, the trajectories of molecules in the system are generated by repeating a process in which the velocity, the coordinates of each molecule and the force acting on each molecule are calculated during infinitesimal period of time, Δt, by integrating the equation of motion for the molecule with respect to Δt. From these trajectories, the average values of properties can be determined.

The equations of motion are integrated with finite difference methods: it is a *N*-body problem and can only be solved numerically. The method is deterministic; once the positions and velocities of each atom are known, the state of the system can be predicted at any time in the future or the past. An outline of the MD simulation method is shown in Figure 10.1.

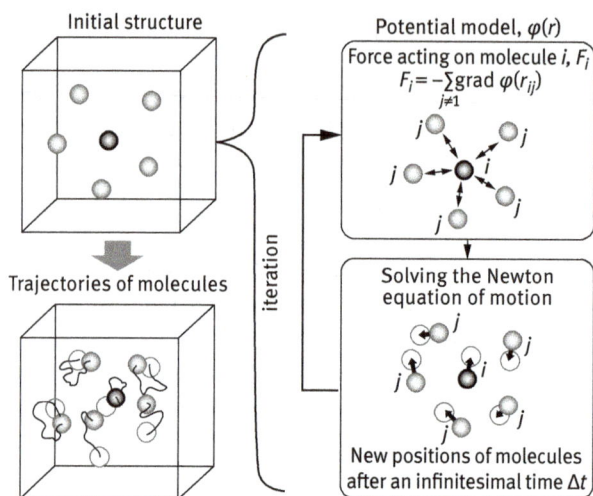

Figure 10.1: Overview of the MD simulation method. (Reprinted with permission from Nada et al. [27]. Copyright 2016 Elsevier Ltd. All rights reserved.)

During the simulation, the configuration space as well as the phase space is explored, allowing to extract information on dynamics of the system. The result of a simulation is the evolution of the positions and the velocities of an ensemble of atoms through time at constant energy and volume (i.e., the microcanonical ensemble). It is possible to extend the set of equations to include additional dynamical variables that enable the simulations to correspond to constant temperature with variable energy (the canonical ensemble) and constant pressure/stress (variable sample volume and shape).

Force fields may be utilized in two directions: to calculate energy minimization and to perform MD simulations.

Over the last 30 years, MD simulations have become a widely used method for the investigation of the molecular structure and surface properties of membrane materials and transport properties in various membrane operations, such as gas separation and water treatment processes [5, 11–13, 15, 28–31]. In this chapter, case studies related to gas separation will be introduced.

10.2.3 Periodic boundary conditions

MD simulations are generally performed under periodic boundary conditions, in order to mimic the presence of the bulk, minimizing boundary effects. The systems containing N atoms are replicated in all directions to produce an infinite periodic lattice of identical cells. When a particle moves in the central cell, its periodic image in every other cell moves accordingly. As one molecule leaves the central cell, its image enters from the opposite side without any kind of interactions with the cell boundary [19–22].

10.3 Dense polymeric membranes

The molecular modeling of polymer-based membrane materials generally starts with the construction of typically rectangular packing models [14]. There, the related chain segments of the respective polymer are arranged in a realistic, that is, statistically possible, way (Figure 10.2). The limited lateral dimensions of packing models of just several nanometers make it impossible to simulate complete membranes or other polymer-based samples. Therefore, on the one hand, bulk models are considered that are typically cubic volume elements of a few nanometers side length that represent a part cut out of the interior of a polymer membrane. On the other hand, interface models are utilized, for example, for the interface between a liquid feed

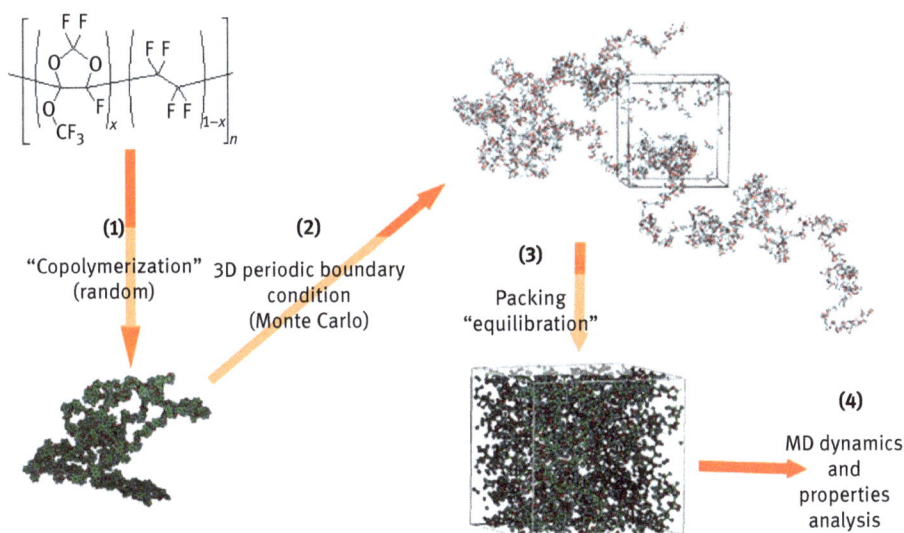

Figure 10.2: Construction procedure of atomistic packing models for Hyflon AD80X. (Reprinted with permission from Jansen et al. [32]. Copyright 2009 American Chemical Society.)

mixture and a membrane surface or between a membrane surface and an inorganic filler [33, 34]. The quality of atomistic packing models is typically validated via comparisons between measured and simulated properties like wide-angle X-ray scattering curves, densities and so on.

10.4 Case study: permeation of gas separation

The permeation of small molecules in amorphous polymers typically follows the solution-diffusion model [35]. According this mechanism, the permeability coefficient (P_A) can be divided into two terms: the solubility coefficient (S_A) and diffusivity coefficient (D_A), following the equation:

$$P_A = S_A \times DA \tag{10.9}$$

Both parameters can be obtained experimentally and in principle also by atomistic simulations [13].

The most common approach to obtain *diffusion coefficients* for gases and vapors is the equilibrium MD. The diffusion coefficient obtained is a self-diffusion coefficient by using the Einstein (eq. 10.10) or by means of the Green–Kubo (eq. 10.11) formulations [22]:

$$D = \lim_{t \to \infty} \frac{\langle [r(0) - r(t)]^2 \rangle}{6t} \tag{10.10}$$

$$D = \frac{1}{3} \int_0^\infty \langle v(0) - v(t) \rangle \tag{10.11}$$

Usually, the diffusion coefficient is calculated from the gradient of the mean-squared displacement as a function of time, although this only applies in the limit of long time (i.e., in equilibrium). The predicted self-diffusivity depends principally on the quality of the force fields used to model the interactions, not only between the penetrant and polymer matrix but also the intramolecular interactions between polymer chains. As temperature is reduced toward and below T_g, penetrant diffusion, in an amorphous polymer matrix, becomes too slow to be predictable by MD simulations. Below T_g, the opening of pathways is constrained by the surrounding polymer, and there is a change in the slope of the variation of diffusion coefficient as a function of temperature. Figure 10.3 depicts the log(msd(t)) of the MD runs as a function of the log simulation time t for oxygen molecules in the perfluoropolymers membrane Hyflon AD60 and in diisopropyldimethyl poly(etheretherketone) DIDM PEEK WC, both glassy polymeric membranes with different amount of free volume. A slope of msd(t) $\approx$ 1, that is, an indication that real Einstein diffusion is reached, can be recognized when the simulation time is greater than 5–10 ns, up to this time anomalous diffusion is dominating.

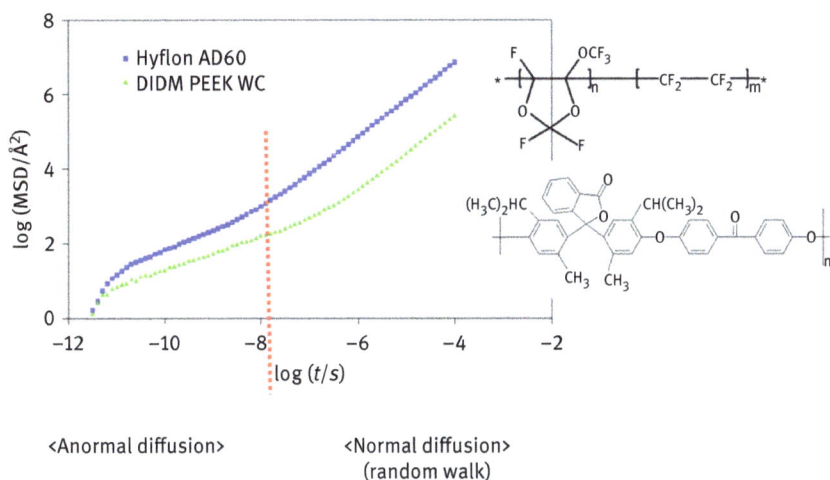

Figure 10.3: Mean-squared displacement of oxygen in Hyflon Ad60 and in DIDM PEEK WC.

Direct MD would require long simulation times for the prediction of diffusivity in low-temperature rubbery polymers and polymer glasses, which are most relevant from the point of view of membrane and barrier material applications [14].

In such cases, diffusion is considered as a sequence of infrequent penetrant jumps (Figure 10.4) and the transition-state theory (TST) of Gusev and Suter [36, 37] and the multidimensional TST approach developed by Greenfield and Theodorou [14] can be utilized. Transport-related diffusion coefficients are less frequently studied by simulation, but several approaches using nonequilibrium MD simulation have also been used [14]. In contrast to a general behavior observed in rubbery polymers, like poly(ether-block-amide) PEBAX [38] and polydimethylsiloxane PDMS [30], for which uniform jump-based mechanism is visible, for glassy systems two types of diffusive motion are available. The inspection of plots of penetrant molecules indicates that there are gas molecules that jump back and forth between two neighboring holes. This finding has been related with the lifetime of temporary channels between different parts of the free volume, and in other glassy polymers, like polyimides, the channel lifetime results much longer than the average residence time of a diffusing molecule in a cavity [3, 6, 8]. Figure 10.4 displays this situation. Here, a nitrogen molecule, in the nitrated PEEK WC model, keeps jumping back and forth between one and the same two voids of the polymer matrix over the second half of the simulation time. A penetrant molecule prefers jumping between two particular holes through channels because this path is energetically much more favorable than moving to a third hole that is separated by a higher energy barrier. For the understanding of small molecule diffusion in stiff chain glassy polymers, it is important to recognize that each pair of jumps of a penetrant molecule back and forth between one and the same two holes basically

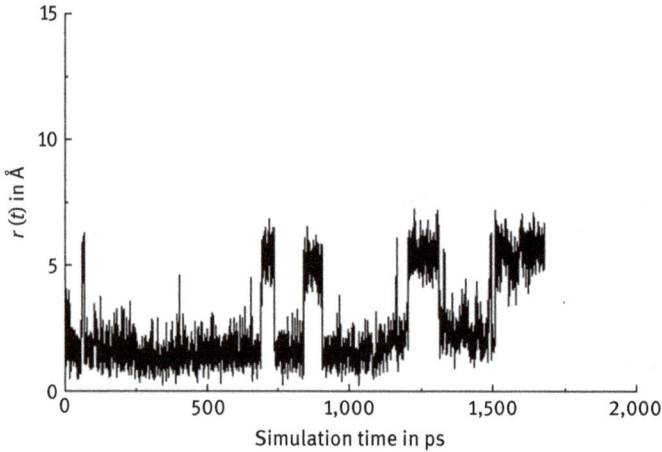

does not contribute to its diffusion through the polymer. The opening and closing of temporary channels is controlled essentially by the movement of the side chain groups of functionalized polymers. The movement of backbone chain is generally more reduced. This kind of behavior is certainly a major reason for the general tendency that constants of diffusion for small molecules in dense amorphous polymers are smaller for the glassy stiff chain case than for the rubbery flexible chain case if similar relative amounts of free volume can be assumed.

A direct visualization of the diffusion of molecules through the matrix is also obtained by investigating their path across the membranes. Figure 10.5 shows the typical trace of hydrogen, oxygen and nitrogen molecules diffusing in poly(ether-etherketone) PEEK WC.

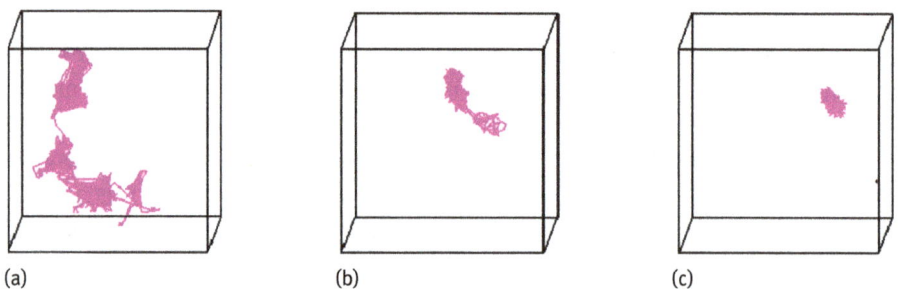

(a)   (b)   (c)

Figure 10.5: Diffusion paths for (a) hydrogen, (b) oxygen and (c) nitrogen molecules.

The figure gives a clear idea of what happens microscopically when gas molecules move into the bulk of polymers.

The predicted self-diffusion coefficients depend principally on the quality of the force fields used to model the interactions, not only between the penetrant and polymer matrix but also by intramolecular interactions between polymer chains. MD simulation of gas diffusion in polymer membranes generates a wealth of information on the mechanism of gas transport, but its use today is limited to high free-volume matrices and small gas molecules due to prohibitive CPU times for diffusion coefficients smaller than 10^{-8} cm^2/s.

Figure 10.6 indicates that the diffusion coefficient results for various gas molecules through thermally rearranged polybenzoxazole (TR-PBO) membranes were compared with experimental data [40]. TR-PBO membranes are diffusivity-enhanced glassy membranes, with outstanding properties determined by the excess free volume in the polymer matrix that can increase and improve molecular transport. The MD diffusion coefficients followed a trend similar to that of the experimental data for TR-PBO membranes: D values decreased with increasing molecular size of gases, indicating their ability to sieve penetrant molecules based on their size. The simulated diffusivities corresponded well with those of experimental ones, apart from CO_2, which was lower as expected due to the fact that the motion was strongly restricted by the immediate environment, and the simulations did not sample large

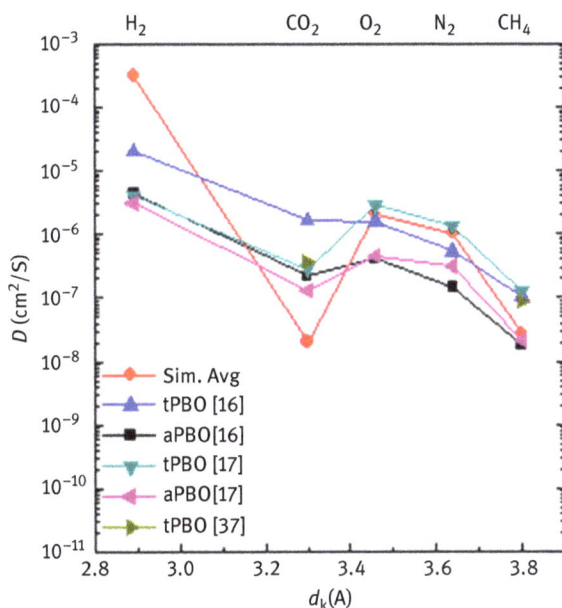

Figure 10.6: Simulated and experimental diffusion coefficient data from the literatures of gases in thermally rearranged polybenzoxazoles (TR-PBO) membranes. (Reprinted with permission from Park et al. [40]. Copyright 2014 American Chemical Society.)

enough time scales to provide reliable information (as previously described in the chapter).

The *solubility* of gases and vapors can be obtained by means of several computational approaches, principally the Widom particle insertion [41] and the grand canonical Monte Carlo (GCMC) [15, 42] methods. The interaction energy of a gas particle inserted within the accessible free volume of the polymer matrix is calculated and the excess thermodynamic potential μ_{excess} can be estimated from the following equation:

$$\mu_{excess} = RT \ln \langle \exp(-E_{int}/kT) \rangle \tag{10.12}$$

The solubility S is then obtained from the following relation:

$$S = \exp(-\mu_{excess}/RT) \tag{10.13}$$

Another approach is the TST of Gusev and Suter [36, 37].

Solubility of a gas molecule is proportional to its condensability, which is represented by a critical temperature in this study. An example is given in Figure 10.7 where the solubility values for different gas molecules in polymer of intrinsic microporosity (PIM) with rigid bridged bicyclic ethanoanthracene (EA) and Tröger's base (TB) PIM-EA-TB are indicated. Here the comparison of PIM-EA-TB data with those of PIM-1 was also shown. Values obtained by using GCMC methods were in reasonable agreement with experimental solubility coefficients up to oxygen then were underestimated. TST method gives the best comparison with larger gases with the well-known deviation for CO_2. The two methods give similar values and trends, independently on the polymers.

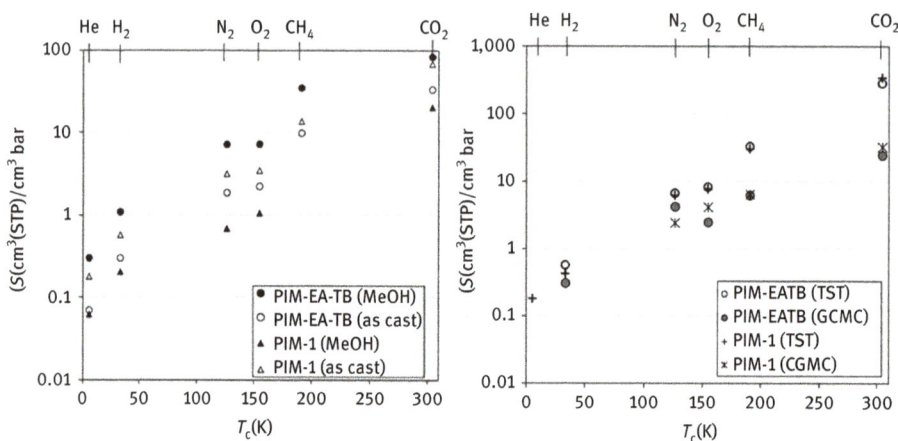

Figure 10.7: Correlation of the solubility coefficients with the critical temperature of penetrants for PIM-EA-TB. From Tocci et al. [43]. Left: experimental values of PIM-EA-TB (190 kDa). Right: data from simulations. Data for PIM-1 are shown for comparison (experimental. From Budd et al [44], TST. From Heuchel et al. [45] and MD values. From Chang et al. [46]). (Reprinted with permission from Tocci et al. [43]. Copyright 2014 American Chemical Society.)

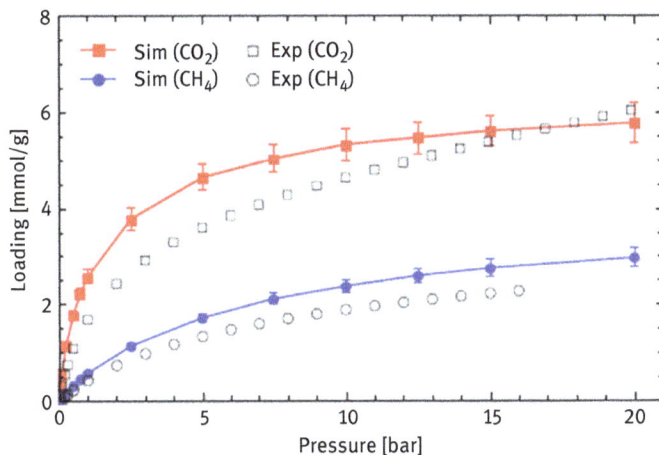

Figure 10.8: Adsorption isotherms of CH_4 and CO_2 at 293 K are in PIM-1. The standard deviation obtained from five independent samples are indicated by the error bars. (Reprinted with permission from Abbott et al. [47]. Copyright 2013 Springer-Verlag Berlin Heidelberg.)

GCMC methods are generally used to calculate *adsorption isotherms*. Figure 10.8 shows CH_4 and CO_2 adsorption isotherms calculated for PIM-1 at 293 K in comparison with experimental data. Generally, the simulated isotherms agree well with the experimental data, mimicking the magnitude of the isotherms and their shape. However, the possible discrepancies can be attributed to the accessibility of the pores and possible swelling of the polymer. The assumption in GCMC simulations is that the polymer matrix is rigid. This is true at low penetrant concentration, being the effect of swelling and dilation negligible.

Indeed, a polymer membrane is expected to undergo reversible elastic dilation with changes in its local structure and free volume during the sorption of concentrated gas molecules.

The CO_2-induced plasticization and swelling can be studied by MD simulations. One of the most widely used strategies is to artificially pack gas molecules with polymers, which corresponds to the swollen state of the polymer, followed by MD simulations. A linear combination of the swollen and nonswollen states isotherms was performed by Hölck et al. (Figure 10.9), allowing the interpolation in order to describe the nonlinear gas sorption in the glassy polymers covering the penetrant pressure range between the reference states showing a good agreement with the experimental results [48].

Figure 10.10 shows the density distributions obtained from the calculation of the sorption energy of CO_2, CH_4, O_2, N_2 and H_2 in PIM-EA-TB and polymer of intrinsic microporosity (PIM) with rigid Tröger's base (TB) and less rigid spirobisindane (SBI) PIM-SBI-TB at 298 K and 1 bar [49]. Gas molecules occupy the voids in the

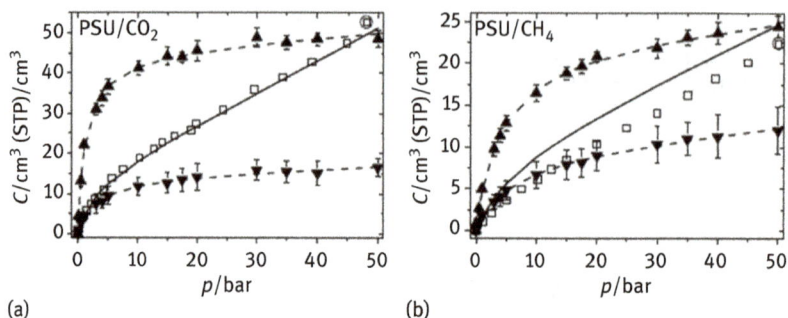

Figure 10.9: Sorption isotherms (308 K, 35 °C) for (a) PSU/CO_2 and (b) PSU/CH_4: calculated for the nonswollen (PSU; ▼) and swollen ((a) PSU80m, (b) PSU35m; ▲) packing models of PSU; experimental data (□); dual mode sorption fits (–) and linearly weighted average of the latter (—). The experimental reference for the swollen state is marked by a circle. (Reprinted with permission from Hölck et al. [48]. Copyright 2012 Elsevier B.V. All rights reserved.)

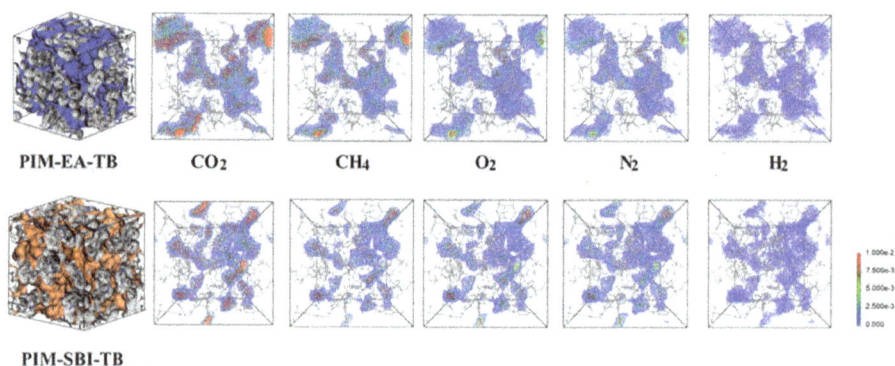

Figure 10.10: Density distributions of CO_2, CH_4, O_2, N_2 and H_2 in PIM-EA-TB and PIM-SBI-TB at 298 K and 1 bar. (Reprinted with permission from Zhou et al. [49]. Copyright 2014 Royal Society of Chemistry.)

membranes. For the same type of gas, PIM-EA-TB exhibits a higher density than PIM-SBI-TB because PIM-EA-TB possesses a larger surface area and fractional free volume (FFV), thus a higher sorptive capacity. Moreover, a relatively higher density was found around the TB units. Therefore, gas sorption was determined not only by the surface area and free volume; furthermore, it was dependent on the affinity for specific units. CO_2 having the strongest affinity for the membranes showed the highest density among the five gases.

The sorption behavior of *mixed gas system*, such as binary mixtures, can be also calculated by GCMC, in order to give a molecular interpretation of the competitive sorption and diffusion of gas molecules in polymeric membranes. An example is

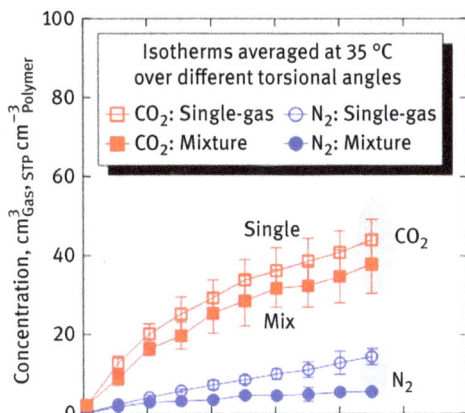

Figure 10.11: Adsorption isotherms in mixture conditions for CO_2 and N_2 (filled symbols) for TR-PBO membranes at 35 °C compared to the theoretical single-gas isotherms (empty symbols). (Adapted from Rizzuto et al. [50]. © 2017 Elsevier B.V. All rights reserved.)

given in Figure 10.11 [50] where the multicomponent adsorption isotherms of CO_2 and N_2 in TR-PBO membranes were simulated.

It was found that the adsorption concentration of both gases increased with the increasing pressure; CO_2 was preferentially adsorbed over nitrogen at the pressure range studied because of its higher solubility. The pressure dependence of the mixture adsorption was similar to that observed for single gas adsorption, with CO_2 showing higher isotherms than N_2. Such behavior was due to the competitive adsorption of one gas over the other, which reduces the ability of the first penetrant to be absorbed in the matrix. More specifically, the adsorption concentration of both gases increased, with increasing pressure. Nevertheless, CO_2 was preferentially adsorbed more than nitrogen over the whole pressure range because of its higher solubility. The pressure dependence of the mixture adsorption was similar to that observed for the single gas adsorption, with CO_2 showing a higher isotherm than N_2.

The transport behavior of polymeric dense membrane models are strongly related to the amount and morphology of the free volume. TR-PBO and the precursor hydroxyl-containing polyimides (HPI) were investigated at the molecular level to focus on how local changes in polymer structure can affect the performance of those membranes in terms of gas permeability and especially selectivity [40].

Figure 10.12 outlines the free volume morphology analysis method used to relate gas permeability and especially selectivity with the shape of the free volume of the HPI compared to TR-PBO membrane materials. It was shown that the higher diffusivity and permeability of TR-PBO was due to the higher amount of interconnected free volume, consisting of larger and/or connected cavities. With the help of an image analysis technique, the authors focused on the shape effect of the free volume

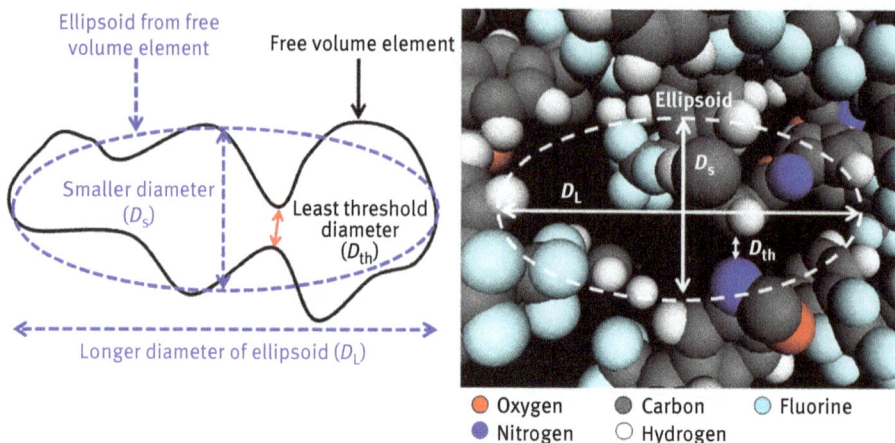

Figure 10.12: Schematic diagram of free volume shape analysis. Free volume shape could range between an ideal spherical profile and an ideal elliptical profile. The critical characteristics include longer diameter (D_L), smaller diameter (D_S), least threshold diameter (D_{th}) and eccentricity of the ellipsoid (D_S/D_L). (Reprinted with permission from Park et al. [40]. Copyright 2014 American Chemical Society.)

elements of both HPI and TR-PBO membranes. HPI possessed a high fraction of spherical free volume elements based on eccentricity values, whereas TR-PBO ones had a higher fraction of elongated free volume elements. From this analysis, the smallest bottleneck in free volume elements was defined from a least threshold diameter and the authors were able to conclude that TR-PBO has elongated pore space and narrow bottlenecks, which supported the hourglass-shaped speculation made in the experimental work [51]. In particular, the bottleneck diameters of the TR-PBO models were wider than those of the HPI models, and this is advantageous for the diffusion of large gas molecules. On the other hand, HPI can have better selectivity for large gas molecules, owing to the narrower and sharply decreased bottleneck diameters in free volume elements.

10.5 Conclusion

The optimal design of membrane units for gas separation processes is by its nature very complex due to different phenomena taking place at different scales. Nevertheless, atomistic molecular modeling techniques have proven to be a very useful tool for the investigation of the structure and dynamics of dense amorphous membrane polymers and of transport processes in these materials. Significant accomplishments have been made in the past decades, leading to the comprehension of different complex problems, and to significant progress in the design of new materials. In this chapter, we have illustrated through some case studies the use and limits of detailed atomistic simulations of transport phenomena in polymeric membranes. In future, innovative utilization of molecular simulations can lead to significantly improved material performance. Moreover, the need to develop a multiscale simulation methodology becomes evident even for this case where for estimating the diffusivity of gases through membrane amorphous cell models is difficult to capture using MD simulations. A real multidisciplinary approach is needed to deal with the different facets that are frontiers of active research in many interconnected fields like materials discovery, membrane fabrication process, optimal fluid flow/distribution inside the membrane module and membrane unit process simulation just to mention some of these.

References

[1]   E. Drioli, A. I. Stankiewicz, F. Macedonio, Membrane engineering in process intensification–An overview. J. Membr. Sci. 2011; 380: 1–8.

[2]   Baschnagel, Jörg. et al. 2000. Bridging the gap between atomistic and coarse-grained models of polymers: Status and perspectives. Retrieved (http://link.springer.com/chapter/10.1007/3-540-46778-5.

[3]   Gubbins, Keith E., Ying-Chun Liu, Joshua D Moore., and Jeremy C. Palmer. "The role of molecular modeling in confined systems: Impact and prospects." Phys. Chem. Chem. Phys : PCCP. 2011; 13 (1): 58–85.

[4]   Karger, Jorg, Douglas M. Ruthven, and Doros N. Theodorou. *.Diffusion in Nanoporous Materials*. 2012.

[5]   Maginn, E. J. and J. R. Elliott. "Historical perspective and current outlook for molecular dynamics as a chemical engineering tool." Ind. Eng. Chem. Res. 49 (7): 3059–78

[6]   Sautet, Philippe and R. A. Van Santen. *Computational Methods in Catalysis and Materials Science: An Introduction for Scientists and Engineers*. 2009.

[7]   Tafipolsky, Maxim, Saeed Amirjalayer, and Rochus Schmid. "Atomistic theoretical models for nanoporous hybrid materials." Microporous and Mesoporous Mater. 2010; 129 (3): 304–18.

[8]   Ma, L. K. et al. "An inverse approach to characterize anisotropic thermal conductivities of a dry fibrous preform composite." J. Reinf. Plast. Compos. 2013; 32 (24): 1916–27.

[9]   Albertus, Paul. et al. . "Identifying capacity limitations in the Li/Oxygen battery using experiments and modeling." J. Electochem. Soc. 2011; 158 (3): A343–51.

[10]   Ferrese, A., P. Albertus, J. Christensen, and J. Newman. "Lithium redistribution in Lithium-Metal batteries." J. Electrochem. Soc. 2012; 159: 10.

[11]   K.E. Gubbins, J.D. Moore. "Molecular modeling of matter: Impact and prospects in engineering". Ind. Eng. Chem. Res. 2010; 49: 3026–3046.

[12]   Membrane Operations: innovative separations and transformations, ed. E. Drioli and L. Giorno, Wiley-VCH Verlag GmbH & Co. KGaA, Weinheim, Germany, 2009 – D.Hofmann, E. Tocci, Chapter 1. Molecular modeling. A tool for the knowledge-based design of polymer-based membrane materials, pp.253–255 and references therein.

[13]   Materials Science of Membranes for Gas and Vapor Separation, ed. Y. Yampolskii, I. Pinnau and B. Freeman, John Wiley & Sons Ltd, England, 2006 – D. N. Theodorou, Chapter 2. Principles of molecular simulation of gas transport in polymers, pp. 49–89 and references therein; J. R. fried,

Chapter 3. Molecular simulation of gas and vapor in highly permeable polymers and references therein.

[14]   Simulation methods for polymers, ed. M. Kotelyanskii and D.N. Theodorou, Marcel Dekker, Inc., USA, 2004 – V. A. Harmandaris and V. G. Mavrantzas, Chapter 6. Molecular dynamics simulations of polymers, pp. 177–222; M. L. Greenfield, Chapter 14. Sorption and Diffusion of small molecules using Transition-State theory, pp. 425–490 and references therein

[15]   Monte Carlo and Molecular Dynamics Simulations in Polymer Science, ed. K. Binder. Oxford university Press, Inc, UK, 1995.

[16]   D. N. Theodorou. "Progress and outlook in monte carlo simulations". Ind. Eng. Chem. Res. 2010; 49: 3047–3058.

[17]   Charpentier, J.C. Among the trends for a modern chemical engineering, the third paradigm: The time and length multiscale approach as an efficient tool for process intensification and product design and engineering. Chem. Eng. Res. Des. 2010; 88(3): 248–254

[18]   Dirac, P.A.M. The Principles of Quantum Mechanics. IV Oxford University Press, Oxford, 1981.

[19]   Frenkel, D. Smit. B. Understanding Molecular Simulation: From Algorithms to Applications. Academic Press: San Diego, CA, 2001.

[20]   Leach, A. R. Molecular Modelling: Principles and Applications. Pearson Education, Harlow, U.K, 2001.

[21]   M. P. Allen and D. J. Tildesley. "Computer Simulation of Liquids". Clarendon Press, Oxford, 1989.

[22]   J.M. Haile. "Molecular Dynamics Simulation". Wiley, Chichester, 1992.

[23]   S. Plimpton. J. Comput. Phys. 1995; 117: 1–19.

[24]   J. C. Phillips, R. Braun, W. Wang, J. Gumbart, E. Tajkhorshid, E. Villa, C. Chipot, R. D. Skeel, L. Kale and K. Schulten. J. Comput. Chem. 2005; 26: 1781–1802.

[25]   M. J. Abraham, T. Murtola, R. Schulz, S. Pall, J. C. Smith, B. Hess and E. Lindahl. SoftwareX. 2015; 1–2: 19–25.

[26]   W. Smith and I. T. Todorov. A short description of DL_POLY. Mol. Simul. 2006; 32: 935–943.

[27]   Nada, H., Miura, H., Kawano, J., Irisawa, T. Observing crystal growth processes in computer simulations. Prog. Cryst. Growth CH. 2016; 62: 404–407.

[28]   Hannah Ebro, Young MiKima, Joon Ha Kim. Molecular dynamics simulations in membrane-based water treatment processes: A systematic overview. J. Membr. Sci. 2013; 438: 112–125.

[29]   Grit Kupgan, Lauren J. Abbott, Kyle E. Hart, and Coray M. Colina Modeling Amorphous Microporous Polymers for CO2 Capture and Separations. Chem. Rev. 2018; 118(11): 5488–5538.

[30]   Hofmann D, Fritz L, Ulbrich J, Schepers C, B hning M. Detailed-atomistic molecular modeling of small molecule diffusion and solution processes in polymeric membrane materials. Macromol Theory Simul. 2000: 9: 293–327,and references therein.

[31]   V N Burganos Modeling and Simulation of Membrane Structure and Transport Properties Comprehensive Membrane Science and Engineering Ed. E. Drioli, L. Giorno, E. Fontananuova. 2015.

[32]   Jansen, J.C., Macchione, M., Tocci, E., De Lorenzo, L., Yampolskii,Y.P., Sanfirova, O., Shantarovich, V.P., Heuchel, M., Hofmann, D., Drioli, E. Comparative Study of Different Probing Techniques for the Analysis of the Free Volume Distribution in Amorphous Glassy Perfluoropolymers, Macromol. 2009; 42, 7589–7604.

[33]   Dashti, A., Asghari, M., Dehghani, M., Rezakazemi, M., Mohammadi, A.H., Bhatia, S.K. Molecular dynamics, grand canonical Monte Carlo and expert simulations and modeling of water–acetic acid pervaporation using polyvinyl alcohol/tetraethyl orthosilicates membrane. J. Mol. Liq. 2018; 265: 53–631.

[34]   Cho, Y.H.,. Kim, H.W., Lee, H.D., Shin, J.E., Yoo, B.M., Park, H.B. Water and ion sorption, diffusion, and transport in graphene oxide membranes revisited. J. Membr. Sci. 2017; 544: 425–435.

[35] Meares, P. (1966). On the mechanism of desalination by reversed osmotic flow through cellulose acetate membranes. European Polymer Journal, 2(3), 241–254.

[36] Gusev, A. A., Arizzi, S., Suter, U. W. J. Chem. Phys. 1993; 99: 2221.

[37] Gusev, A. A.; Suter, U. W. J. Chem. Phys. 1993; 99: 2228.

[38] Tocci, E., Gugliuzza, A., De Lorenzo, L., Macchione, M., De Luca, G., Drioli, E. Transport properties of a co-poly(amide-12-b-ethylene oxide) membrane: A comparative study between experimental and molecular modelling results. J. Membr. Sci. 2008; 323: 316–327.

[39] Tocci, E., Bellacchio, E., Russo, N., Drioli, E. Diffusion of gases in PEEKs membranes: molecular dynamics simulations. J. Membr. Sci. 2002; 206: 389–398.

[40] Park C.H., Tocci E., Kim S., Kumar A., Lee Y.M, Drioli E. A simulation study on OH-containing polyimide (HPI) and thermally rearranged polybenzoxazoles (TR-PBO): Relationship between gas transport properties and free volume orphology. J. Phys. Chem. B. 2014; 118: 2746–2757.

[41] Widom, B. Some topics in the theory of fluids. J Chem Phys. 1963; 39: 2808–2812.

[42] Razmus, D.M., Hall, C.K. Prediction of gas adsorption in 5 A° zeolites using Monte Carlo simulation. AIChE J. 1991; 37: 769–779.

[43] Tocci, E., De Lorenzo, L., Bernardo, P., Clarizia, G., Bazzarelli, F., Mckeown, N.B., Carta, M., Malpass-Evans, R., Friess, K., Pilnáček, K., Lanč, M., Yampolskii, Y.P., Strarannikova, L., Shantarovich, V., Mauri, M., Jansen, J.C. Molecular. Modelling and gas permeation properties of a polymer of intrinsic microporosity composed of ethanoanthracene and tröger's base units. Macromol. 2014;47 (22): 7900–7916.

[44] Budd, P. M., McKeown, N. B., Ghanem, B. S., Msayib, K. J., Fritsch, D., Starannikova, L., Belov, N., Sanfirova, O., Yampolskii, Y. P., Shantarovich, V. J. Membr. Sci. 2008; 325: 851–860.

[45] Heuchel, M., Fritsch, D., Budd, P. M., McKeown, N. B.,Hofmann, D. J. Membr. Sci. 2008; 318: 84–99.

[46] Chang, K.-S., Tung, K.-L., Lin, Y.-F., Lin, H.-Y. RSC Adv. 2013; 3: 10403–10413.

[47] Abbott, L. J., Hart, K. E., Colina, C. M. Polymatic: A generalized simulated polymerization algorithm for amorphous polymers. Theor. Chem. Acc. 2013; 132: 1334.

[48] Hölck, O., Böhning, M., Heuchel, M., Siegert, M. R., Hofmann, D. Gas sorption isotherms in swelling glassy polymers – detailed atomistic simulations. J. Membr. Sci. 2013; 428: 523–532.

[49] Zhou, J., Zhu, X., Hu, J., Liu, H., Hu, Y., Jiang, J. Mechanistic insight into highly efficient gas permeation and separation in a shape-persistent ladder polymer membrane. Phys. Chem. Chem. Phys. 2014; 16: 6075–6083.

[50] Rizzuto, C., Caravella, A., Brunetti, A., Park, C.H., Lee, Y.M., Drioli, E., Barbieri, G., Tocci, E. Sorption and Diffusion of CO_2/N_2 in gas mixture in thermally-rearranged polymeric membranes: A molecular investigation. J. Membr. Sci. 2017; 528: 135–146.

[51] Park, H. B., Jung, C. H., Lee, Y. M., Hill, A. J., Pas, S. J., Mudie, S. T., Van Wagner, E., Freeman, B. D., Cookson, D. J. Polymers with cavities tuned for fast selective transport of small molecules and ions. Science. 2007; 318: 254–258.

Index

https://doi.org/10.1515/9783110281392-011

www.ingramcontent.com/pod-product-compliance
Lightning Source LLC
Chambersburg PA
CBHW080913220326
41598CB00034B/5562